DESCRIPTIVE ANALYSIS OF

NAVAL TURBINE PROPULSION PLANTS

C. N. PAYNE

Commander, U. S. Navy

UNITED STATES NAVAL INSTITUTE

Annapolis, Maryland

Fifth Printing, 1970

Library of Congress Catalogue Card Number 58-7047

ISBN 0-87021-139-0

Printed in the United States of America

by

George Banta Company, Inc.,
Menasha, Wisconsin

PREFACE

The author has prepared this book for use by Midshipmen of the United States Naval Academy. It replaces, in their curriculum, the textbooks *Naval Turbines* and *Naval Auxiliary Machinery.*

The purpose of this publication is three-fold: *first,* to provide the basis for the study of a naval steam propulsion plant which will utilize the student's knowledge of the basic principles of thermodynamics and fluid mechanics; *second,* to re-orient the instruction of Midshipmen in order to emphasize the reasons for the use of each component in the propulsion plant, and to investigate, at least briefly, how the component performs its function; and *third,* to modernize the instruction of Midshipmen through the study of machinery components in use in the latest ships.

The book is organized to provide a view of the propulsion system as an integrated unit, and then to examine in some detail the function and operation of each component. Simple relations, upon which the operation and functions of the components are based, are given wherever practicable. Finally, a simplified heat balance serves to re-assemble the components into a working power plant.

Many illustrations from *Naval Turbines* and *Naval Auxiliary Machinery* have been used. Acknowledgement is due to LTJG J. C. Lochridge, USNR, for the preparation of new illustrations and to CDR. R. G. Mills, USN, for review and criticism of the manuscript. The helpful suggestions of Professor R. M. Johnston and Associate Professor W. F. Eckley are also acknowledged.

R. B. Madden
Captain, United States Navy
Head of the Department of Marine Engineering
United States Naval Academy

June, 1957
Annapolis, Md.

TABLE OF CONTENTS

DESCRIPTIVE ANALYSIS OF

NAVAL TURBINE PROPULSION PLANTS

CHAPTER I
INTRODUCTION

100. FUNDAMENTAL CONSIDERATIONS.

The operation of all engines which are designed to produce mechanical work from thermal energy is based on the fundamental set of principles known as *the First and Second Laws of Thermodynamics.* They are briefly discussed here to provide a basis for understanding some of the broader aspects of engine performance, and consequent design arrangements. These laws are deceptively simple in their statement, but they have such extensive ramifications that their proper application is essential in nearly all fields of engineering.

The First Law of Thermodynamics is a form of the Law of Conservation of Energy. It may be simply stated as follows: *When a quantity of heat energy disappears, an equivalent quantity of mechanical, electrical or chemical energy appears.* We are concerned in this study only with the conversion of thermal to mechanical energy. This law resulted from work done by *Mayer*, *Joule*, and *Helmholtz*.

The Second Law of Thermodynamics is concerned with the proportion of available heat energy which can be converted into mechanical energy, i.e., the largest fraction of a given quantity of heat energy which can be made to disappear and then reappear in the form of mechanical energy. *Clausius*' original statement of this law was as follows: The transformation of heat to work is always dependent on a temperature drop and the transfer of heat to a lower temperature level. The law has been restated in numerous forms. Perhaps the most useful form to the engineer is as follows: *It is impossible to construct an engine that will deliver mechanical work derived solely from the cooling of a single heat source if no heat is rejected to a reservoir at a lower temperature.* A moment's reflection on these statements will establish that the requirements for producing mechanical energy from thermal energy are first, a source of heat; second, a device or engine for converting a part of the thermal energy to mechanical work; third, a reservoir (often called a "heat sink") at a lower temperature than the source, to which heat can flow; and fourth, a working fluid which provides the means of transporting the thermal energy from one part of the system to another. It should also be apparent from the Second Law that no matter what kind of engine is utilized, some of the heat energy can never be converted to work and must be rejected to the sink. Consideration of this law explains why the enormous thermal energy in the atmosphere, or in the ocean, has not been tapped. Incidentally, this law has never been proved in the rigorous sense, and has been subjected to doubt and criticism by such able minds as Max Planck and many others, *but no one has ever demonstrated its failure.*

101. THE CARNOT CYCLE.

An important consequence of these laws was developed by *Sadi Carnot*, who showed that no engine which produces mechanical work from heat can be more efficient than the ideal engine operating on the Carnot cycle, whose efficiency is given as

$$\eta = \frac{\text{work output}}{\text{heat input}} = \frac{T_1 - T_2}{T_1} = 1 - \frac{T_2}{T_1}.$$

In the equation, T_1 is the mean absolute temperature of heat addition to the working fluid from the source; T_2 is the mean absolute temperature of heat rejection from the working fluid to the sink. Note that this expression is independent of *the type of engine, the kind of working fluid, and the source of heat* (whether from chemical combustion or nuclear fission). η then represents the maximum fraction of heat generated which can be converted to work in any cycle for which T_1 and T_2 have been established.

As an example, let us compute the ideal amount of work that can be obtained from one pound of fuel oil in a steam plant whose mean temperature of heat addition is 625°F and which rejects heat to a condenser at 85°F. Each pound of fuel contains about 18,000 BTU of thermal energy. Assume that all of this energy is transferred to the working fluid. Converting the temperature to absolute (degrees Rankine), and substituting in the Carnot equation:

$$T_1 = 625° + 460° = 1085°R$$

$$T_2 = 85° + 460° = 545°R$$

$$\eta = 1 - \frac{545}{1085} = 1 - .502 = .498.$$

Therefore, the maximum amount of heat which can be extracted from each pound of fuel *and converted to work* is $.498 \times 18{,}000 = 8964$ BTU. The remainder of the heat energy in each pound of fuel is lost to the sea through the condenser, the sea constituting the low temperature reservoir, or sink.

It must be realized that this performance is *ideal,* and assumes no losses of any kind except that intrinsic in the Second Law. In actual practice, many losses occur. Some of these are incomplete fuel combustion; loss of heat in the boiler stack gases; radiation, conduction and convection losses from the boiler, piping, and turbine; friction and other mechanical losses in the turbine; and friction losses in the reduction gearing and shafting. These losses combine to yield a considerably smaller fraction of useful work than that given above. Their magnitudes will be discussed later. As a matter of practice, marine power plant performance is usually expressed as the ratio of the pounds of fuel (supplied to the cycle) to the shaft work available from the cycle, i.e., pounds of fuel per shaft horse-power hour, abbreviated as lbs/SHP-hr. Therefore, using the following conversions:

$$1 \text{ BTU} = 778 \text{ ft-lbs}$$

$$33{,}000 \text{ ft-lbs/min} = 1 \text{ HP}$$

the performance of the above cycle may be expressed as:

.284 pounds fuel per shaft horsepower-hour.

102. THERMODYNAMIC CYCLE CONSIDERATIONS.

A *thermodynamic cycle*, as used in the following discussion, comprises the series of events and processes which the working fluid undergoes from the time heat is added until work is extracted and the fluid is discarded, or returned to the source of heat. Although many working fluids have been used in thermodynamic cycles, this discussion will be confined to the steam cycles, in which the fluid is water in its various physical states. Many cycle arrangements have been used since the inception of the steam engine. A few of the important ones will be discussed to give a clear understanding of why the arrangement presently used has been adopted.

The simplest cycle, or *open cycle*, meets the four requirements discussed in article 100. It consists of a *boiler* (source of heat) to liberate heat by the combustion of the fuel and transfer it to the working fluid in the form of stored thermal energy, an *engine* in which the thermal energy of the *working fluid* is converted into mechanical work, and a *heat sink* at a lower temperature than the boiler. This sink, in an open cycle, is usually the atmosphere. See Figure 1-1. Since the temperature, T_1, may be increased by boiling the water under pressure, and since a pressure difference is needed to transfer the working fluid from the boiler through the engine, the boiler pressure is maintained above that of the atmosphere.

To avoid shutting the engine down when the boiler runs dry, the working fluid must be continually supplied to the boiler. For this purpose, a feed pump must be added which will permit continuous operation. Raising water from a low pressure to the high pressure necessary to introduce it into a boiler requires work. The power necessary to drive the feed pump amounts to a fraction of that developed by the engine, so that the net work available from the cycle is less than the work available from the steam generated in the boiler. The work necessary for the feed pump is commonly developed by a separate engine which obtains steam from the same boiler as the main engine. This engine is known as an *auxiliary* engine. See Figure 1-2. Other auxiliary engines drive fuel oil pumps, forced draft blowers, lube oil pumps, etc.

In the open steam cycle, the exhaust temperature, T_2, is limited by the boiling point of water corresponding to atmospheric pressure. Furthermore, the open cycle is wasteful of water, since the exhaust steam goes to the atmosphere. In order to conserve fresh water, a condenser is employed to condense the steam as it leaves the engine. The latent heat is removed from the steam by means of a circulating water system, condensing the working fluid, which is then returned to the boiler. Thus the heat is still re-

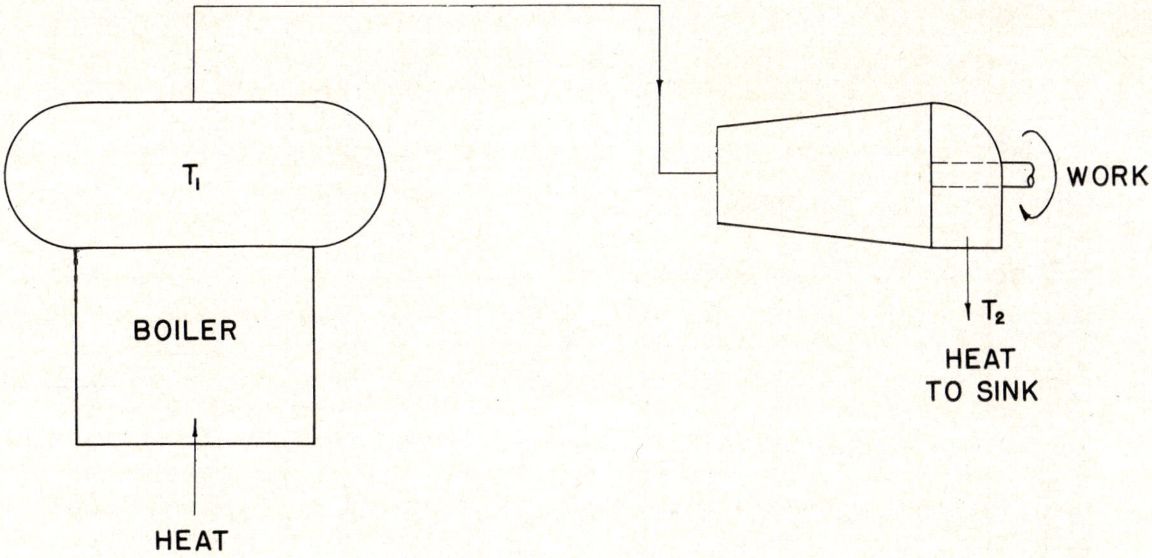

FIG. 1-1. Simple Open Steam Cycle.

jected, but the working fluid is not lost. The addition of the condenser, and its connection back to the boiler, forms the simple *closed cycle.*

One other major advantage of a condenser is realized by placing it under a vacuum, which is easily done. Since the temperature at which steam condenses (equal to the temperature at which water boils) decreases considerably with decreased pressure (see Figure 1-3), the utilization of a condenser under a vacuum lowers T_2. This increases the Carnot efficiency and makes it possible to extract more work from the steam. Figure 1-4 is a schematic diagram of a simple closed cycle.

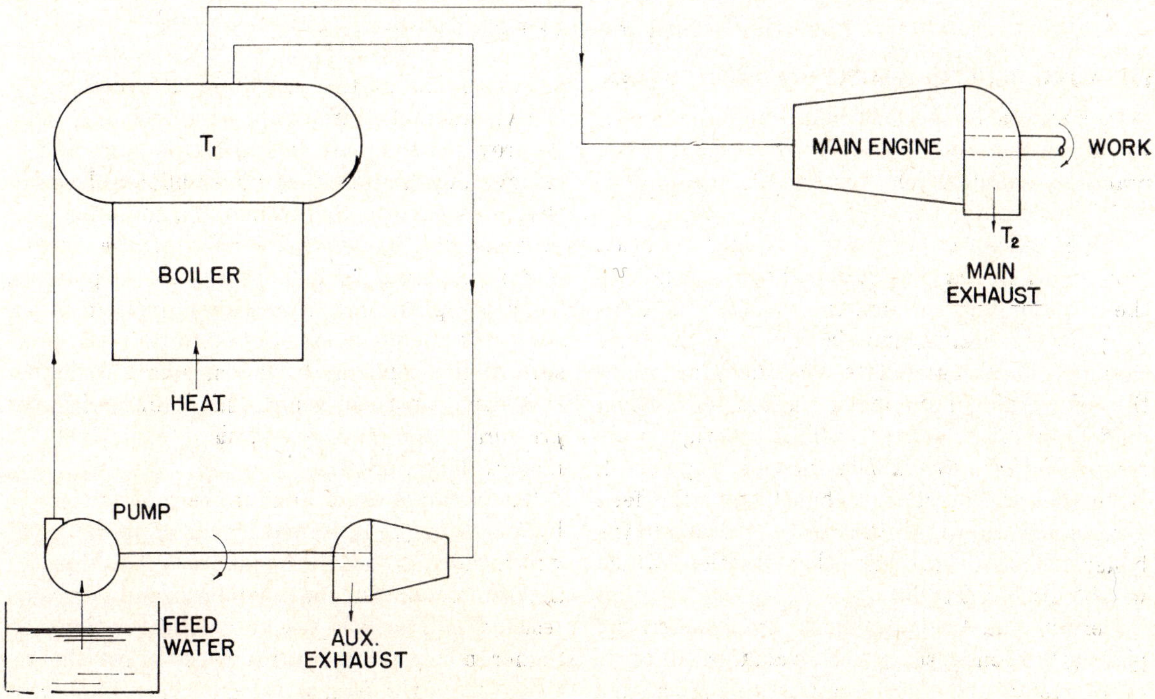

FIG. 1-2. Open Steam Cycle with Auxiliary Engine.

Pressure Closed System is what we have on the ship.

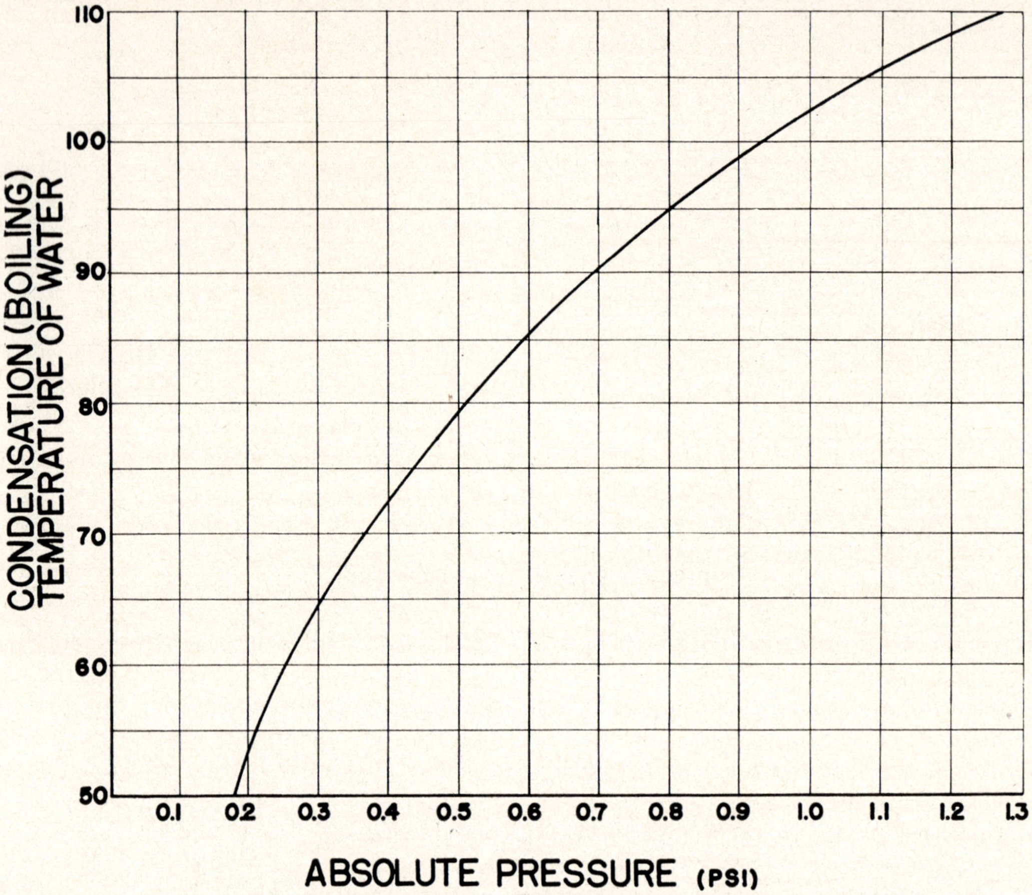

FIG. 1-3. Variation of Boiling Point with Pressure.

103. AUXILIARY EQUIPMENT IN POWER CYCLES.

In addition to the feed pump and condenser, which are the bare necessities for a closed cycle, practical considerations require the use of several other units in the cycle.

When the power output of the plant is varied (commonly by varying the weight rate of flow of the working fluid through the turbine) the total quantity of working fluid circulating in the cycle changes. To accommodate this changing quantity of working fluid in circulation, a working fluid reservoir is necessary. This function is accomplished by a *feed tank* as shown in Figure 1-4. Note that the auxiliary exhaust from the feed pump is also led to this tank to be returned to the boiler. The feed tank has other functions which will be discussed below.

Certain practical problems are created by placing the condenser under a vacuum. All parts of the system under vacuum must be kept as free as possible from air leakage, to prevent air from accumulating in the condenser and destroying the vacuum. An air pump, or air ejector, must be provided to create the initial vacuum, and to remove any air or other non-condensable gases that do leak into the system. To minimize this problem of leakage the portion of the system under vacuum must be kept as small as possible. To accomplish this, a condensate pump is installed at the condenser outlet to raise the pressure of the condensed exhaust steam to higher than atmospheric pressure. This characterizes the *pressure closed feed system* used exclusively in modern naval practice.

The attainment of greater thermal efficiencies by increasing T_1 is limited by the temperature to which the materials in the plant can be subjected and still retain adequate strength and corrosion resistance. The increase in chemical activity at higher boiler temperatures necessitates the removal from the feed water of oxygen and other dissolved gases which tend to react chemically with the materials used in the boiler to the detriment of the strength of these materials. This removal is accomplished by a process of deaeration in the feed tank, to be described in detail later. Therefore, the feed tank is commonly known as the deaerating feed tank (D.F.T.).

The function of the main feed pump has been discussed. In addition, on many shipboard installations, another pump is necessary for reasons which will be discussed later. This pump transfers feed water from the D.F.T. to the main feed pump, and is known as the *main feed booster pump*, or commonly, as the "booster pump."

A final piece of apparatus found in naval pressure closed feed systems is a heat exchanger, called the *economizer*. It is located in the boiler uptakes just above the boiler tube array. Feed water flows through the economizer tubes before entering the boiler drum, absorbing heat from the stack gases. The economizer is usually considered a part of the boiler, and is discussed in detail in *Naval Boilers*.

104. SUMMARY OF THE STEAM CYCLE.

To summarize the important units in a modern steam cycle, the following are listed in order of steam or feed water flow (see Figure 1-4):

1. Boiler (with its economizer)
2. Engine
3. Condenser (with its air ejector)
4. Condensate Pump
5. D.F.T.
6. Main Feed Booster Pump
7. Main Feed Pump

105. IMPROVEMENT OF STEAM CYCLES.

Various additional processes and arrangements have been used in steam cycles in order to improve efficiency, improve operating characteristics, and decrease weight. Some of the more common of these are discussed.

It will be recalled that when heat is added to water to produce steam, the steam will be produced at a constant temperature as long as the steam remains in contact with the water during

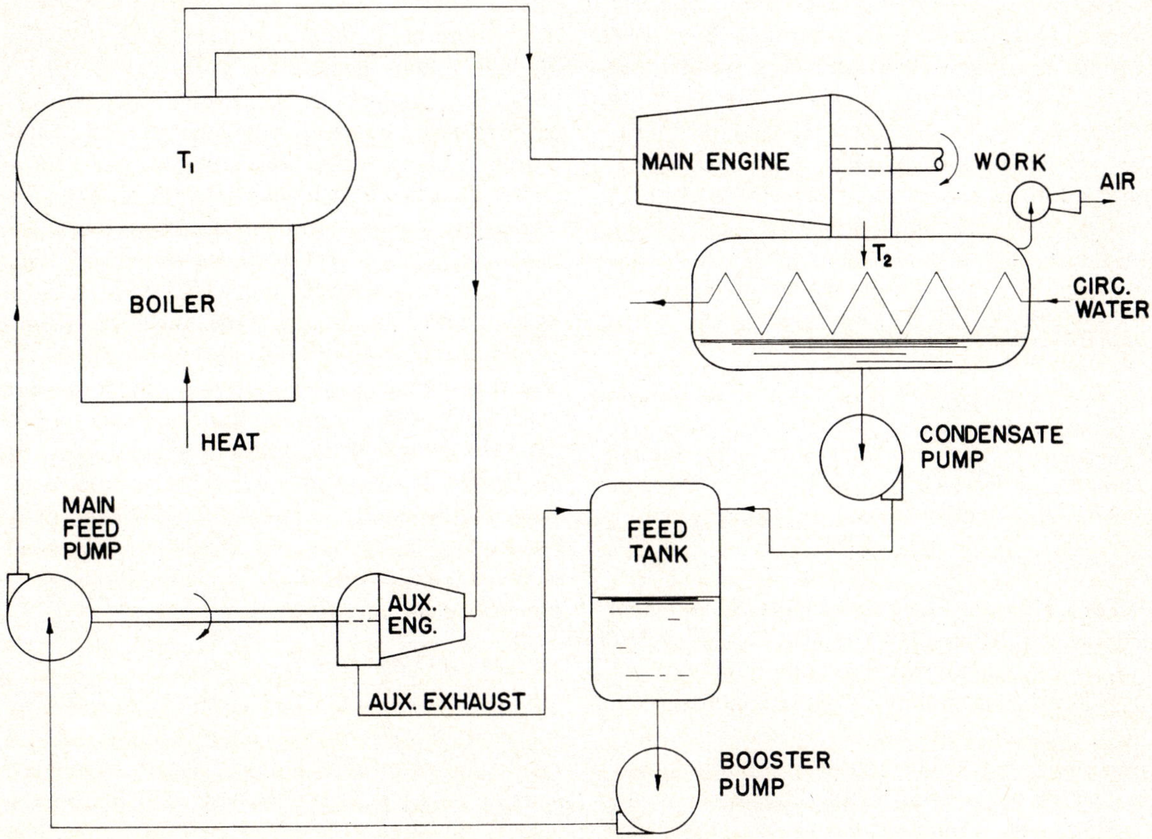

FIG. 1-4. Simple Closed Steam Cycle.

the heat addition process. The steam so produced is said to be *saturated.* As soon as this steam begins to expand in the turbine, the conversion of heat to work begins and the temperature drops, causing partial condensation. The droplets of condensed water so formed, carried through the turbine in increasing quantity at high velocity, have a considerable erosive effect, and affect the efficiency of the turbine. In modern plants the saturated steam from the boiler is superheated to a sufficiently high temperature, before being introduced into the turbine, that the moisture content at the turbine outlet is not greater than 12% to 15%. *Superheat is an increase in temperature above the boiling temperature for a given pressure.*

Other advantages result from the use of superheat. First, a portion of the heat absorbed by the steam is at a temperature above the boiling or saturation temperature. This increases the mean temperature of heat addition, T_1, and results in some increase in the overall thermal efficiency of the cycle. Second, more energy is contained in a pound of superheated steam than saturated steam. Therefore, more work can be performed during expansion. Thus, for a given required power, less superheated steam is required and the engine size can be decreased. The comparison below will illustrate these points. Computations are omitted for clarity.

Compare the ideal thermal efficiency and theoretical work available in each pound of steam for two engines, one operating at 600 psi saturated, and one operating at 600 psi with 350°F of superheat. Both engines exhaust saturated steam to a condenser pressure of 1 psia.

	SATURATED	SUPERHEATED
Heat added in boiler	1133.6 BTU	1355.7 BTU
Heat rejected in condenser	737.1 BTU	851.9 BTU
Theoretical heat available for conversion to work	396.5 BTU	503.8 BTU
Theoretical efficiency	35.2%	37.2%

For ideal cycle efficiency, thermodynamic considerations require that the heat to the cycle be received by the working fluid at a temperature as near as possible to that of the source of supply of heat. This indicates that it is uneconomical to inject relatively cool condensate directly into the boiler, where the temperature level is the highest in the cycle. Therefore, if the condensate temperature can be raised in steps, from sources of heat lower than the boiler temperature, the cycle efficiency will increase. This is accomplished by *regenerative feed heating,* or *extraction,* as it is commonly called. In this scheme, a part of the steam is extracted or bled from the turbine at one or more points, after a part of its thermal energy has been converted into useful work, and led to heat exchangers in the feed circuit, where it gives up thermal energy to raise the feed temperature. After the bled steam is cooled (and condensed) it is mixed with the feed water and returned to the boiler. Figure 1-5 indicates a scheme for three-stage extraction. Five or six stages is about the practical limit in any steam plant. When extraction is used in modern naval practice, only one stage is used. The D.F.T. serves as the heat exchanger for single stage extraction in naval practice, and the two feed heaters indicated in Figure 1-5 are not used. Even in naval installations where no steam is bled from the main turbine, the condensation in the D.F.T. of exhaust steam from the main feed pump and other auxiliaries heats the feed, and is similar in effect to extraction.

Still another scheme for improving efficiency involves adding heat to the steam again after it has done some work in the turbine. This requires extracting the steam at some point in the turbine, piping it back to the boiler where more heat is added, and then returning it to the turbine immediately downstream from the extraction point and allowing it to complete its expansion. This process is called *reheat* and when properly used results in two advantages. First, the average temperature of heat addition to the cycle is increased, and this increases cycle efficiency. Second, more work per pound of working fluid can be extracted from the steam, while maintaining final moisture content at an acceptable level. Reheat plants involve considerably increased plant complexity and cost, and they are complicated to operate at varying turbine speeds. They are not used in modern naval practice.

106. MISCELLANEOUS.

In the foregoing, elementary principles of thermodynamics have been considered as applied to a steam power plant, and in particular, a steam power plant using a steam turbine as the energy conversion device. These principles hold regardless of whether the energy is supplied to the cycle from a conventional boiler, or from a nuclear reactor. It is emphasized that the basic laws apply to *any* plant which produces work from the conversion of thermal energy to mechanical energy, and that the Carnot cycle efficiency limits the performance of all such plants. Gasoline engines, diesel engines, steam reciprocating engines, gas turbines, jets, and turbo-jets all follow these principles, though detailed considerations may vary. The steam turbine is the most suitable energy conversion device for marine propulsion at high power levels, though at moderate to low powers, diesel engines are extensively used. Some fuel rate comparisons are given in Figure 1-6. Steam reciprocating engines are relatively cheap, reliable and easy to construct, but are impractical at higher powers because of the large reciprocating masses required by the low pressure cylinders to handle the quantities of steam involved. In addition, the use of highly superheated steam, necessary for good thermodynamic efficiency, is impractical in reciprocating engines, chiefly because the lubrication between piston and cylinder depends upon some moisture content in the steam. Therefore, the remainder of this book will be devoted to the marine steam turbine plant. The student will have opportunity to study internal combustion types in subsequent courses.

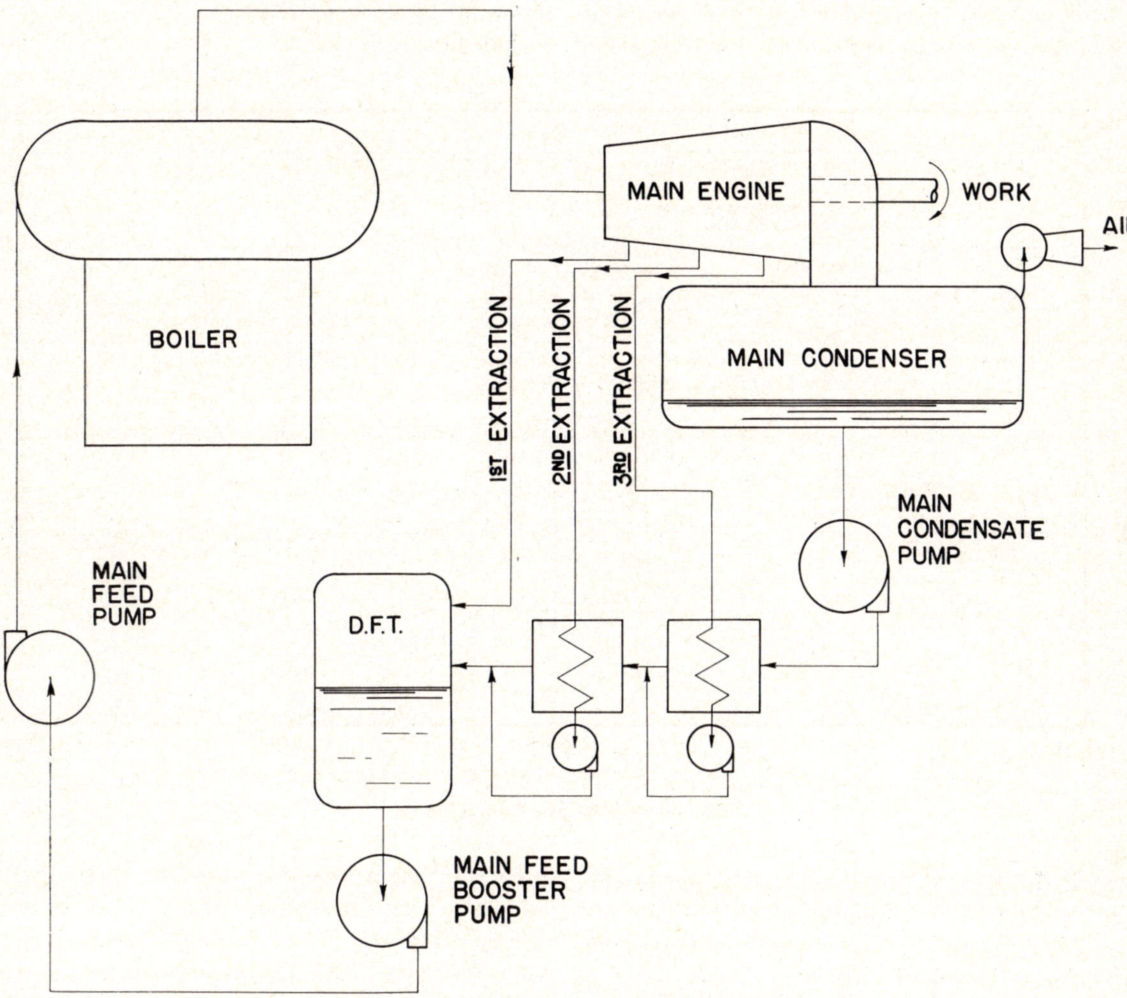

FIG. 1-5. Steam Cycle with Three Extraction Stages.

107. DEFINITIONS.

The terms defined below are standard terms as

used in *General Specifications for Ships of the United States Navy*. They will be used in subsequent discussion and their meaning should be kept clearly in mind.

A propulsion plant consists of the propulsion unit, boilers, propulsion auxiliaries and associated equipment, piping systems, and electrical systems that are required to drive one propeller.

plant(s) or independent auxiliary plants. Fire pumps, bilge pumps, fuel and lubricating oil transfer pumps, anchor engines, and steering engines are examples.

A machinery plant consists of all the above units required to propel the ship, and provide power for all of the various services and operating equipment aboard it.

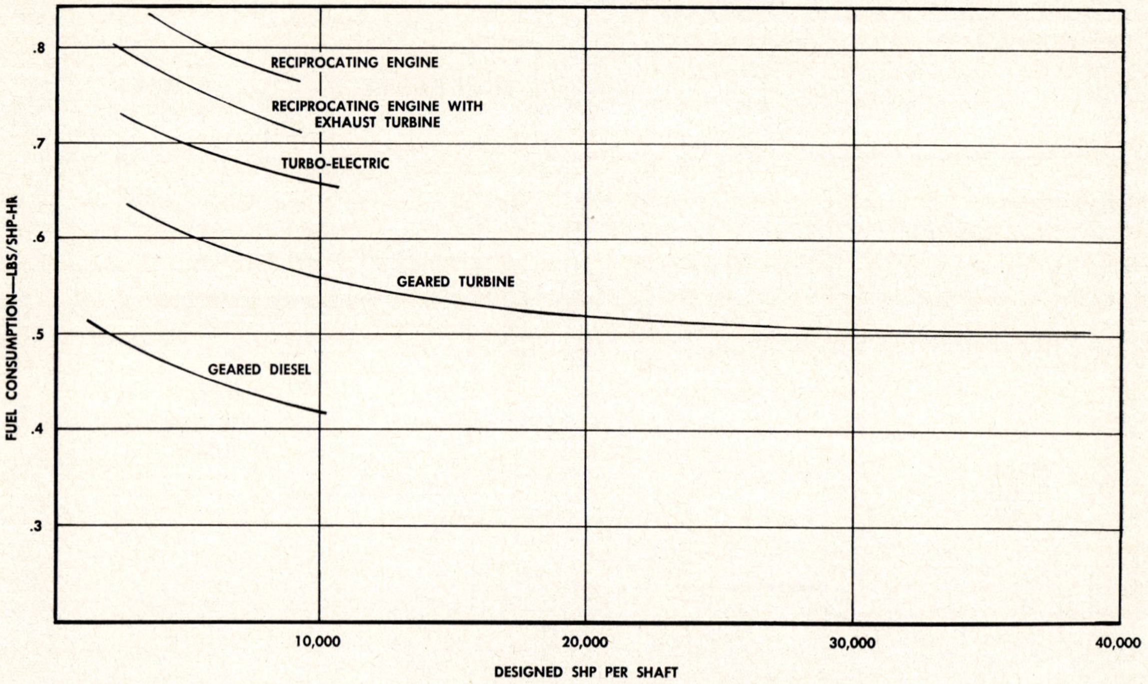

FIG. 1-6. Fuel Rate Comparison.

A propulsion unit consists of the machinery and equipment, including their controls, that are mechanically, electrically, or hydraulically connected to a propulsion shaft.

Propulsion auxiliaries are those auxiliaries, directly associated with a propulsion plant, which perform functions essential to the operation of the propulsion plant. Examples are the condenser, the condensate pump, the deaerating feed tank, the main feed booster pump, etc.

Independent auxiliary plants are those complete plants that are provided to perform specific functions other than those associated with propulsion. Such plants include refrigeration plants, distilling plants, air conditioning plants, ship service electric plants, etc.

Independent auxiliaries are those auxiliary units not directly associated with the propulsion

In addition to the above standard definitions two more terms hereafter will be used with precise meaning to avoid confusion. A *turbine* is considered to be a single component consisting of a single rotor enclosed within one shell. It may be characterized by appropriate descriptive terms, such as "high-pressure," "low-pressure," "single flow," etc. When two or more turbines are mounted on the same shaft, or on shafts which are geared together, the assembly will be called a *turbine unit.*

108. DESIGN CONSIDERATIONS.

Three important considerations in the design of naval machinery plants are reliability, self-sufficiency, and flexibility. A brief discussion of these factors follows:

Reliability. The principle of reliability is paramount. No compromise of this principle is made with any other requirement. This principle results in frequent duplication of those units essential to the operation of the plant. The requirement for reliability is sometimes responsible for seemingly slow acceptance in naval steam plants of certain equipments, controls, or automatic devices which simplify normal operations, but which may add to maintenance complexity and to vulnerability under battle conditions. When such apparatus is proven in reliability, after rigorous service testing, it is incorporated in naval design. Despite this necessary rigorous trial process, naval engineering designers have always been in the forefront in the conception, design, and development of reliable marine steam plant improvements.

Self-sufficiency. Where propulsion plants are installed in separate machinery spaces, each plant is independent and self-sufficient so that maloperation of one plant will not affect operation of any other plant. Where two propulsion units are installed in the same machinery space, the units are independent with respect to systems such as lubricating oil, cooling water, and condensate. Other equipment or components may serve both units. In either case, full flooding from damage of any space through which the propeller shaft passes must not prevent emergency operation of shaft bearings in that space for propulsion shafting originating in another space.

Flexibility. System flexibility incorporating cross connections between propulsion plants and bypasses around components is provided in all modern machinery plants. Such flexibility provides the possibility of operating a given propulsion unit with main and auxiliary steam from a boiler or boilers which are a part of another propulsion plant.

In summary, it can be appreciated that the requirements of self-sufficiency and flexibility are in support of the principle of reliability. Not only must individual components be reliable in their operation, but the engineering plant as a whole must be reliable to the extent of providing propulsive power for the ship under conditions of severe damage. The self-sufficiency of one propulsion plant permits it to continue in operation when its companion plant or plants are inoperative. Flexibility among the propulsion plants permits linking undamaged portions of one plant to those of others to assure continued operation (for instance, steam supplied from No. 1 plant may drive the turbine of No. 2 plant).

CHAPTER II

THE NAVAL STEAM CYCLE

200. PURPOSE.

It is the purpose of this chapter to examine a typical propulsion plant as a whole. This will enable the student to obtain a clear picture of the entire plant prior to detailed study of the various units within the plant. The arrangement herein described is that in a modern destroyer. With few relatively minor variations, this arrangement is closely similar to that installed in any modern steam propelled ship. The plant consists essentially of the machinery necessary to the main steam cycle (as summarized in article 104.), an auxiliary steam system, an auxiliary exhaust system, and the connections necessary to interrelate these systems. Other important systems which will be discussed later are the gland sealing steam system, the lubricating oil system and the high and low pressure drain systems.

201. THE MAIN STEAM CYCLE.

The main steam cycle is here considered as consisting of four parts, or systems, three of which will be considered in detail in this text. These are the *expansion system*, the *condensate system*, and the *boiler feed system.* The latter two systems, grouped together, are often referred to as the *main feed* system. The fourth system in the cycle is the *generation system.* It is thoroughly covered in the course on naval boilers and will receive only passing attention at this point.

Figure 2-1 is a diagram of the main steam cycle indicating these systems. Summarizing, they are:

1. Expansion system
   - A. Main steam piping, valves and fittings
   - B. Turbine control valves
   - C. Main propulsion turbines
2. Main feed system
   - A. Condensate system
     - (1) Main condenser
     - (2) Main condensate pump
     - (3) Various heat exhangers
     - (4) Heating and deaerating portions of D.F.T.
   - B. Boiler feed system
     - (1) Storage portion of D.F.T.
     - (2) Main feed booster pumps
     - (3) Main and emergency feed pumps
     - (4) Economizer
   - C. Piping and valves to reserve feed tanks (Connected at various places to both the condensate and boiler feed systems.)

The expansion system includes that part of the main steam cycle beginning at the main steam boiler outlet, and ending at the inter-connection between the turbine unit and condenser. This includes the main steam piping (sometimes referred to as the main steam system) and the valves therein, the turbine control valves, and the turbines themselves. In this part of the cycle, steam is delivered to the turbine unit, its flow is controlled, and the thermal energy therein is converted into mechanical energy.

The main feed system is that portion of the main steam cycle beginning at the steam inlet to the condenser, including the condenser, condensate pump, various heat exchangers to be described later, the D.F.T., the feed booster pump, the main and emergency feed pumps. The emergency and reserve feed tanks and their inter-connection to the above units will be considered later. The main feed system is subdivided into two principal systems, the condensate system and the boiler feed system. See Figure 2-1.

The condensate system is that portion of the main feed system which begins at the condenser steam inlet and ends with that portion of the D.F.T. which accomplishes heating and deaerating. In this system, the exhaust steam is converted to liquid condensate by rejection of heat to circulating sea water; the pressure is raised by the condensate pump; heat flows into the condensate in the various heat exchangers mentioned above and in the D.F.T.; and dissolved air is removed from the condensate. The final deaerating process converts the condensate to boiler feed water. This part of the cycle is labeled in Figure 2-1.

The boiler feed system comprises the storage part of the D.F.T., the feed booster pump and the main and emergency feed pumps, and the economizer. In this portion of the system the pressure of the boiler feed is raised sufficiently to

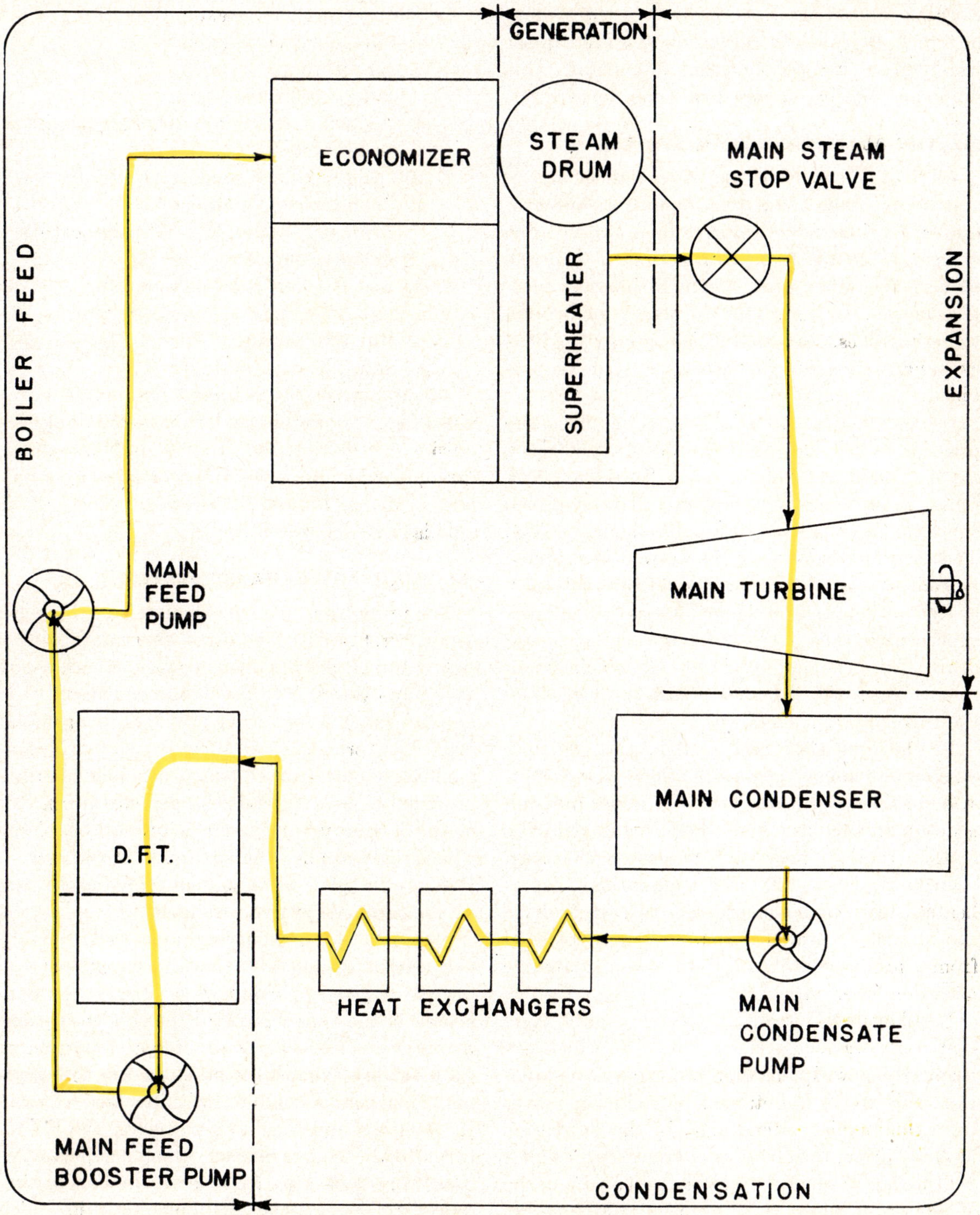

FIG. 2-1. Main Steam Cycle.

inject it into the boiler drum, and thermal energy is added in the economizer from the boiler stack gases.

The generation system in the main steam cycle includes the boiler, the superheater, and the desuperheater, all of which are ordinarily in one integral unit. It is in this portion of the system that the largest quantity of heat is added to the working fluid. This is the portion of the system in which the nuclear power plant differs most widely from the conventional fuel plant. This difference will be discussed in some detail later.

202. THE AUXILIARY STEAM SYSTEM.

All of the propulsion auxiliaries named above, and many other auxiliaries, require a source of power or prime mover to drive them; or a source of heat to enable them to function. The prime movers are either small steam turbines or electric motors. To distribute the steam for driving these turbines, and to furnish energy to heat exchangers, an auxiliary steam system is provided.

In past modern installations, it has been the practice to provide saturated steam at 600 psig for the principal auxiliary loops. Subsidiary systems, requiring lower pressures, are supplied from this principal loop through reducing valves. The saturated steam for the system is supplied by routing a portion of the steam from the superheater outlet of the boiler back through desuperheater coils located in the boiler steam drum. This scheme provides a continuous steam flow through the superheater tubes, even when the turbines are shut down.

In the first 1200 psig system, the principal auxiliary steam system is still at 600 psig. In this system, the superheated steam at boiler pressure is desuperheated as before, then led to a reducing station where its pressure is reduced to 600 psig.

In more recent 1200 psig installations, superheated steam at boiler pressure is used to drive the auxiliary turbines. This steam is obtained from a branch connection to the main steam line near the boiler outlet.

In all systems, steam at lower pressures is required for various services, some of which are itemized below. In the first two types of installation, this steam is obtained by reducing valves from the auxiliary steam line. In the third type of installation, the steam is obtained by desuperheating a portion of the superheated steam in the steam drum desuperheating coils, and then reducing the pressure to that desired.

The steam system that supplies the major auxiliaries is termed the *auxiliary steam system.* Systems at lower pressures are identified by their pressures, as the *150 lb. auxiliary steam system.* Three of these systems, together with their uses, are:

1. 150 psig auxiliary steam system for
   a. Fuel oil service heater.
   b. Air ejectors.
   c. Lube oil purifier heater.
   d. Reducing valve for 100 psig auxiliary steam system.
2. 100 psig auxiliary steam system for
   a. Laundry and tailor shop.
   b. Reducing valve for 50 psig auxiliary steam system.
3. 50 psig auxiliary steam system for
   a. Galley cooking.
   b. Hot water system.
   c. Ship's heating system.

To summarize, the auxiliary steam system provides steam to power the driving turbines of propulsion auxiliaries and it also provides steam through reducing valves to lower pressure auxiliary systems. Figure 2-2 shows a skeletonized auxiliary steam system.

203. AUXILIARY EXHAUST SYSTEM.

The propulsion turbine exhausts into the main condenser. Similarly the turbo-generator, being a sizable unit, exhausts into its own condenser (but also may exhaust into the main condenser). It is impracticable, however, to provide an individual condenser for each small turbine which drives an auxiliary pump. Consequently, a piping circuit is provided to collect the discharged steam from all of the auxiliary turbines in a propulsion plant. This circuit is called the *auxiliary exhaust system.* Pressure in this system is maintained at 15 psig by an automatic valve arrangement.

Steam in the auxiliary exhaust system at 15 psig retains considerable thermal energy which is utilized in two important ways. First, the steam is used in the D.F.T., to heat the condensate and to purge it of dissolved air. A major portion of the auxiliary exhaust steam is used in this manner. This portion combines with the feed water in the heating and deaerating process and so returns to the boiler. Second, a smaller portion is used to provide a part of the thermal energy required to operate the distilling plant, in which fresh water is obtained from sea water. This portion is condensed, and finds its way back to the boiler through the low pressure drain system, discussed later.

When more steam is delivered to the auxiliary exhaust system than is required for these two functions, the pressure in the system commences

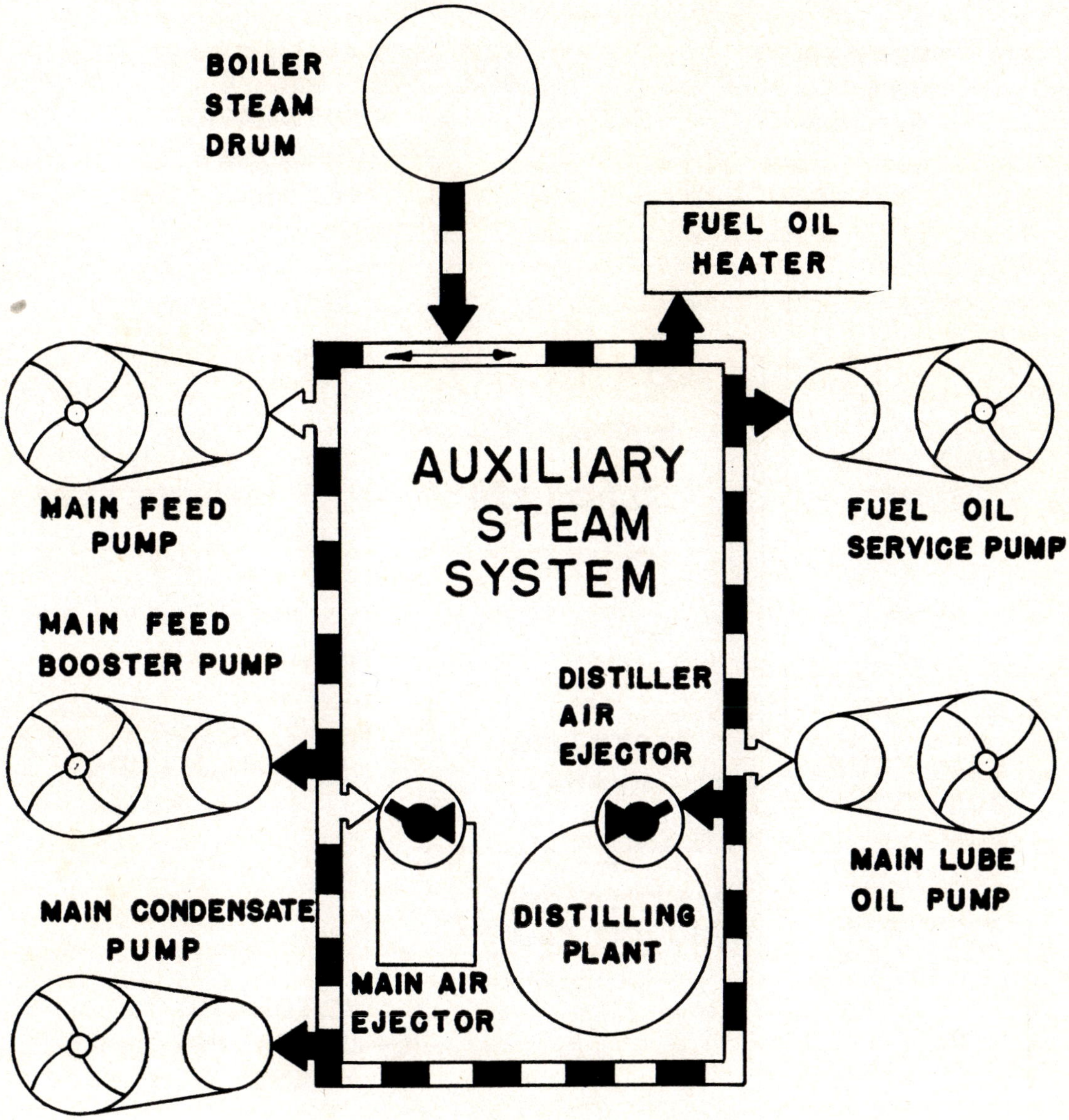

FIG. 2-2. Auxiliary Steam System.

to rise above 15 psig. This rise in pressure causes an *automatic unloading valve* to open, allowing steam to flow from the auxiliary exhaust system to the main condenser. This valve closes when the pressure falls back to 15 psig. The steam that flows to the main condenser loses its thermal energy to circulating sea water and is condensed. Note that the thermal energy in the portions of auxiliary steam used in the D.F.T. and distilling plant performs useful functions. However, the steam unloaded to the condenser loses its thermal energy to the sea. Therefore it is desirable to control the supply of auxiliary exhaust steam discharged into the system to balance the sum of the requirements of the D.F.T. and the distiller. This is done by proper choice of the number of steam driven pumps and electric driven pumps in use under various plant load conditions.

Under some load conditions, there may be insufficient steam in the auxiliary exhaust system to maintain the required pressure, even when no electric pumps are in use. In such circumstances provision is made for augmenting the pressure by admitting steam directly to this system from the auxiliary steam system, through a pressure reducing valve.

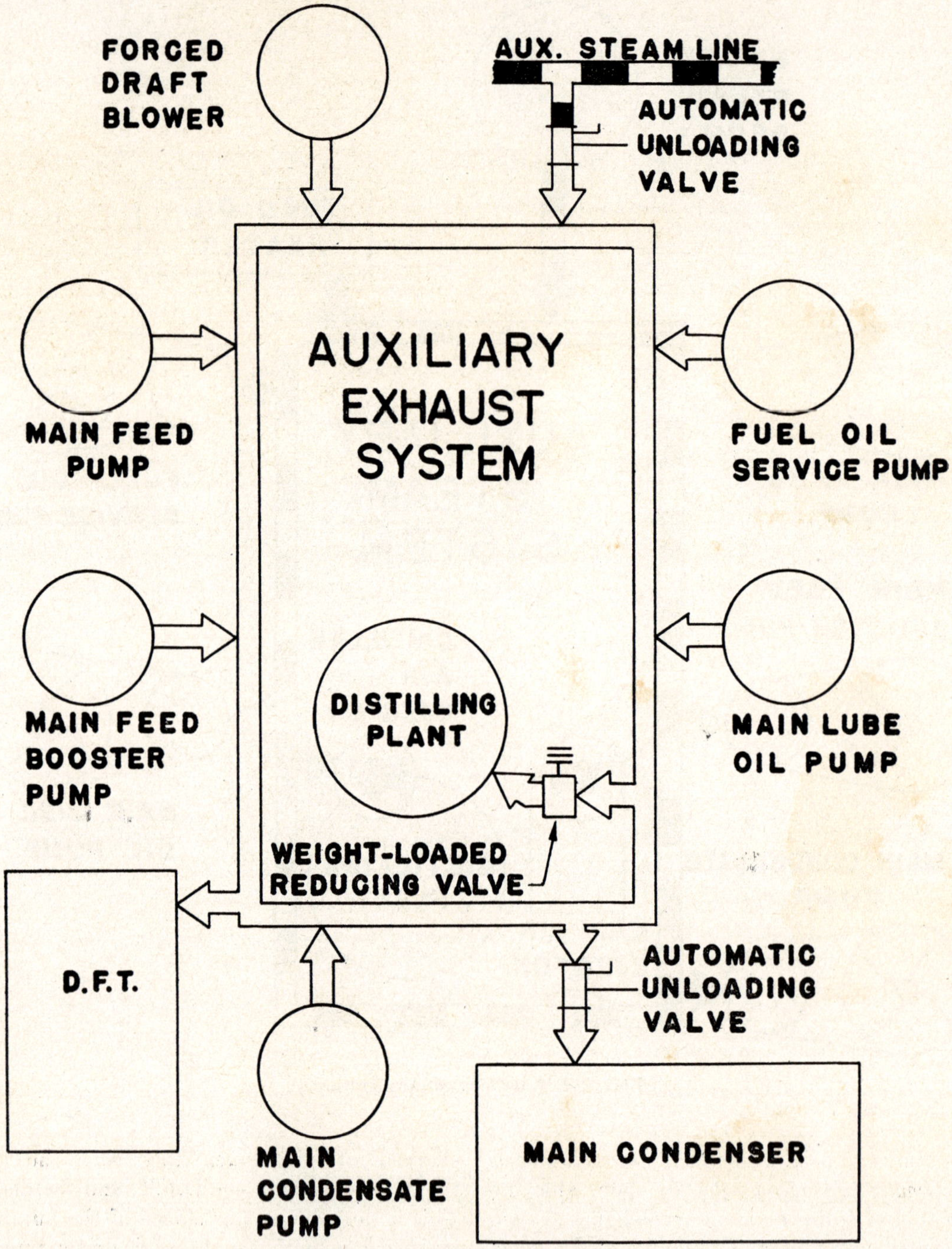

FIG. 2-3. Auxiliary Exhaust System.

Figure 2-3 shows the auxiliary exhaust system.

204. SUMMARY.

To recapitulate this chapter, the main steam cycle, with its four systems, has been discussed; the source and use of the auxiliary steam system has been explained; and the auxiliary exhaust system has been described. A diagram showing the integration of these systems for one propulsion plant is given in Plate I.

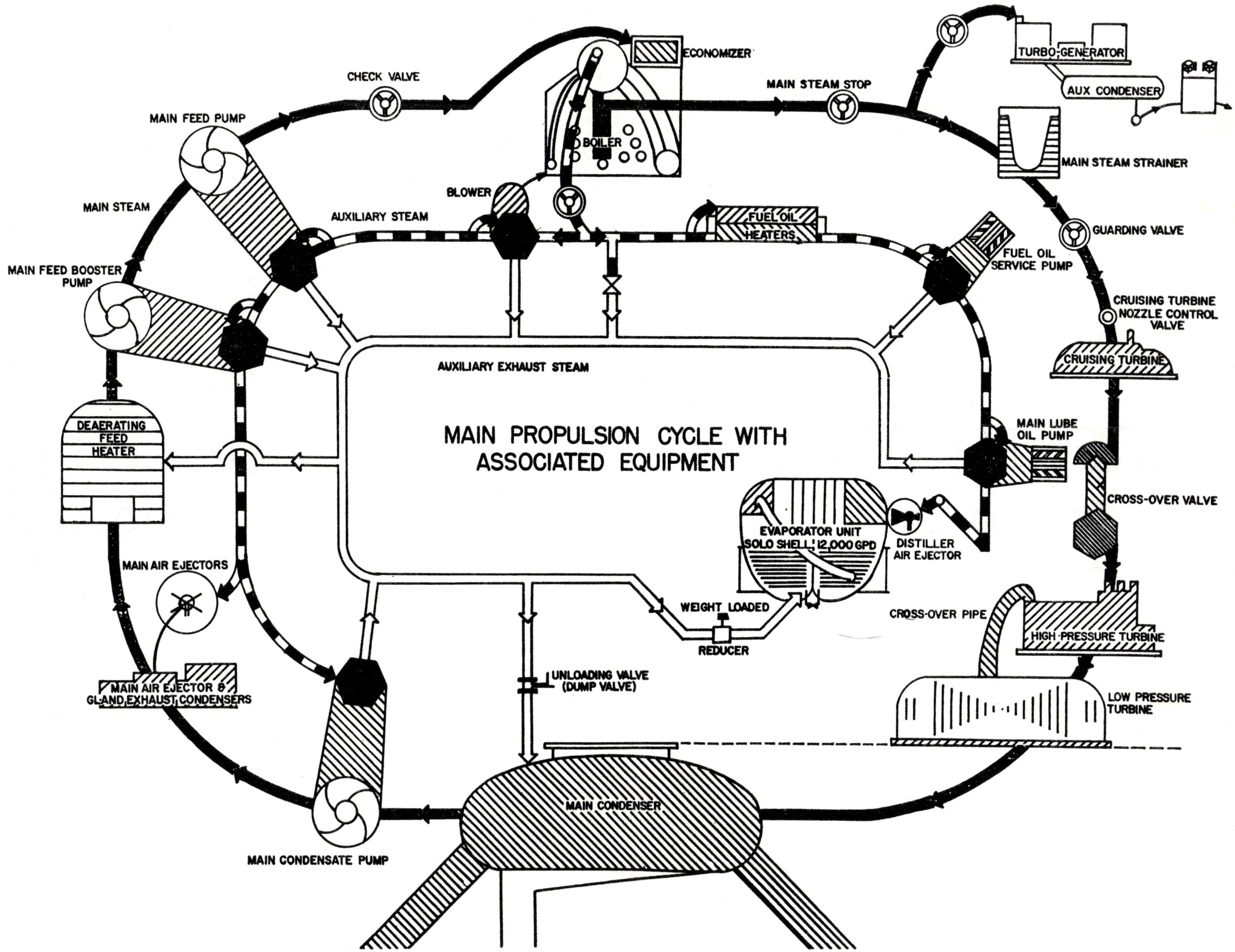

PLATE I. Main Propulsion Cycle with Associated Equipment.

CHAPTER III

BASIC CONSIDERATIONS OF FLUID FLOW AND THERMODYNAMICS

300. GENERAL.

In this chapter basic relationships of fluid flow and thermodynamics will be discussed. For the student who has studied elementary courses in these subjects, the recollection of these relationships will not be difficult. For their reference, and for others who have not studied these subjects, this chapter is included. No attempt here is made to develop fully the principles. The student who desires a more thorough treatment should refer to standard textbooks in the various subjects.

301. PRINCIPLES OF FLUID FLOW.

Three basic physical laws applied to the flow of fluids, plus some empirical relations (derived from quantitative observation) will be discussed in connection with fluid flow principles. The three basic governing laws are: (1) The law of Conservation of Mass (2) The law of Conservation of Energy and (3) The law of Conservation of Momentum.

302. CONSERVATION OF MASS, CONTINUITY.

Continuity of fluid flow in a channel is obtained when the same mass of fluid flows past every point in the channel (whether it is an open or closed channel) in the same increment of time. That is, between any two points simultaneously under observation, there is no accumulation or depreciation of the quantity of fluid anywhere between the two points.

Consider the flow depicted in Figure 3-1. Fluid flows past an observation point, section 1, into a system whose configuration is unknown. It flows out through an observation point at section 2. The cross-sectional area (A) of the channel (or pipe) at sections 1 and 2 are measurable. Also the specific weight (weight per unit volume—w) of the fluid and its velocity (V) are determinable at the two sections. The volume, Q, passing each section per unit time is found by multiplying the cross-section area by the velocity,

$$A \text{ ft}^2 \times V \text{ ft/sec} = Q \text{ ft}^3\text{/sec}.$$

The weight rate of flow, W, is determined by multiplying this volume by the specific weight of the fluid.

$$W = wAV \left(\frac{\text{lbs}}{\text{ft}^3} \times \text{ft}^2 \times \frac{\text{ft}}{\text{sec}} = \frac{\text{lbs}}{\text{sec}} \right).$$

Under these circumstances, the weight rate of flow is equal at all sections and the law of Conservation of Mass is fulfilled. Thus

$$W = w_1 A_1 V_1 = w_2 A_2 V_2. \quad (1)$$

This equation is the mathematical expression of continuity of flow. When the fluid is incompressible, there is no change in density during flow, or w_1 is equal to w_2, and the volume rate of flow is constant at all sections.

$$Q = A_1 V_1 = A_2 V_2. \quad (2)$$

Even when dealing with compressible fluids, changes in density are frequently small and may be neglected. Under such circumstances equation (2) is valid.

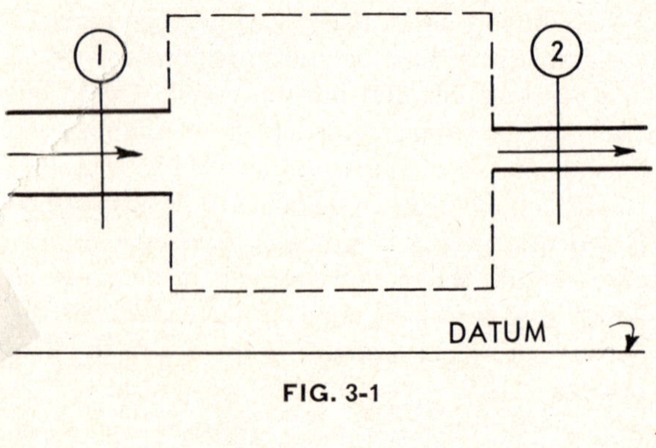

FIG. 3-1

303. CONSERVATION OF ENERGY, BERNOULLI'S EQUATION.

Ignoring thermal energy, the energy contained in a moving fluid may be of two basic types: potential energy and kinetic energy. The potential energy consists of two components, *potential head* and *pressure head.*

A potential head exists whenever a body of fluid is located above a given datum. A pound of water ten feet above a datum plane has the capacity to perform 10 ft/lbs of work in falling to

the datum plane, therefore it contains 10 ft/lbs of potential energy with respect to that datum plane. Since this energy, expressed on a unit weight basis, has the dimensions, ft-lbs/lb, the pound units are conventionally cancelled or ignored, and the energy is expressed in feet of *potential head,* where head refers simply to the height above the established datum plane. Thus a head, Z, of ten feet indicates a potential energy of 10 ft-lbs/lb.

A closed container filled with liquid may be put under pressure by connecting a pipe to the container and filling it with liquid to some level above the top of the container (Figure 3-2). The pressure at section 1 of the pipe is equal to the pressure at the top of the container, else there would be flow. The volume of fluid resting on section 1 is hA, if A is the cross-section area of the pipe. Its weight is whA, and its pressure, or weight per unit area is

$$p = \frac{whA}{A} = wh \text{ lbs/ft}^2. \tag{3}$$

It is common practice to express pressure in pounds per square inch, specific weight in pounds per cubic foot, and h in feet. Noting this, and solving equation (3) for h,

$$h = \frac{144p}{w} \text{ ft}, \tag{4}$$

which defines h as *pressure head.* Like the potential head, this quantity could be expressed in

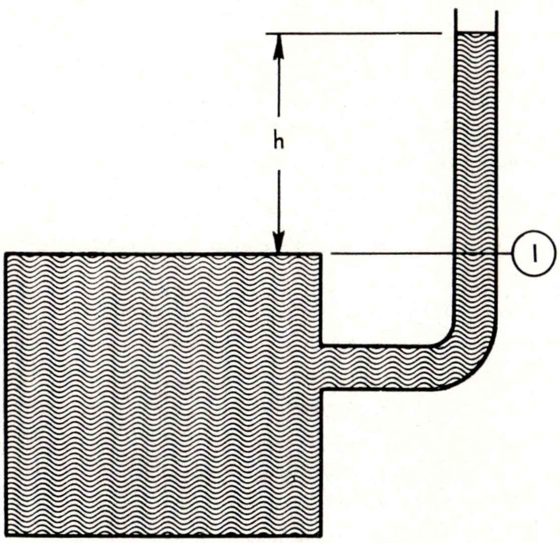

FIG. 3-2

units of ft-lbs/lb. Thus the pressure, no matter how produced, denotes a capacity for creating a potential head, and therefore a capacity for doing work, and it is thus a form of potential energy.

Kinetic energy is a property of a body in motion. The kinetic energy of a weight W, moving at a velocity V is

$$KE = \frac{WV^2}{2g} \text{ ft-lbs}$$

where g is the gravitational constant, expressed conventionally in ft/sec². A weight of one pound of fluid in motion then, has the kinetic energy expressed by

$$KE = \frac{V^2}{2g} \frac{\text{ft-lbs}}{\text{lb}}. \tag{5}$$

It follows that the total mechanical energy of a flowing fluid is the sum of the potential head and the pressure head (which together measure the potential energy) and the kinetic energy.

$$E = Z + \frac{144p}{w} + \frac{V^2}{2g}. \tag{6}$$

If no energy is added to or taken from a fluid passing through a system in steady flow, the law of Conservation of Energy states that the energy of each mass unit of the fluid remains constant. Thus, for any two observable sections in the system, denoted by the subscripts 1 and 2 (Fig. 3-1),

$$Z_1 + \frac{144p_1}{w} + \frac{V_1^2}{2g} = Z_2 + \frac{144p_2}{w} + \frac{V_2^2}{2g}. \tag{7}$$

Equation (7) is applicable to incompressible flow. Therefore the specific weight is the same on both sides and it is not identified by a subscript denoting where it was measured. Equation (7) is the well known *Bernoulli Equation.*

It is important to note the difference between the terms compressible and incompressible *fluid* and compressible and incompressible *flow.* Generally liquids are assumed to be incompressible and all of the above equations apply to them. Gases, on the other hand, are compressible. However, in many instances, processes involving gases fulfill the conditions for incompressible flow. This is true if there is no change in the density, or specific weight, of the gas in the flow process.

304. CONSERVATION OF MOMENTUM—THE MOMENTUM EQUATION.

The momentum relation is derived from the Newtonian relationship

$$F = ma$$

where

F = force

m = mass

a = acceleration.

Since acceleration is time rate of change of velocity, the relationship may be written

$$F = m\frac{dV}{dt} = \frac{m}{\Delta t}(V_2 - V_1). \qquad (8)$$

In equation (8), F is the force required to produce a change of velocity of a mass, m, of fluid from V_1 to V_2 in a time interval Δt. The ratio, $m/\Delta t$, may be interpreted as the *mass rate of flow*, and when converted into the weight rate of flow, the equation takes the form

$$F = \frac{W}{g}(V_2 - V_1) \qquad (9)$$

where

W = weight rate of flow, lbs/sec

g = gravitational constant, ft/sec^2

V_2 = final velocity, ft/sec

V_1 = initial velocity, ft/sec

F = force *on the stream* required to change its velocity, lbs.

The term F in Equation (9) is the force exerted *on the stream*. The force, R, which a stream exerts on any object is equal and opposite to F. Thus

$$R = -F = \frac{W}{g}(V_1 - V_2). \qquad (10)$$

It is important to note that V_1 and V_2 in equations (9) and (10) are *vector* quantities having both magnitude and direction. Their difference is a vector difference, and must be used as such in the equations. Two examples follow to illustrate this.

Example 31: A stream of water strikes a vane as shown in the figure and it is deflected through an angle of 30°. The stream velocity is 50 ft/sec at both entrance and exit and its cross sectional area is 4 in^2. What is the magnitude and direction of the force produced on the vane?

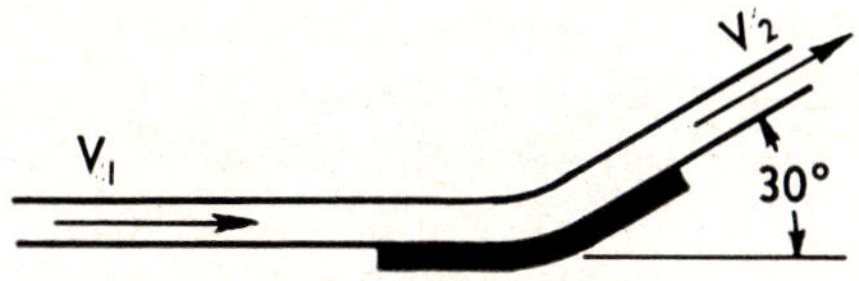

Solution: There is no change in the *magnitude* of the stream velocity. The difference, $V_1 - V_2$, may be found vectorially as indicated in the vector polygon. Solving the polygon either analytically or graphically gives the magnitude of

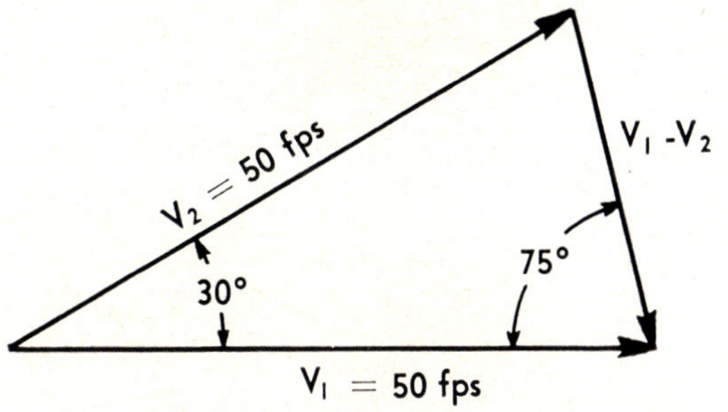

$V_1 - V_2$ as 25.8 fps and its direction as 105° from the original direction of V_1. Therefore

$$R = \frac{W}{g}(V_1 - V_2)$$

$$= \frac{Qw}{g}(V_1 - V_2)$$

$$= 50 \times \frac{4}{144} \times \frac{62.4}{32.2} \times 25.8$$

$$= 69.4 \text{ lbs, in the direction of } \Delta V.$$

Example 32(a): Water flows through a 3″ fire hose at 10 fps, entering the nozzle on the end of the hose at this velocity. The nozzle exit diameter is $1\frac{1}{4}$ in. What is the magnitude and direction of the force on the nozzle due solely to the change in velocity of the fluid?

Solution: Applying the continuity equation to determine the exit velocity,

$$A_1V_1 = A_2V_2$$

$$\frac{\pi}{4}\frac{(3)^2}{144} \times 10 = \frac{\pi}{4}\frac{(1\frac{1}{4})^2}{144}V_2$$

or

$$V_2 = \left(\frac{3}{1.25}\right)^2(10) = 57.6 \text{ fps.}$$

Since there is no change in the direction of the fluid between the entering and leaving section, the resultant force is determined by use of the magnitude difference of $V_1 - V_2$, and is opposite to the direction of V_1, the direction of flow.

$$R = \frac{W}{g}(V_1 - V_2)$$

$$= \frac{Qw}{g}(V_1 - V_2)$$

$$= 10 \times \frac{\pi}{4}\frac{(3)^2}{144} \times \frac{62.4}{32.2} \times (10 - 57.6)$$

$$= \underline{\underline{-45.3}} \text{ lbs (opposite to the direction of flow)}$$

Example 32(b): Whenever a change in fluid velocity occurs, a change in fluid pressure also occurs, according to Bernoulli's equation. The above nozzle discharges to atmospheric pressure, 15 psia. Find the force on the nozzle caused by the pressure change across the nozzle, and find the total force on the nozzle.

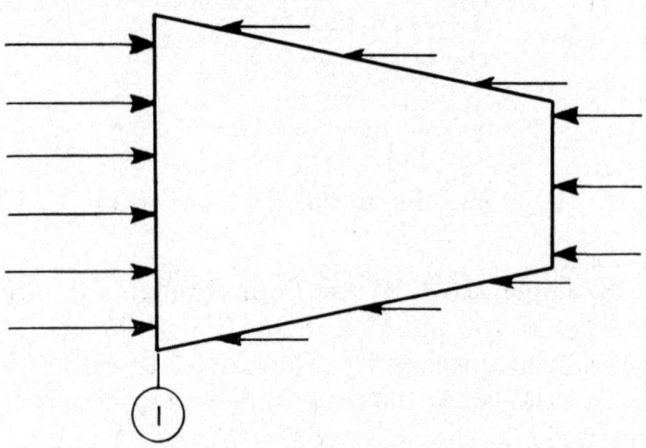

Solution: The fire main pressure acts on the cross section of the nozzle at the entrance. Atmospheric pressure acts on the outside of the nozzle, including the exit area. The net pressure force is equal to the cross section area at 1 multiplied by the difference between the entrance and exit pressures. Since the exit pressure is atmospheric, the entrance pressure can be found by Bernoulli's equation.

$$144\frac{p_1}{w} + \frac{V_1^2}{2g} = 144\frac{p_2}{w} + \frac{V_2^2}{2g}$$

whence

$$p_1 = p_2 + \frac{(V_2^2 - V_1^2)w}{288g}$$

$$= 15 + \frac{(57.6^2 - 10^2)62.4}{288 \times 32.2}$$

$$= 15 + 21.6 = \underline{\underline{36.6}} \text{ psia.}$$

Then the net pressure force is

$$F = A_1(p_1 - p_2) = 21.6(3)^2\frac{\pi}{4}$$

$$= 152.7 \text{ lbs.}$$

The total force on the nozzle is the algebraic sum of the above force and the force found in part (a). Choosing the plus sign in the direction of flow, the total force is

$$F_T = 152.7 - 45.3 = \underline{\underline{107.4}} \text{ lbs}$$

in the direction of flow. This is the force exerted on the threads holding the nozzle on the end of the hose.

305. FLUID FRICTION IN PIPES.

There are two modes of fluid flow. These are termed *laminar* flow and *turbulent* flow. In laminar flow, layers of fluid may be imagined sliding smoothly over adjacent layers, or *lamina*. In turbulent flow, cross currents and eddies are set up, superimposed upon the average motion of the fluid, which destroy the lamina.

In the flow of an incompressible fluid completely filling a pipe, the mode of flow is determined by the velocity of the fluid, the diameter of the pipe and the kinematic viscosity of the fluid. Dimensional analysis, confirmed by experiment, shows that these variables may be grouped to form a dimensionless constant, the *Reynold's Number*, as follows

$$N_R = \frac{VD}{\nu} = \rho\frac{VD}{\mu}$$

where

N_R = Reynolds number
V = velocity, fps
D = pipe diameter, ft
ν = kinematic viscosity, ft²/sec
ρ = mass density = w/g = lb-sec²/ft⁴
μ = dynamic viscosity = lb-sec/ft².

The dynamic viscosity of water varies from 3.74×10^{-5} lb-sec/ft^2 at 32°F to 0.64×10^{-5} lb-sec/ft^2 at 200°F.

Experiment shows that laminar flow of an incompressible fluid filling a pipe occurs at values of N_R less than about 2000. Above this value, turbulent flow begins to develop.

The loss due to friction in incompressible flow may be considered as a loss in pressure head. Equation (7) may be written to include this loss as follows

$$Z_1+144\frac{p_1}{w}+\frac{V_1^2}{2g}=Z_2+144\frac{p_2}{w}+\frac{V_2^2}{2g}+H_L \qquad (11)$$

where H_L has now been included to represent the loss in head due to friction. However, a means for evaluating H_L must be developed.

It is reasonable to assume that the head loss will vary:

(a) Directly as the length of pipe, since a longer pipe has more surface area.

(b) Inversely as the diameter of the pipe, since a pipe with a larger diameter has a smaller ratio of pipe surface to quantity of fluid carried.

(c) Directly with the velocity of flow, or velocity head which is a measure of flow velocity.

In general, then, head loss may be expressed by the following relation

$$H_L=f\frac{l}{D}\cdot\frac{V^2}{2g} \qquad (12)$$

where f is a proportionality constant, l is pipe length in feet, D is diameter in feet, and the other symbols have their previous meaning and units.

The value of f is found to depend on the Reynolds number and the degree of pipe roughness. The latter is expressed by the ratio of the average height of surface projections on the pipe-fluid interface to the diameter of the pipe. In the laminar flow region, f depends on the Reynolds number alone, but for N_R values above 2000, the relative roughness, e/D, also becomes a determining variable. Figure 3-3 is a plot of the relationships which have been determined by careful measurement, principally by Nikuradse. The absolute roughness, or value of e, of piping materials varies from 150×10^{-6} ft for steel pipe to 5×10^{-6} ft for drawn brass tubing. Fouling of salt water mains by marine growth may increase the roughness to the order of .01 ft or higher.

Example 33: A $\frac{5}{8}$ in drawn brass condenser tube 10 ft long carries sea water at 75°F at the rate of 9 fps. Find the head loss due to friction. μ is 1.5×10^{-5} lb-sec/ft^2 at 75°F.

Solution: Check the Reynolds number,

$$N_R=\frac{\rho VD}{\mu}=\frac{wVD}{g\mu}\,.$$

For sea water, the specific weight is 64 lbs/ft^3.

$$N_R=\frac{64\times9\times5/8}{32.2\times1.5\times10^{-5}\times12}=62{,}000$$

The relative roughness is

$$\frac{e}{D}=\frac{5\times10^{-6}\times12}{5/8}=.000096.$$

From Figure 3-3, with e/D as above and N_R equal 62,000 (approximately), f is found to be .021. Then

$$H_L=f\left(\frac{l}{D}\right)\left(\frac{V^2}{2g}\right)$$

$$=.021\left(\frac{10\times12}{5/8}\right)\left(\frac{81}{64.4}\right)=\underline{\underline{5.07\text{ ft.}}}$$

This is equivalent to a pressure drop through the tube of

$$\frac{5.07\times64}{144}=\underline{\underline{2.25\text{ psi.}}}$$

306. FRICTION LOSSES IN VALVES AND FITTINGS.

The passage of fluid through a partially closed valve, a pipe bend, or other type of fitting results in a pressure drop which may be expressed as some fraction of the velocity head, since it varies with the velocity of flow. Thus H_L for these fittings may be expressed as follows

$$H_L=K\frac{V^2}{2g}\,.$$

Values of K for various fittings as compiled by the Crane Company are given on following page.

307. FURTHER RELATIONS.

Other relationships of fluid mechanics will be discussed when required in particular applications. Those relationships having to do with the

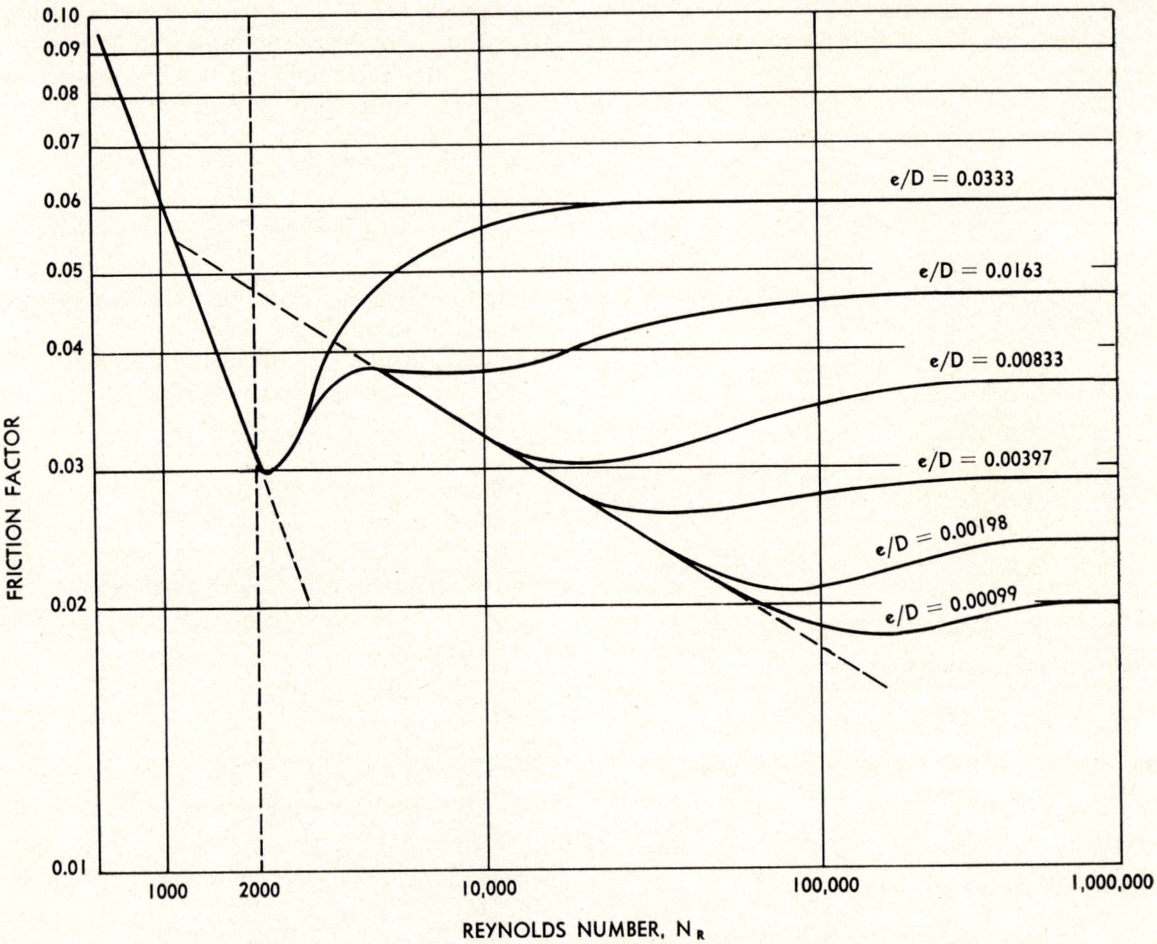

FIG. 3-3. Friction Factor vs Reynolds Number.

conversion of the potential and kinetic energy into useful work are considered as special cases of thermodynamic relationships and are discussed in the following articles.

Table 3-1
Loss Coefficients for Valves and Fittings

Valve or Fitting	K
Globe Valve, open	10
Angle Valve, open	5
Gate Valve, open	0.19
Gate Valve, ¼ closed	1.15
Gate Valve, ½ closed	5.6
Gate Valve, ¾ closed	24.0
U-bend	2.2
Standard tee	1.8
Standard elbow	0.9
Medium sweep elbow	0.75
Long sweep elbow	0.60
45-degree elbow	0.42

308. THERMODYNAMIC CONSIDERATIONS.

The science of thermodynamics is concerned primarily with the conversion of thermal energy into useful work. Certain definitions are necessary in the following discussion. These are:

(a) *Work*—mechanical energy in transition. Work is evidenced by a force moving through a distance. Work, as such, may not be stored, but may be converted into another form of transitory energy, *heat;* or into *kinetic energy;* or into a form of stored mechanical or thermal energy.

(b) *Heat*—thermal energy in transition. For heat to flow, a temperature difference must exist. Heat may not be stored as such, but it may be converted into another form of transitory energy, *work;* or into a form of stored thermal energy.

(c) *Internal Energy*—stored thermal energy. When heat or work is done on a substance, the molecular kinetic energy may be increased, as evidenced by an increase in temperature of the substance, resulting in an increase in internal energy.

Other types of energy have been discussed in Art. 303 and summarized in Equation (6).

309. THERMAL AND MECHANICAL ENERGY.

From Equation (6), and the above definitions, we can include the consideration of thermal energy, and express the energy associated with a unit weight of fluid at any instant as

$$E=\frac{Z}{J}+\frac{p}{wJ}+\frac{V^2}{2gJ}+u. \tag{13}$$

The constant J, Joule's constant, is used to express the relationship between thermal and mechanical energy. J equals 778 ft-lbs/BTU. The term, u, has been added to Bernoulli's equation to account for stored thermal or internal energy. It is common practice in thermodynamics to use specific volume, v (ft^3/lb), instead of specific weight, w (lb/ft^3), in the term involving pressure. This reciprocal will be used hereafter.

310. THE GENERAL ENERGY EQUATION.

When a fluid flows from one point to another, heat may be transferred to or from the fluid and external work may be done on or by the fluid. The energy associated with the fluid at the second point must be the net remaining after the heat and work are transferred. Denoting the heat transferred as Q, and considering it a positive quantity when *added to* the fluid; denoting the external work as w, and considering it as a positive quantity when *done by* the fluid; and considering flow through some system from point 1 to point 2, the following equation can be written:

$$\frac{Z_1}{J}+\frac{p_1v_1}{J}+\frac{V_1^2}{2gJ}+u_1+Q_{12}-w_{12}$$

$$=\frac{Z_2}{J}+\frac{p_2v_2}{J}+\frac{V_2^2}{2gJ}+u_2. \tag{14}$$

Equation (14) is called the *General Energy Equation*. In the absence of chemical, electrical, magnetic, and capillary forces, it expresses the energy relation in a fluid before and after a mechanical and/or thermodynamical process. It holds for any fluid, real or imaginary. The science of thermodynamics is concerned with, among other things, the application of this equation to particular cases, and the evaluation of its terms for these cases. A brief discussion of evaluation of the terms follows.

311. PROPERTIES OF A FLUID.

A *property* of a fluid is defined as any observable characteristic. Three common and well known thermodynamic fluid properties are pressure, temperature and density (or specific volume). Other properties are viscosity, internal energy, enthalpy and entropy. For a pure fluid (a chemically and mechanically homogeneous fluid) any two thermodynamic properties are sufficient to determine all other thermodynamic properties. Thus, it may be said that a pure fluid has a fixed thermodynamic *state* when any two *independent* thermodynamic properties are fixed. Note that work and heat are not properties. Potential and kinetic energy may be thought of as mechanical properties, if desired, but they are completely independent of thermodynamic properties, except as Equation (14) relates them.

Since the term, pv/J, and the term, u, in Equation (14) are both composed of thermodynamic properties, it is often convenient to combine them into another property, *enthalpy*, especially in processes involving flow of the thermodynamic fluid. Thus

$$h=u+\frac{pv}{J}$$

where

h = enthalpy, BTU/lb
u = internal energy, BTU/lb
pv/J = "pressure energy," BTU/lb.

One other fluid property is important in the use of the general energy equation. This property has the name *entropy*. The reader is referred to any standard text on thermodynamics for a complete discussion of its meaning and derivation. Its usefulness in this book lies in the following fact: *when the kinetic energy of the fluid is increased in an ideal reversible process, or when work is done by the fluid in a similar process; either being accomplished at the sole expense of stored fluid energy, the entropy of the working substance remains constant.* On the other hand, a process involving only the addition or rejection of heat to or from the fluid is *always* accompanied by a change in entropy.

This will permit analysis of nozzles and turbines in the subsequent discussion by first considering the thermodynamic processes as isentropic, and then considering the factors which cause these processes to depart from the ideal.

Equation (14) may be rewritten conveniently, by combining the u and pv terms into the single h term, as

$$\frac{Z_1}{J}+h_1+\frac{V_1^2}{2gJ}+Q_{12}-w_{12}=\frac{Z_2}{J}+h_2+\frac{V_2^2}{2gJ}, \quad (15)$$

where all energy terms are expressed in thermal units on a per pound basis (BTU/lb).

312. THE EVALUATION OF HEAT AND WORK.

The evaluation of the quantity of energy added to or abstracted from a fluid in the form of heat is conveniently done if the fluid is maintained at either constant pressure or constant volume during the process. For water in the range of pressures and temperatures found in most portions of a conventional naval power plant, the specific heat may be assumed as 1.0 BTU/lb-°F with little error. At higher pressures and temperatures, appropriate values are available in *Thermodynamic Properties of Steam* by Keenen and Keyes, Figures 4 and 5, or in other data sources. Given the appropriate specific heat, the relation is

$$Q=C(T_2-T_1) \quad (16)$$

where

C = specific heat for condition of process
T_2 = final fluid temperature
T_1 = initial fluid temperature.

Equation (16) must be used with care, since the value of the specific heat may change with temperature, especially in the case of gases, but sufficient accuracy is obtainable if the temperature difference is not too large.

For the flow processes involved in later discussion in this book, the general energy relation as given in equation (15) will ordinarily be sufficient to evaluate the work involved in the process. In this equation it will be necessary to refer to *Thermodynamic Properties of Steam* by Keenan and Keyes, or the Mollier Chart, in order to determine the value of h_1 and h_2. For the student unfamiliar with these tables, Chapter IV of *Thermodynamics* by Johnston, Brockett, and Bock is recommended

CHAPTER IV

THE EXPANSION SYSTEM

400. PURPOSE AND SCOPE.

In this chapter the expansion system will be considered in some detail. It consists of the main steam piping, valves, drains, and the propulsion unit. The latter is made up of two or more turbines, a means for coupling them to the shaft, their lubrication systems, and their steam seal systems. Simple illustrative examples are used to clarify the principles.

PIPING

401. MAIN STEAM PIPING SYSTEMS.

The main steam piping system on a ship is obviously of vital importance to its satisfactory and safe operation. Aside from the boiler itself, it is the only major portion of the propulsion plant which is subjected at all times during operation of the ship to the maximum temperature and pressure of the plant. For this reason the materials, workmanship and design of this system must be the best obtainable.

In addition to the requirement that the design of the pipe permit it to withstand the temperatures and pressures which it must encounter, the system as a whole must be so designed as to afford satisfactory mechanical flexibility. The necessity for this arises from forces encountered under service conditions caused by thermal expansion, shock, vibration, and the working of the ship itself. This piping flexibility must be sufficient to prevent (a) failure of the system by overstressing of the piping materials or the supports, (b) leakage of joints caused by excessive bending moments which tend to part the joint faces and (c) distortion of connected equipment caused by excessive forces and moments transmitted by the piping run.

Flexibility in main steam piping is obtained by use of bends, loops, and offsets in the piping. Where the piping passes through a bulkhead means must be provided to allow it to move axially and radially without distortion of the bulkhead or piping, and without impairing the bulkhead's watertight integrity. This is done by means of a bulkhead expansion joint. One type is shown in Figure 4-1. With this arrangement, it will be noted that motion of either piping or bulkhead will be taken up in the corrugated sleeve without being transmitted to the other. At the same time a watertight joint is provided.

A straight run of piping connecting two units, and under no stress when cold, will tend to expand and force the units farther apart when at its operating temperature. If sufficient rigidity is built into the foundations of the two units so that distortion of their position is negligible under these circumstances, the piping will be subjected to compressive stresses. These stresses, when added to the stresses resulting from the pressure of the fluid within the pipe, increased the total operating stress the design must withstand. To minimize these, it is a common practice to install piping so that it is slightly shorter than necessary at room temperature. Joints are made up by pulling or springing. This places the piping under initial tensile stress which is relieved when the piping warms up to operating temperature. This is known as "cold-spring" or "cold-pull."

402. PIPE SIZE AND MATERIAL.

The size of the main steam piping is determined by considerations of the pressure drop from boiler to turbine for the required weight rate of steam flow at full power, and of the total weight of piping. These considerations conflict, since a smaller pressure drop requires larger pipe size, while less weight dictates smaller pipe size. Therefore, the optimum size is a compromise. It generally turns out to be a size which produces a pressure drop from the boiler to the turbine at full power which is in the order of 10% of the boiler pressure. Thus a plant having a boiler pressure of 1200 psig will have a turbine throttle pressure at full power of about 1080 psig.

Main steam piping in modern high pressure, high temperature plants is of the seamless type. Lap or butt welded pipe is not permitted. The

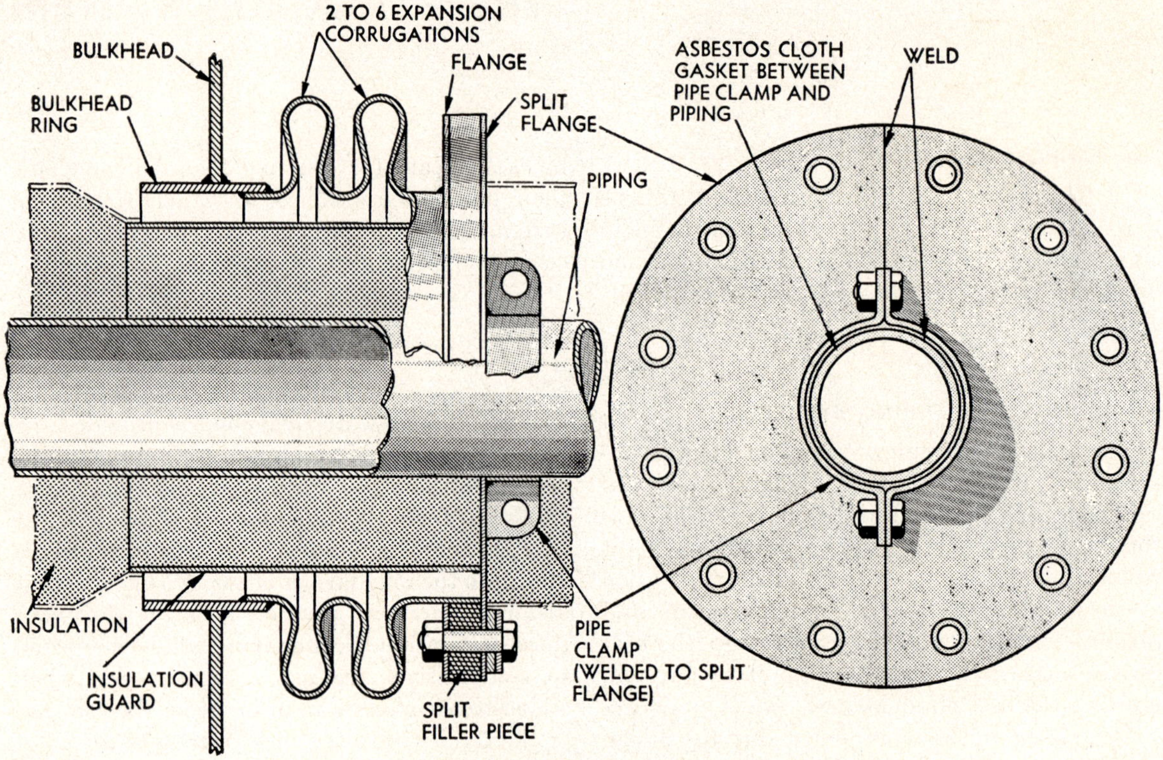

FIG. 4-1. Bulkhead Expansion Joint.

materials are carbon-molybdenum steel, chrome-molybdenum steel or corrosion resisting steel. The metallurgical problem involved is one of "creep"—the plastic deformation of a material at high temperatures under stresses which at ordinary temperatures would be well within the elastic limit. This consideration dictates the specification of maximum allowable fiber stress (the combined stress at any point) for different temperatures. Table 4-1 is an example of such specifications.

The joints are generally welded, except when connected to units which must be capable of being removed for overhaul. Such connections are carefully designed bolted flange fittings. To prevent loss of heat, and to protect personnel, insulation of the asbestos or fiber glass variety forms a covering for the main steam piping system, the thickness being from $3\frac{1}{2}''$ to $4\frac{1}{2}''$ depending on the pipe size and steam temperature.

Table 4-1

Maximum Allowable Fiber Stresses, Psi

Material	Temperature, Degrees F				
	700	800	900	1,000	1,050
Carbon-Moly, Seamless	13,750	13,450	←Not Used→		
Chrome-Moly, Seamless					
1% Chrome; ½% Moly	15,000	14,750	13,100	←Not Used→	
2¼% Chrome; 1% Moly	15,000	15,000	13,100	7,800	5,800
Corrosion Resisting Steel, (CRS) Seamless	17,000	16,750	16,000	14,000	12,200

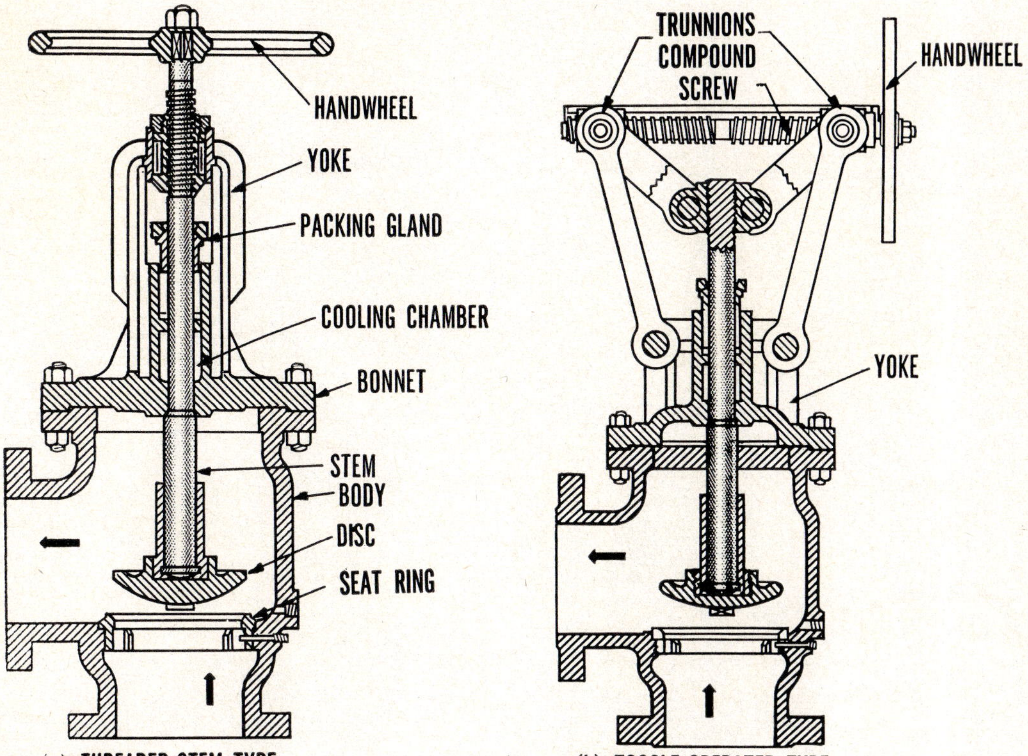

FIG. 4-2. Steam Stop Valves.

403. VALVES.

Two main types of valves are used in main steam piping. These are the globe valve and the gate valve. Many other types are used in various systems, and will be discussed individually with their systems where their importance warrants. Two arrangements of the globe valve are shown in Figure 4-2. Note the cooling chamber, characteristic of high temperature valves. This space around the stem fills with condensate or cooler steam, protecting the packing from direct contact with the superheated steam. Also notice the beveled shoulder on the stem, which seats against the bonnet when the valve is fully open, sealing off steam from leaking around the stem. This feature allows the valve to be packed while under pressure. The seat ring is replaceable for ease of maintenance of the valve. The globe valve has the advantage of simplicity of design and maintenance. However, it has a much greater resistance to flow than does the gate valve, when both are fully opened. In fact, for main steam line sizes, the angle type globe valve shown in Figure 4-2 has as much as ten to fifteen times the resistance to flow as a fully opened gate valve. When the flow is throttled by partial opening of the valve, the globe type loses its disadvantage.

The gate valve is shown in Figure 4-3. Gate valves in main steam lines are invariably used either fully closed or fully opened. The action of high temperature steam around the sharp-edged seating surfaces would cut out grooves across the seats, known as "wire drawing" if the valve were partially closed. Also, impact of steam flowing against the partially open gate can cause chattering, resulting in damage to the seating surfaces. Since the seating surfaces must be exactly plane and parallel, repair of the surfaces is somewhat more difficult than is the case for the globe valve. Both the body and disc seat rings are replaceable.

404. TRAPS AND DRAINS.

When the valves in a main steam line are secured after operation, the steam remaining in the line cools and condenses. On resuming operation, a portion of the steam that initially flows into the line also condenses as it strikes the cold pipe surfaces. As steam moves into the line, slugs of this

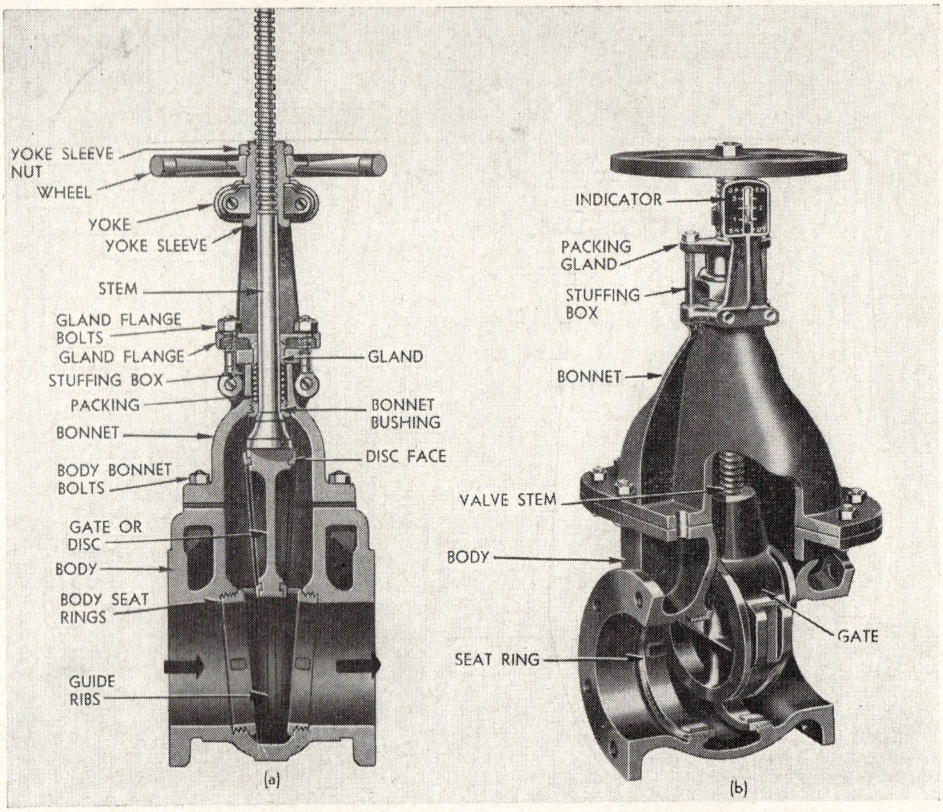

FIG. 4-3. (a) Gate Valve, Rising Stem, Outside Screw Type. (b) Gate Valve, Non-rising Stem, Inside Screw Type.

water are carried along with great force and can result in damage to the line and connected equipment. Therefore small drain lines to collect water are connected to low points and pockets in the main line and its valves. These drain lines lead to "steam traps," devices which open to discharge liquid, but close to prevent the passage of the steam. Since this liquid, saturated or nearly so at steam line pressures, may flash into steam when released by the trap, the drains are piped to the deaerating feed tank, where the thermal energy is utilized to assist in heating the boiler feed.

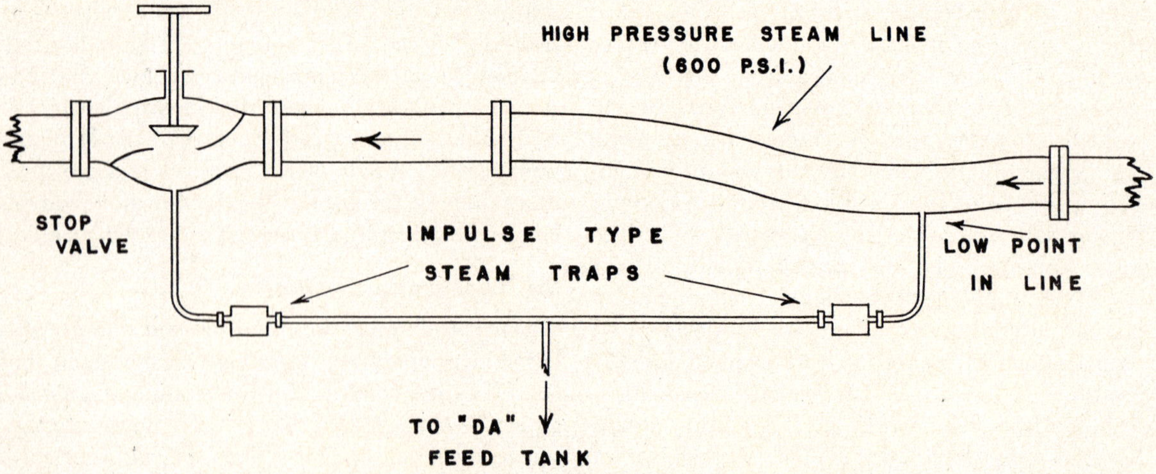

FIG. 4-4. Main Steam Piping Drains.

Small by-pass lines with valves are provided around the main valves to permit warming up the whole system without opening the main valves. The above discussion is illustrated in Figure 4-4. The main steam line must be warmed up slowly and evenly to prevent rapid temperature changes which may cause uneven expansion resulting in leakage or overstressing.

In addition to valves, drains and traps, strainers are incorporated in the main steam system to prevent particles of scale or other foreign solid matter from being carried into the turbine where expensive damage may result.

TURBINES

405. TURBINE PRINCIPLES.

The steam turbine is a device for the conversion of thermal energy in the steam to useful work. As seen before, it cannot convert *all* of the thermal energy, since the exhaust from the turbine must always contain thermal energy not available for conversion. The steam turbine accomplishes the conversion in two steps: (1) the conversion of thermal energy into kinetic energy, and (2) the conversion of kinetic energy into mechanical energy. These processes do not necessarily occur separately, but to give a clearer understanding of them, they will be discussed separately.

406. CONVERSION OF THERMAL ENERGY TO KINETIC ENERGY—THE IDEAL NOZZLE.

Since the nozzle is a device through which a fluid flows, the continuity equation must be satisfied. This equation simply states that the mass passing any section of the device in a given time is equal to the mass passing any other section in the same time. If w = weight rate of flow in lb per sec then

$$w_0 = w_1 = w_2 = w_3, \qquad (1)$$

where the subscripts denote arbitrary sections of the nozzle. But $w = AV/v$, where

A = cross section of channel, sq ft
V = fluid velocity, ft per sec
v = specific volume of fluid, cu ft per lb

Then, from Equation (1),

$$\frac{A_0V_0}{v_0} = \frac{A_1V_1}{v_1} = \frac{A_2V_2}{v_2}, \text{ etc.} \qquad (2)$$

Also, the law of conservation of energy applies in the form of the First Law of Thermodynamics. This is expressed by the general energy equation:

Entering Fluid

Potential Energy + Kinetic Energy + Thermal Energy + Heat added from 0 to 1

Leaving Fluid

= Potential Energy + Kinetic Energy + Thermal Energy + External work done from 0 to 1

$$\frac{z_0}{J} + \frac{V_0^2}{2gJ} + h_0 + q_{01} = \frac{z_1}{J} + \frac{V_1^2}{2gJ} + h_1 + \frac{wk_{01}}{J} \qquad (3)$$

where

z = height above datum, ft
J = mechanical equivalent of heat, 778 ft lb per BTU
V = velocity of steam, ft per sec
h = enthalpy of steam, BTU/lb
q_{01} = heat added in process, BTU/lb
wk_{01}/J = work done in process, BTU/lb.

Equation (3) then, expresses the equivalence of the total energy entering the nozzle for each pound of steam to the total energy leaving. In the nozzle, no work is done on or by the working substance; heat transferred from the nozzle is negligible; no heat is added; and due to the short length of real nozzles, the difference in elevation of entrance and exit may be neglected, regardless of position. Therefore Equation (3) reduces to:

$$\frac{V_0^2}{2gJ} + h_0 = \frac{V_1^2}{2gJ} + h_1$$

or

$$V_1^2 - V_0^2 = 2gJ(h_0 - h_1). \qquad (4)$$

As a matter of actual practice in a steam turbine, V_0 is so small compared to V_1 that it may be neglected, and Equation (4) further reduces to

$$V_1 = \sqrt{2gJ(h_0 - h_1)}. \qquad (5)$$

407. THE REAL NOZZLE.

Equation (5) is the relation for the process in an ideal nozzle. This is an isentropic process. In a real nozzle, there are always some friction

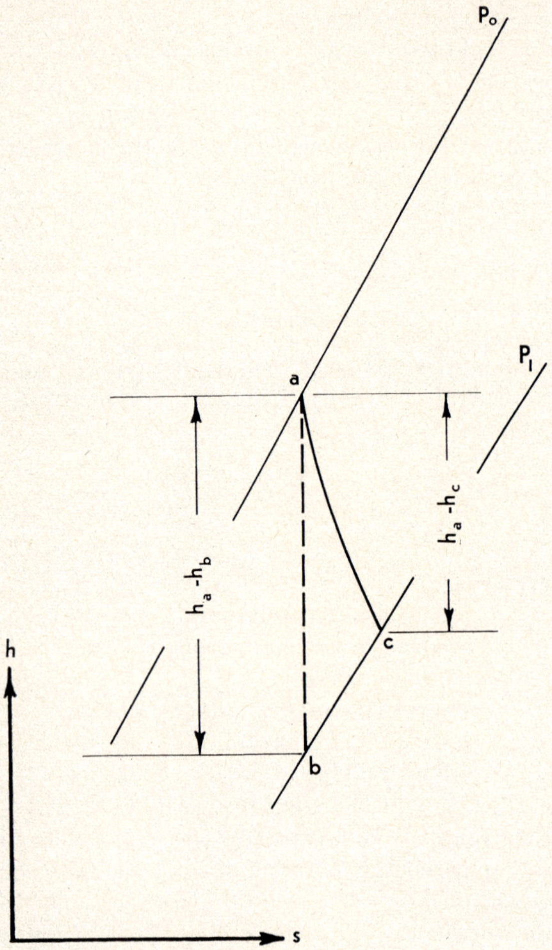

FIG. 4-5. h-s Diagram for Real and Ideal Nozzle.

losses, though they may be kept to a minimum by good design and careful construction. The friction of the steam on the nozzle walls results in reconversion of some of the velocity gain back into thermal energy, so that the ideal drop in enthalpy is not realized in the actual nozzle. Since the change in enthalpy is equivalent to the change in kinetic energy, somewhat less kinetic energy is obtained from a real nozzle than is indicated by Equation (5) for the ideal nozzle. If h_c is the exit enthalpy (see Figure 4-5) from the real nozzle, then the efficiency of the nozzle may be defined as:

$$\eta_n = \frac{\text{actual kinetic energy obtained}}{\text{kinetic energy from an ideal nozzle}}$$

$$= \frac{h_a - h_c}{h_a - h_b}. \qquad (6)$$

Since in steady flow the outlet pressure is fixed by the surroundings of the nozzle exit, this pressure is the same for both the ideal and real nozzles. Figure 4-5 shows the process lines on the *h-s* coordinates.

Point a is established by the entrance pressure and temperature, point b by assuming an isentropic enthalpy drop to the exit pressure, and point c by multiplying $h_a - h_b$ by the efficiency, solving for h_c, and plotting on the exit pressure line. From Equations (5) and (6) it is seen that the velocity can be obtained for the actual nozzle by including the efficiency in Equation (5) as follows:

$$V_{(\text{actual})}s = \sqrt{2gJ\eta_n(h_a - h_b)}$$

or

$$V_{(\text{actual})}s = \sqrt{\eta_n}\sqrt{2gJ(h_a - h_b)} = C_N\sqrt{2gJ(h_a - h_b)}. \qquad (7)$$

where $C_N = \sqrt{\eta_n}$ is defined as the velocity coefficient of the nozzle. For a well designed nozzle C_N may be taken as 0.97 or 0.98. Equation (7) is valid for any nozzle, within the limits of accuracy of C_N, which has only slight variations for well designed nozzles operating under designed conditions.

408. NOZZLE SHAPE.

The shape of the nozzle is governed to an extent by the ratio of the outlet pressure, P_1, to the inlet pressure, P_0. A *critical pressure ratio* exists for all gases which is a function of their specific heats. For dry steam the critical pressure ratio may be taken as 0.58. When P_1/P_0 is greater than 0.58, a converging nozzle is used. See Figure 4-6. For this type nozzle, if P_0 is held constant

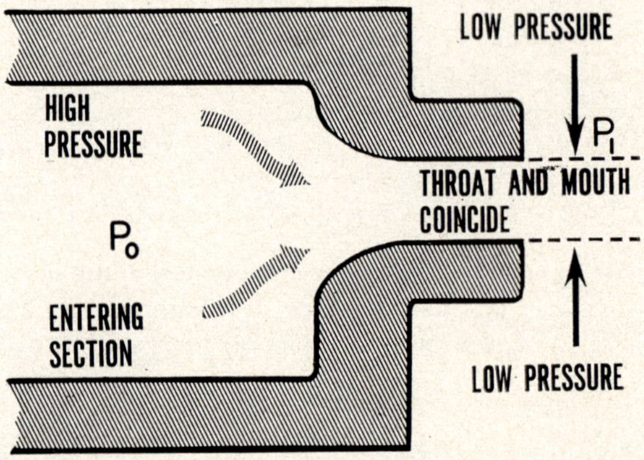

FIG. 4-6. Converging Nozzle.

and P_1 is set at various values from P_0 to 0.58 P_0, it will be found that the exit velocity will increase as the values of P_1 decrease. However, when P_1 drops to 0.58 P_0, it is found that exit velocity reaches the local velocity of sound, and if P_1 is lowered further, no increase in the exit velocity results. This velocity occurs at the smallest cross section of the nozzle, or the *throat*. From Equation (2) the weight rate of flow is

$$W=\frac{A_T V_T}{v_T} \qquad (8)$$

where

A_T = area of the throat, ft²
V_T = velocity at throat (here = V_1) fps
v_T = specific volume at throat (here = v_1)ft³/lb.

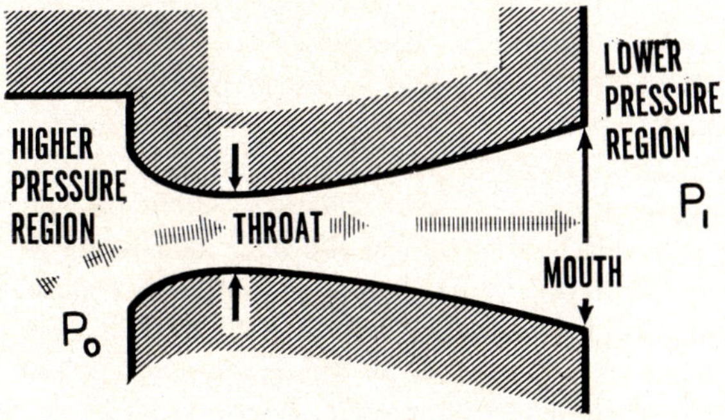

FIG. 4-7. Convergent-Divergent Nozzle.

Consequently, it is seen that the convergent nozzle produces both the maximum exit velocity and the maximum weight rate of flow at the critical pressure ratio and that further lowering of the exit pressure does not alter this flow.

By adding a divergent portion to the nozzle when $P_1/P_0<0.58$, further expansion of the steam can be accomplished with a further gain in velocity, and therefore an increase in kinetic energy is obtainable. A convergent-divergent nozzle is shown in Figure 4-7. In this nozzle when P_1/P_0 is less than 0.58, the pressure at the throat is always 0.58 P_0 regardless of the value of P_1. Therefore, the weight rate of flow is limited by Equation (8), and cannot increase beyond this. However, if the divergent section is properly designed, the velocity continues to increase as the pressure decreases from P_T at the throat to P_1 at the exit. The exit velocity is still determined by Equation (7) as before.

The following example will serve to illustrate the above discussion.

Example 41: A convergent-divergent nozzle expands steam from 350 psia and 700°F to 80 psia. $C_N=.967$. The diameter of the throat is 1 inch. Find:

(a) Outlet velocity
(b) Throat velocity
(c) The weight rate of flow, lbs/sec
(d) The kinetic energy per pound of steam.

Solution:

(a) $h_0=1365.5$ BTU/lb (from steam tables, for $p_0=350$ psia and $T_0=700$)

$h_1=1211.5$ BTU/lb (from steam tables isentropic drop from h_0)

$$V_1=0.967\sqrt{2gJ(h_0-h_1)}$$
$$=0.967\times 223.8\sqrt{1365.5-1211.5}$$
$$=\underline{\underline{2685\text{ fps}}}$$

(b) $p_t=0.58\ p_0=0.58\times 350=203$ psia

$h_0=1365.5$ BTU/lb

$h_t=1292.0$ BTU/lb (from steam tables isentropic drop from h_0)

$$V_t=0.967\times 223.8\sqrt{1365.5-1292.0}$$
$$=\underline{\underline{1856\text{ fps}}}$$

(c) v_t = specific volume corresponding to h_t
$=2.829$ ft³/lb (from steam tables)

$$A_t=\frac{\pi d^2}{4\times 144}=\frac{\pi}{4\times 144}=.00545\text{ ft}^2$$

$$w=\frac{A_t V_t}{v_t}=\frac{0.00545\times 1856}{2.829}$$
$$=\underline{\underline{3.56\text{ lbs/sec}}}$$

(d) $$\text{K.E.}=\frac{V^2}{2gJ}$$

$$=\frac{(2685)^2}{2\times 32.2\times 778}=\underline{\underline{144.0\text{ Btu/lb}}}$$

Check:

$$\text{K.E.}=C_N{}^2(h_0-h_1)$$
$$=(0.967)^2(1365.5-1211.5)$$
$$=\underline{\underline{144.0\text{ Btu/lb}}}$$

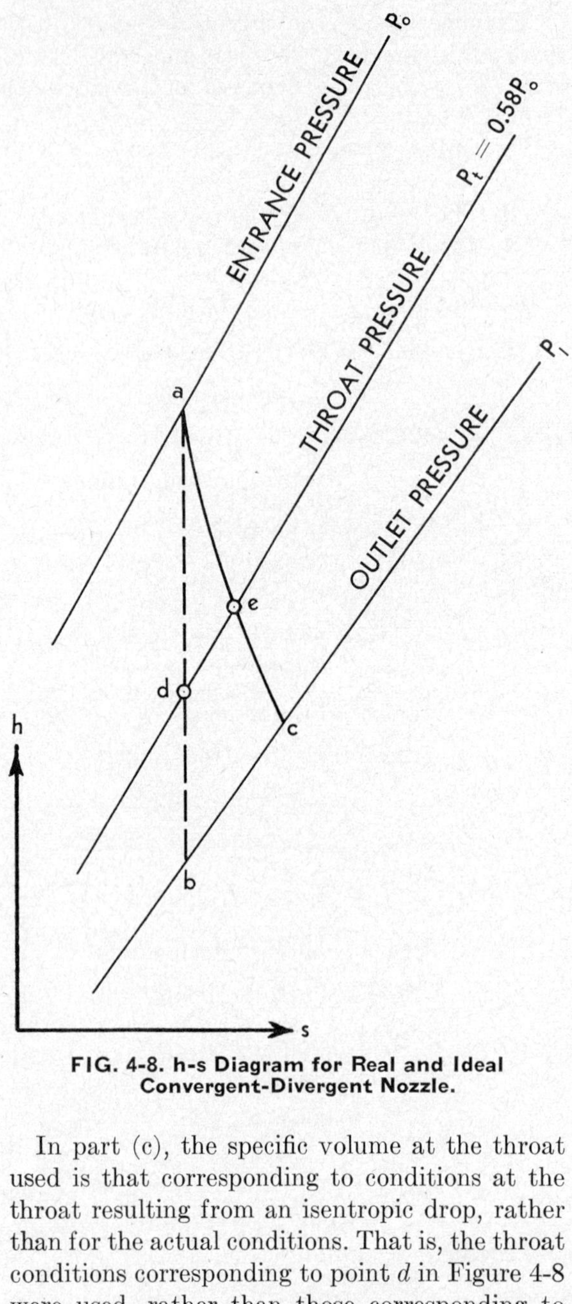

FIG. 4-8. h-s Diagram for Real and Ideal Convergent-Divergent Nozzle.

In part (c), the specific volume at the throat used is that corresponding to conditions at the throat resulting from an isentropic drop, rather than for the actual conditions. That is, the throat conditions corresponding to point *d* in Figure 4-8 were used, rather than those corresponding to point *e*. This simplifies calculations and causes negligible error.

409. APPLICATIONS OF NOZZLES IN TURBINES.

Steam supplied to turbines flows from the line through the controlling valves into a *steam chest.* It passes from the steam chest into groups of nozzles assembled in a circular array to direct it into the first row of moving blades or buckets. This first group of nozzles, in a ring, or block, may be either convergent, or convergent-divergent, depending on the designed pressure drop. For convergent-divergent nozzles, the blocks or rings are generally made from solid blocks, drilled and reamed to the required shape. For converging nozzles, the vane type construction is generally preferred. This arrangement consists of successive vanes or blades spaced around the periphery, shaped so that the space between them gives the desired nozzle cross section. A vane type block may be either built up from individual parts or cast as a unit. Figure 4-9 indicates the construction of these two types.

In impulse turbines, after leaving the first wheel, the steam passes through several more combinations of nozzles and wheel buckets before leaving the turbine. These nozzles are of the vane type, mounted in diaphragms as shown in Figure 4-10. The buckets are for the purpose of changing the direction of the steam only and usually do not function as nozzles. The impulse turbine and its details will be described in the following pages.

In reaction turbines, the moving blading *and* the intervening fixed rows of blading are shaped so as to form convergent nozzles. The fixed blades are attached directly to the casing, and the moving blades are mounted on a single rotor as shown in Figure 4-11.

410. CONVERSION OF KINETIC ENERGY TO MECHANICAL ENERGY.

The foregoing discussion has described how the nozzle is used to convert the thermal energy of the working fluid into kinetic energy. It remains to describe how this kinetic energy in a stream of fluid can be converted into mechanical energy, and with what efficiency this conversion is possible. In order to understand this process, it is necessary to recall the principles of fluid momentum. Starting from Newton's Second Law

$$F = ma.$$

but

$$a = \frac{dv}{dt}.$$

Therefore

$$F = m\frac{dv}{dt}$$

or

$$F\,dt = m\,dv. \tag{9}$$

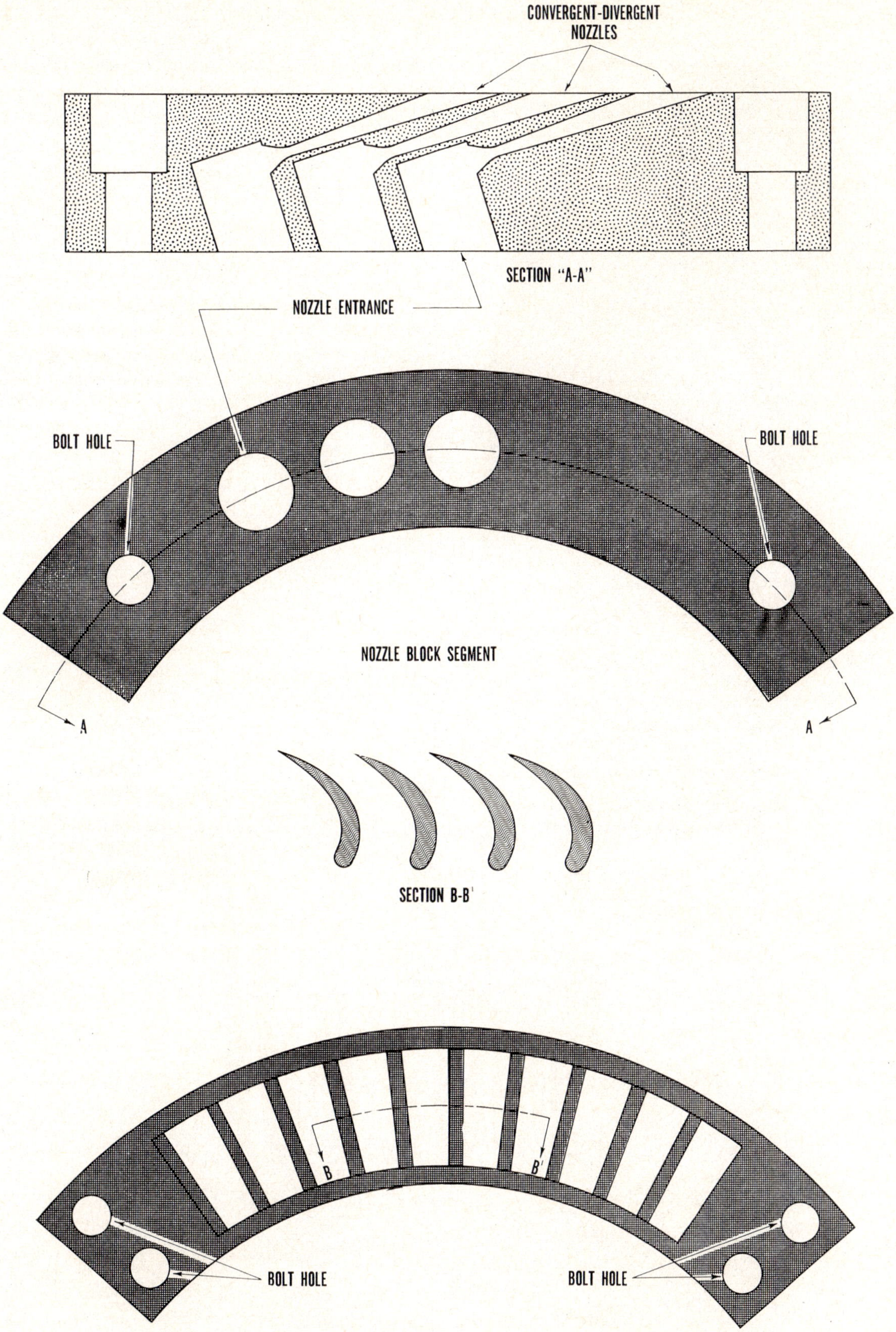

FIG. 4-9. (a) Single Piece Construction of Convergent-Divergent Nozzle Block. (b) Vane Type Construction of Converging Nozzle Block.

FIG. 4-10. Impulse Type Nozzle Diaphragm.

Assuming a constant average value of F over an interval of time Δt, and integrating Equation (9):

$$F\Delta t = m(V_2 - V_1).$$

or

$$F = \frac{m}{\Delta t}(V_2 - V_1).$$

In fluid flow, the term $m/\Delta t$ is the mass rate of flow. But $m = W/g$, where W is weight. Therefore mass rate of flow can also be expressed as $W/g\Delta t$. For convenience let $W/\Delta t = w$, where w is now the weight rate of flow. Then

$$F = \frac{w}{g}(V_2 - V_1) \qquad (10)$$

where

F = lbs of force

w = weight rate of flow, lbs/sec

V_1 = initial velocity, ft/sec

V_2 = final velocity, ft/sec.

Equation (10) shows that a force F is required to change the velocity of a stream of fluid flowing at the rate w from V_1 to V_2. Recalling that ve-

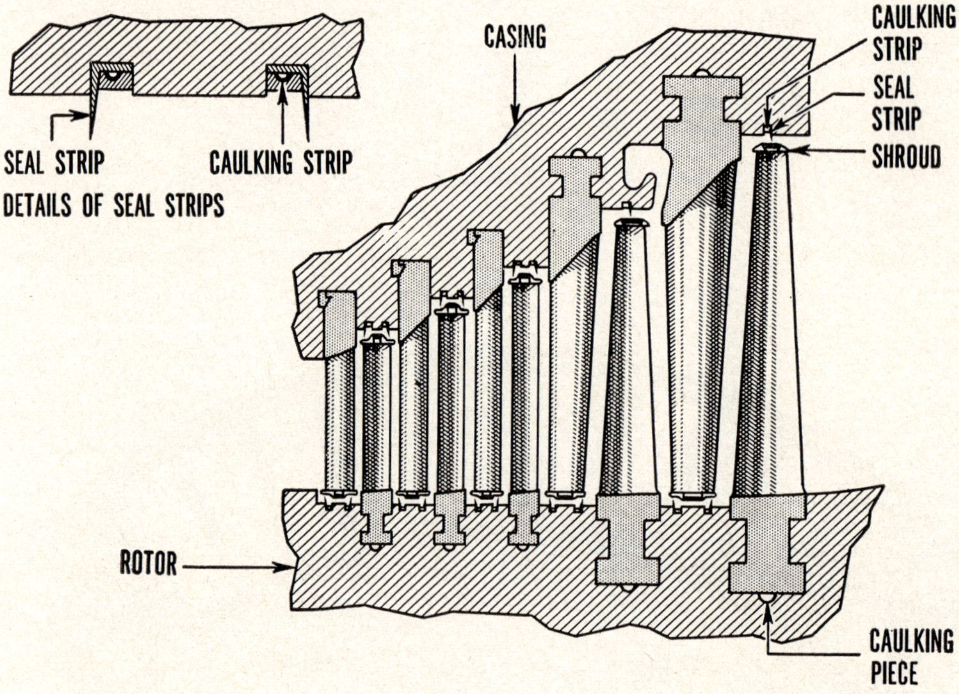

FIG. 4-11. Typical Reaction Blading.

locity is a vector quantity having direction as well as magnitude, *the change in velocity can be brought about by changing the direction of flow of the fluid, or by changing the magnitude of its velocity, or both.*

The arrangement for producing kinetic energy from thermal energy, and then producing a force (and work) from the kinetic energy is called a turbine *stage*. Stages in all turbine types of interest to us always include at least one fixed nozzle (or more arranged in parallel) to convert thermal energy to kinetic energy. This nozzle, or nozzle row, is followed by at least one row of moving blading, and in some cases, by additional rows of fixed and moving blading, as will be discussed later. As long as such fixed blading does not contain nozzle shaped passages, there is no significant pressure change across it, and it is a part of the same stage as its associated moving rows. A turbine stage is defined, therefore, as *a fixed nozzle or nozzle row plus all of the moving and fixed blading which follows it prior to the next fixed nozzle row.*

There are two major classifications of turbines stages which will be discussed below. These are based on the mechanics of how the kinetic energy of the steam is converted into work on the turbine blading. They are the *impulse* stage and the *reaction* stage.

411. THE ZERO ANGLE IMPULSE STAGE.

We shall first consider the ideal case where only a change in direction of the stream occurs. This ideal system is made up of a nozzle and bucket such as is shown in Figure 4-12. The stream leaves the nozzle with velocity V_1, strikes the blade and has a velocity of V_2 after leaving the blade. Then, from Equation (10), the force on the stream is $w(V_2-V_1)/g$. This force is exerted by the moving blade, and therefore, the stream exerts an equal and opposite force, R, on the blade. It is this continuous force acting on the blade which causes it to move, and to produce work and power. Its expression is

$$R=-F=\frac{w}{g}(V_1-V_2). \qquad (11)$$

When the blade moves with a velocity, V_b, the power is equal to RV_b.

Note that if V_b equals V_1, the stream cannot overtake the blade and therefore it is not deflected. Also, if the blade is stationary, V_b is zero, and the power is therefore zero. The blade velocity for power production, then, lies between 0 and V_1.

To an observer standing on the blade, the stream strikes the blade with a relative velocity V_1-V_b. The blade reverses the direction of the stream relative to it; but, if we neglect friction, it does not change the magnitude of the relative velocity. *It is this fact which characterizes the blade as an impulse blade.* Therefore, the velocity of the stream leaving the blade (still viewed from the observer on the blade) is $-(V_1-V_b)$. But the absolute leaving velocity is the sum of the relative leaving velocity and the velocity of the blade, or

$$V_2=-(V_1-V_b)+V_b=-V_1+2V_b. \qquad (12)$$

The maximum quantity of work done on the blade in Figure 4-12 occurs when all of the kinetic

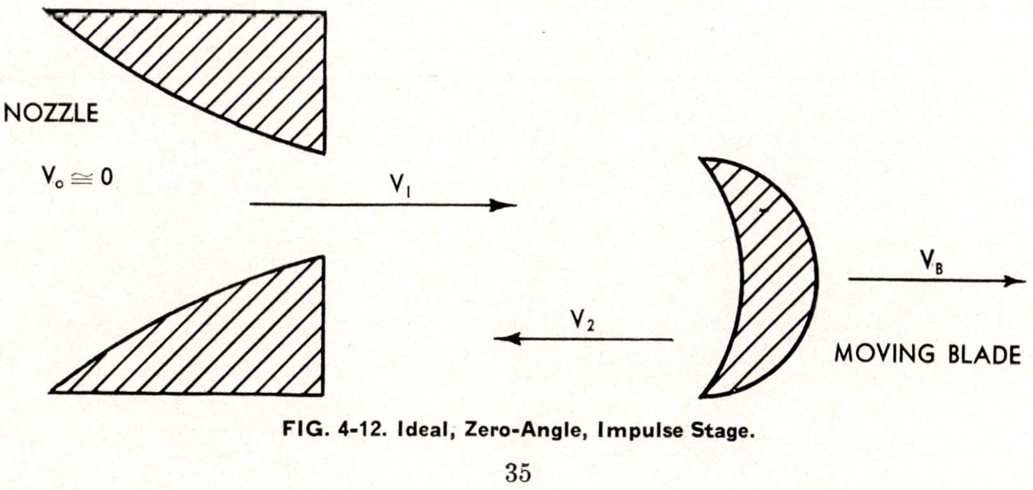

FIG. 4-12. Ideal, Zero-Angle, Impulse Stage.

energy in the stream is converted to work, that is, when the kinetic energy of the stream leaving the blade is zero. This implies that the absolute velocity of the stream leaving the blade must be zero, which occurs only when $V_2=0$.

From Equation (12), when $V_2=0$,

$$V_1=2V_b$$

or

$$V_b=\tfrac{1}{2}V_1. \qquad (12a)$$

Therefore, in the zero-angle impulse stage, the maximum work occurs when the blade speed is one-half the stream speed. In practice this type turbine is approached by the tangential flow turbine, used to power auxiliary machinery. It will be discussed in more detail later.

412. FINITE ANGLE IMPULSE STAGE—VELOCITY ANALYSIS.

Most impulse turbines have a steam entrance arrangement similar to that shown in Figure 4-13. The velocity relationships are readily studied by the velocity vector diagram shown in Figure 4-14. In the figure, α is the nozzle angle and V_1 is the velocity of the jet leaving the nozzle. The blade velocity, or wheel speed, is V_b. V_{r1} is the relative velocity of the steam entering the bucket and is

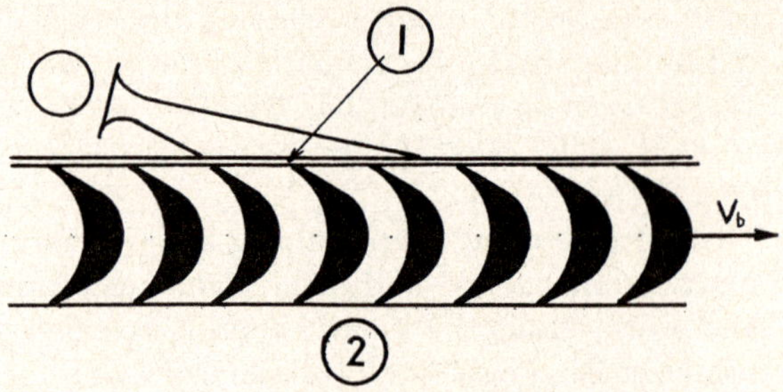

FIG. 4-13. Simple Impulse Stage.

obtained by the vectoral subtraction V_1-V_b. The angle β_1 is the steam inlet angle, and it is seen that it is also fixed by the above vectoral subtraction. The steam is deflected by the bucket through an angle $(\beta_1+\beta_2)$. β_2 is the steam exit angle and is fixed by the bucket design. In the impulse turbine, if the bucket were frictionless, the relative velocity at the bucket exit would be identical in magnitude to the relative entrance velocity, V_{r1}. The effect of friction is expressed by the following relation

$$V_{r2}=C_bV_{r1} \qquad (13)$$

where the constant, C_b, is the bucket velocity coefficient. This coefficient is somewhat smaller and more variable than the similar coefficient C_N discussed previously, but a fair average value ap-

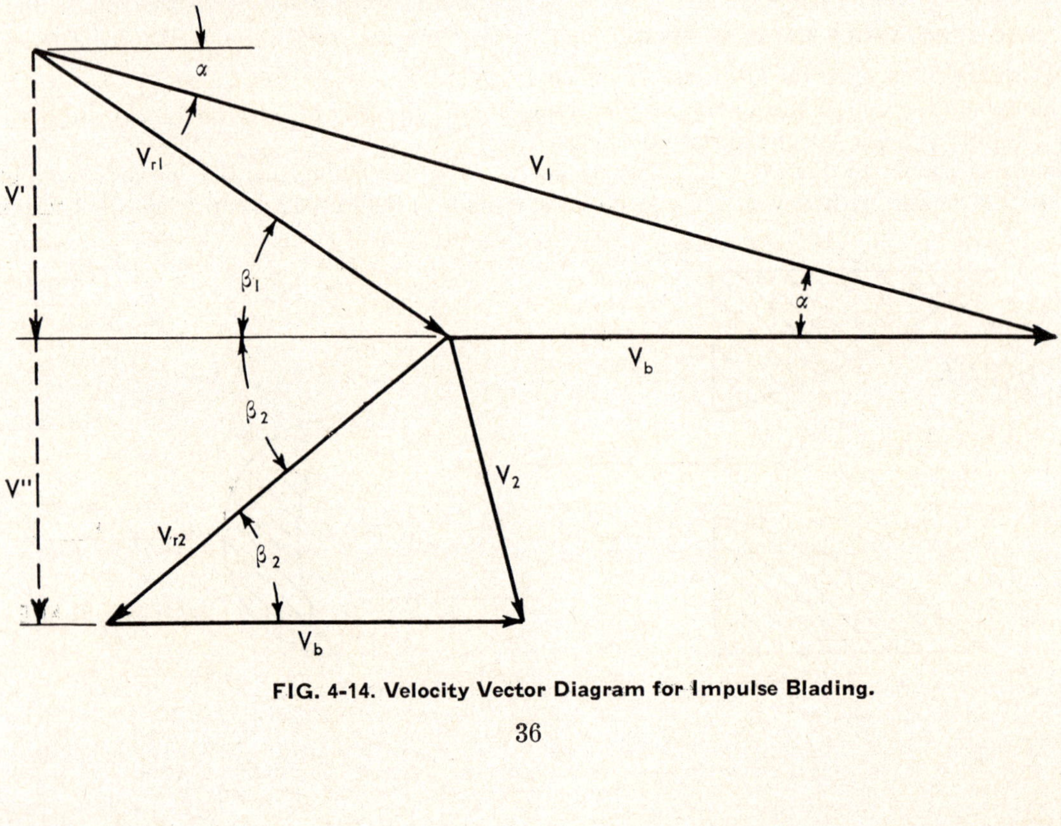

FIG. 4-14. Velocity Vector Diagram for Impulse Blading.

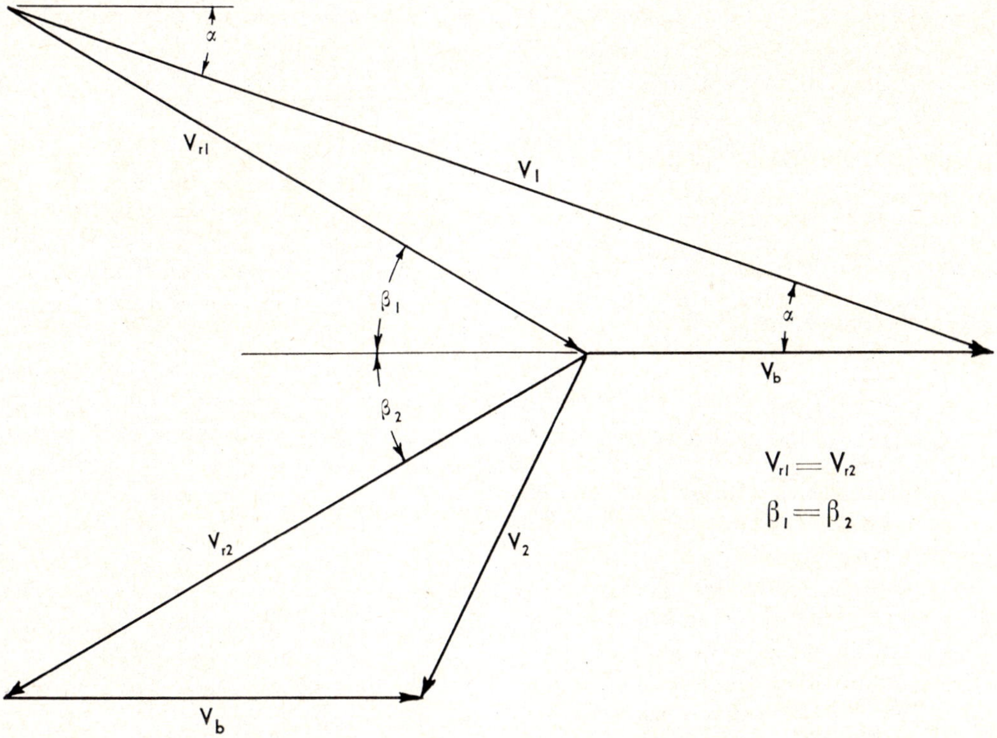

FIG. 4-15. Velocity Vector Diagram for Frictionless Impulse Stage.

pears to be in the range of 0.8 to 0.9. Having obtained V_{r2} from a knowledge of the angle β_2 and Equation (13), the absolute exit velocity of the steam, V_2, can be found by adding the velocity of the steam relative to the blade (V_{r2}) to the velocity of the blade V_b.

The dotted vectors, V' and V'', represent the velocity of axial flow through the turbine. These two axial flow components are practically equal in a pure impulse stage, that is, a stage in which there is no pressure drop across the moving blades. If the stage were a frictionless impulse stage, there would be no friction loss in the buckets, and V_{r2} would equal V_{r1}. Also, V' would exactly equal V'', and β_1 would be identical with β_2. Under these conditions a typical diagram would be as shown in Figure 4-15.

The exit velocity of the steam, V_2, represents a loss to the stage, since it represents kinetic energy not utilized in the stage. From Figure 4-15, it can be seen that this loss is a m nimum when the vector representing V_2 is perpendicular to the vector V_b. It is also readily seen that this occurs when

$$2V_b = V_1 \cos \alpha$$

or

$$\frac{V_b}{V_1} = \frac{1}{2} \cos \alpha. \qquad (14)$$

Note the similarity of this expression with Equation (12a). Cos α is .98+, since α is seldom greater than 15°. Although the derivation of a similar expression for the impulse stage with friction is complex, it turns out that Equation (14) holds very nearly true for the actual stage, i.e., the leaving loss is a minimum when the blade speed is approximately one-half the steam speed. V_b/V_1 is known as the velocity ratio. Whatever the direction of V_2 (which deviates from the perpendicular when V_b/V_1 deviates appreciably from $\frac{1}{2}$), its perpendicular component represents the velocity of flow of the steam through the turbine at any particular stage, and therefore is necessary to obtain continuity of flow.

413. IMPULSE STAGE WORK.

Applying an energy balance to the flow entering and leaving the buckets, an expression for work may be found. The total energy in the steam entering the blades is its enthalpy plus its entering kinetic energy. Leaving the stage, the energy is the sum of the enthalpy plus the leaving kinetic energy. Also, work is done on the buckets. Thus, on a per pound of steam flow basis

$$h_1+\frac{V_1^2}{2gJ}=h_2+\frac{V_2^2}{2gJ}+W_b$$

where W_b is the bucket work in BTU/lb. Then

$$W_b=(h_1-h_2)+\frac{V_1^2-V_2^2}{2gJ}. \qquad (15)$$

Another expression relating initial and final enthalpy to the relative velocities may be found by considering the flow from the point of view of an observer riding on the blades. He sees steam approaching with an enthalpy h_1 and a velocity V_{r1}, and leaving with an enthalpy h_2 and a velocity V_{r2}. From his point of view, there is no work done since the blades are stationary with respect to him. He would write the energy balance as follows:

$$h_1+\frac{V_{r1}^2}{2gJ}=h_2+\frac{V_{r2}^2}{2gJ}. \qquad (16)$$

Eliminating h_1 and h_2 from Equations (15) and (16) gives a useful expression for bucket work, or stage work, in terms of velocities only.

$$W_b=\frac{V_1^2-V_{r1}^2+V_{r2}^2-V_2^2}{2g}. \qquad (17)$$

Note that the J is now omitted, giving the results in ft-lb/lb of steam flow.

Equations (15) and (17) apply equally well to the real stage or to the frictionless stage. The available kinetic energy to the buckets is $V_1^2/2g$. Dividing the bucket work by this quantity gives the bucket efficiency, commonly called *diagram efficiency*.

$$\eta_b=\frac{V_1^2-V_{r1}^2+V_{r2}^2-V_2^2}{V_1^2}. \qquad (18)$$

Note in the numerator of Equation (17) that $V_1^2/2g$ is diminished by the leaving loss, $V_2^2/2g$ and by the loss due to bucket friction, $(V_{r2}^2-V_{r1}^2)/2g$. Therefore Equations (17) and (18) apply to real stages with no further allowance necessary for bucket friction loss or leaving loss.

414. IMPULSE STAGE LOSSES.

The shaft work of a given stage is less than the bucket work produced by the stage because of additional internal losses. The principal ones are:

(a) *Windage loss*—The disc carrying the buckets is rotating in an atmosphere of steam. Friction of the steam on the non-working portion of the disc sets up turbulence which consumes power.

(b) *Suction loss*—In many impulse stages the nozzles do not completely fill the entire wheel periphery. Such stages are known as *partial admission* stages. These occur in high pressure stages where full admission would dictate impractically short buckets, due to relatively small required flow area. When partial admission is used, a pumping action takes place in the buckets during the inactive portion of the arc, tending to pump steam from the upstream side of the disc to the exit side. This loss is proportional to the size of the inactive arc.

(c) *Nozzle end loss*—In a partial admission stage, when the bucket reaches the beginning of the admission arc, the relatively slow-moving steam therein must be accelerated again by the nozzle jet. This loss may amount to about 0.5% of the stage work.

(d) *Diaphragm packing loss*—It is possible for a small amount of steam to leak from one stage to the next without passing through the nozzle. This steam passes through the diaphragm packing which will be described later. This portion of the steam does no work.

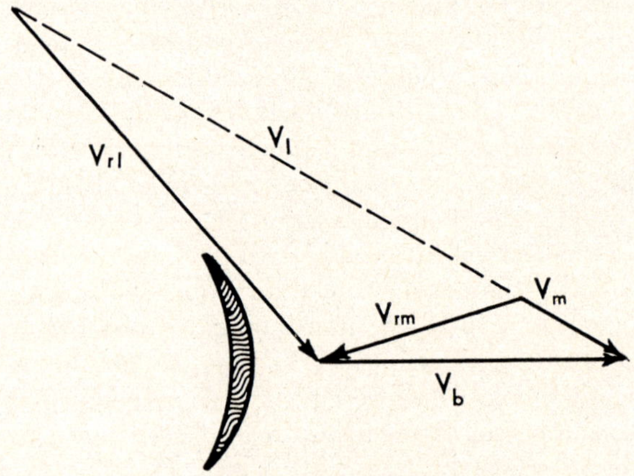

FIG. 4-16. Velocity Vector Diagram Showing Moisture Effect.

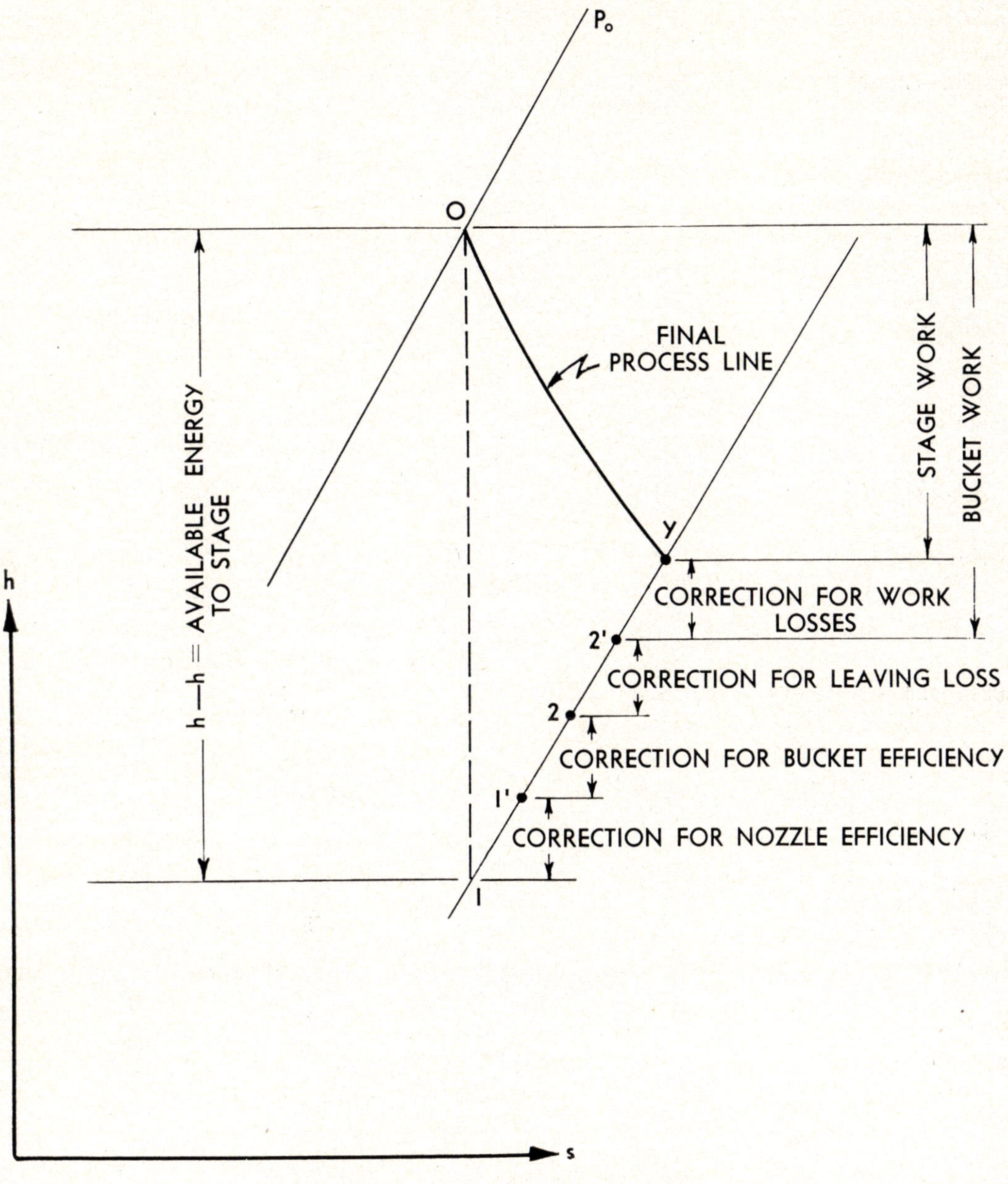

FIG. 4-17. h-s Diagram for Impulse Stage.

(e) *Moisture loss*—At some point in the turbine, the steam reaches its saturation pressure and continues to expand with the accompanying formation of moisture droplets. Energy from the vapor is required to accelerate these droplets, which are estimated to travel at about 10% of the velocity of the steam. Due to this slow speed, the relative velocity of the droplets to the blade actually causes a braking effect on the blade, resulting in an additional loss of power. This is explained by Figure 4-16. Note that V_{r1} strikes the face of the blade with a component in the direction of blade motion, while V_{rm} strikes the back of the blade, opposing the motion.

The total of these losses amounts to about 3% to 10% of the bucket work. Therefore the stage efficiency, defined as shaft work over energy available to the stage is about 90% to 97% of the nozzle-bucket efficiency.

Figure 4-17 shows an exaggerated *h-s* diagram

for the complete stage. Note from it the following relationships:

Available energy to stage:

$$h_0 - h_1$$

Entering steam velocity:

$$V_1 = C_N\sqrt{2gJ(h_0 - h_1)}$$
$$= \sqrt{2gJ(h_0 - h_1')}$$

Nozzle loss:

$$h_1' - h_1$$

Bucket loss:

$$h_2 - h_1' \left(= \frac{-V_{r1}^2 + V_{r2}^2}{2gJ} \right)$$

Leaving loss:

$$h_2' - h_2 \left(= \frac{V_2^2}{2gJ} \right)$$

Nozzle-bucket efficiency:

$$\frac{h_0 - h_2'}{h_0 - h_1}$$

Bucket work:

$$h_2' - h_1 \left(= \frac{V_1^2 - V_{r1}^2 + V_{r2}^2 - V_2^2}{2gJ} \right)$$

Stage work:

$$h_0 - h_x$$

Stage efficiency:

$$\frac{h_0 - h_x}{h_0 - h_1}$$

Example 42: The second stage inlet conditions on the H.P. impulse turbine of the DL-4 at 20 kts. are 68.2 psia and 392°F. The exit conditions are 55 psia and 360°F. (a) Find the stage efficiency and the stage work per pound of steam. (b) The nozzle angle is 15°, the blade velocity at 20 knots is 400 fps, the blade exit angle is 24°, $C_N = 0.975$ and $C_b = 0.85$. Construct the velocity diagram and find the bucket work. (c) Find the leaving loss and the work losses. (d) Find the nozzle loss and bucket loss, and check whether the sum of all losses plus stage work equal the available energy to the stage.

Solution:

(a) $h_0 = 1228.3$ (from steam tables for entrance conditions)

$s_0 = s_1 = 1.6935$ (from steam tables for entrance conditions)

$h_1 = 1209.1$ (isentropic drop from 68.2 psia to 55 psia)

$h_x = 1214.3$ (from steam tables for exit conditions)

$$\text{Stage efficiency} = \frac{h_0 - h_x}{h_0 - h_1} = \frac{14.0}{19.2} = \underline{\underline{72.9\%}}$$

$$\text{Stage work} = h_0 - h_x = \underline{\underline{14.0}} \text{ BTU/lb}$$

(b) $V_1 = C_N\sqrt{2gJ(h_0 - h_1)} = 0.975\sqrt{50100(19.2)}$

$= 0.975 \times 981 = 957$ fps

$V_{r1} = \sqrt{V_1^2 + V_b^2 - 2V_1V_b \cos 15°}$

(See diagram)

$= \sqrt{(957)^2 + (400)^2 - (2)(957)(400) \cos 15°}$

$= 580$ fps

$V_{r2} = C_bV_{r1} = 0.85 \times 580 = 493$ fps

$V_2 = \sqrt{V_{r2}^2 + V_b^2 - 2V_{r2}V_b \cos 24°}$

$= \sqrt{(493)^2 + (400)^2 - 2(493)(400) \cos 24°}$

(See diagram)

$= 207$ fps

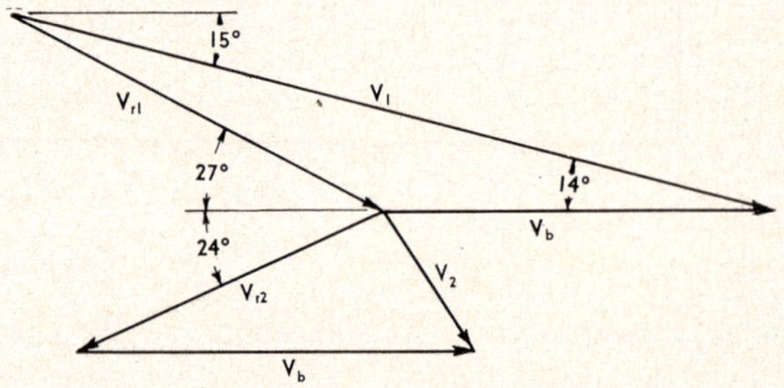

$$\text{Bucket work} = \frac{V_1^2 - V_{r1}^2 + V_{r2}^2 - V_2^2}{2gJ}$$

$$= \frac{(957)^2 - (580)^2 + (493)^2 - (207)^2}{2gJ}$$

$$= \underline{\underline{15.56}} \text{ BTU/lb}$$

(c) Leaving Loss $= \frac{V_2^2}{2gJ} = \frac{(207)^2}{64.4 \times 778}$

$= .85$ BTU/lb

Work Loss = Bucket Work − Stage Work (See Fig. 4-17)

$=$ Bucket Work $-(h_0 - h_x)$

$= 15.56 - 14.0 = 1.56$ BTU/lb

(d) Nozzle efficiency $= C_N^2 = (0.975)^2 = 0.951$

Nozzle loss factor $= 1 - 0.951 = .049$

Nozzle loss $= (.049)(h_0 - h_1) = (.049)(19.2)$

0.94 BTU/lb

Bucket loss:

Energy of relative steam entering $= \frac{V_{r1}^2}{2gJ}$

Energy of relative steam leaving $= \frac{V_{r2}^2}{2gJ}$

Bucket loss $= \frac{(580)^2 - (493)^2}{2gJ}$

$= 1.86$ BTU/lb

Check: From computation—

Work + Work Loss + Leaving Loss

+ Nozzle Loss + Bucket Loss

$= 14.0 + 1.56 + 0.85 + 0.94 + 1.86$

$= 19.21$ BTU/lb

From $h-s$ chart, or steam tables—

available energy to stage, $h_0 - h_1$

$= 19.2$ BTU/lb

415. THE REACTION STAGE.

From the foregoing description of the impulse stage, it is apparent that the process of conversion of thermal to kinetic energy takes place only in the nozzles, and that the sole function of the moving buckets is to convert the kinetic energy into mechanical work. It is entirely feasible, however, to construct a stage in which the change from thermal to kinetic energy occurs in both the nozzles and the moving blades. This is done by shaping the moving blade passages so that they behave as nozzles also. Such blades, in addition to changing the *direction* of the steam relative to the blade, also operate to change its *magnitude* relative to the blade. This scheme of operation characterizes *the reaction stage.* It is possible, in fact, to construct a turbine in which *all* of the thermal energy is changed to kinetic energy and then into mechanical energy in the moving blades alone. Hero's turbine was essentially such a machine. It is common practice, however, to so design the reaction stage that in normal power ranges of the turbine, about one-half of the thermal energy is converted into kinetic energy in the nozzles, or fixed blades, and one-half in the moving blades. This type stage is called a *pure reaction* stage, and the percent of thermal energy converted in the moving blades is termed the *degree of reaction.* Thus a conventional reaction stage is commonly called a 50% reaction stage.

416. THE ZERO ANGLE REACTION STAGE.

The analysis of the reaction stage is almost the same as that of the impulse stage. Let us consider a zero angle reaction stage in which the total enthalpy drop is $h_0 - h_1$. Assume that this drop produced the V_1 used in the discussion of the zero angle impulse stage in Art. 411. In the pure reaction stage, however, the drop across the nozzle is $\frac{1}{2}(h_0 - h_1)$. Then the velocity produced by the reaction nozzle, V_n, is found as follows:

$$\frac{V_n}{V_1} = \frac{\sqrt{\frac{1}{2}(h_0 - h_1)}}{\sqrt{(h_0 - h_1)}} = \sqrt{\tfrac{1}{2}}$$

or

$$V_n = V_1/\sqrt{2} = .707\ V_1. \qquad (19)$$

In Figure 4-18 this stream approaches the blade with a relative velocity of $.707\ V_1 - V_b$. The blade reverses the stream, but also accelerates it by a quantity equal to the acceleration of the steam in the nozzle, so that the relative velocity of the

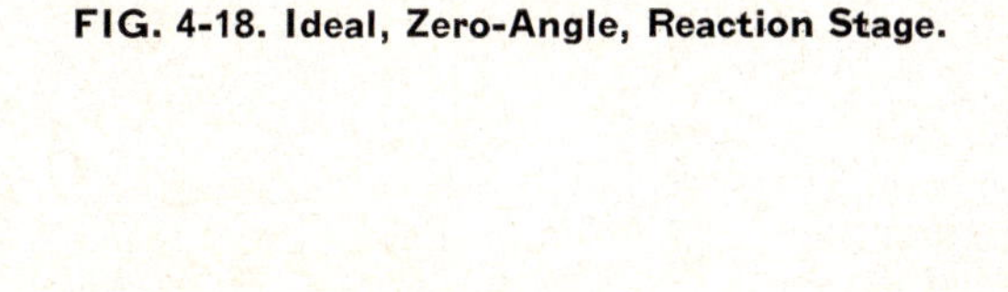

FIG. 4-18. Ideal, Zero-Angle, Reaction Stage.

stream leaving the blade is

$$-(V_n - V_b) - V_n$$
$$= V_b - 2V_n.$$

The absolute velocity of the stream leaving the blade is the sum of the velocity with respect to the blade plus the velocity of the blade, or

$$V_2 = V_b - 2V_n + V_b = 2V_b - 2V_n.$$

Maximum efficiency of the stage occurs when $V_2 = 0$, so that no kinetic energy is lost to the stage. Then

$$2V_b - 2V_n = 0$$

or

$$V_b = V_n. \qquad (20)$$

Equation (20) indicates that in this reaction stage, the blade speed and the speed of the steam coming from the nozzle are equal under conditions of minimum loss of kinetic energy from the stage. Compared to a zero angle impulse turbine, in which the nozzle velocity produced is exactly the same, this indicates that the blade speed of a reaction turbine is twice that of an impulse turbine under ideal conditions. However, a more valid comparison can be made between the two turbines when the total thermal energy conversion is the same. In this case, the nozzle velocity of the reaction turbine is $1/\sqrt{2}$ times that of the impulse turbine (Equation 18). Then for the impulse turbine, as before

$$V_b = 0.5V_1$$

and for the reaction turbine

$$V_b = V_n = V_1/\sqrt{2} = .707\ V_1.$$

417. THE FINITE ANGLE REACTION STAGE.

Assume that the enthalpy of the steam entering the nozzle of a reaction stage is h_0 and that of the steam leaving the stage is h_1, and that no losses occur in the stage so that the entropy is equal at entrance and exit. Then the total enthalpy drop is $h_0 - h_1$, and some fraction of this is converted into kinetic energy in the nozzles, and the remaining fraction is converted in the buckets. Let X equal the fraction converted in the buckets. Then $1 - X$ is the fraction converted in the nozzles. For a pure reaction stage, $X = 0.5$, so that the amount of thermal energy converted in

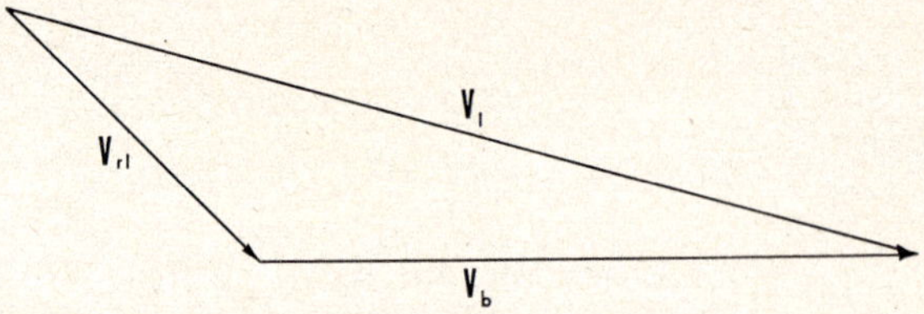

FIG. 4-19. Vector Diagram for Entering Steam, Reaction Stage.

the nozzle is equal to that converted in the buckets. Then the velocity, V_1, leaving the nozzle is

$$V_1 = \sqrt{2gJ(.5)(h_0 - h_1)}. \qquad (21)$$

With a given blade speed and nozzle angle, V_b and α, the first part of the vector diagram can be constructed, giving V_{r1}, as shown in Figure 4-19. The vector V_{r2} must now be found. Its direction is determined by the blade exit angle β_2, which is known. Its magnitude may be found by applying energy balance across the moving blades.

$V_{r1}^2/2gJ$ = K.E. entering blades, BTU/lb

Let h_a = enthalpy of steam entering the blades, BTU/lb

$V_{r2}^2/2gJ$ = K.E. leaving blades, BTU/lb

h_1 = enthalpy of steam leaving the blades, BTU/lb

Then

$$\frac{V_{r1}^2}{2gJ} + h_a = \frac{V_{r2}^2}{2gJ} + h_1$$

or

$$V_{r2} = \sqrt{2gJ(h_a - h_1) + V_{r1}^2}.$$

But $(h_a - h_1) = X(h_0 - h_1)$ from previous discussion, and when $X = .5$,

$$V_{r2} = \sqrt{2gJ(.5)(h_0 - h_1) + V_{r1}^2}.$$

The first term under the radical is identical to that in Equation (21), so that

$$V_{r2} = \sqrt{V_1^2 + V_{r1}^2}. \qquad (22)$$

If the nozzle angle α is small (in the order of 14°) then for reasonable values of blade speed,

$$V_{r1} \approx V_1 \sin 14°$$
$$\approx 0.242\ V_1.$$

Therefore Equation (22) becomes

$$V_{r2} = \sqrt{V_1^2 + (0.242)^2 V_1^2}$$
$$= \sqrt{V_1^2 + .0585V_1^2}$$
$$= \sqrt{1.0585V_1^2} = 1.03\ V_1$$
$$\approx V_1. \qquad (23)$$

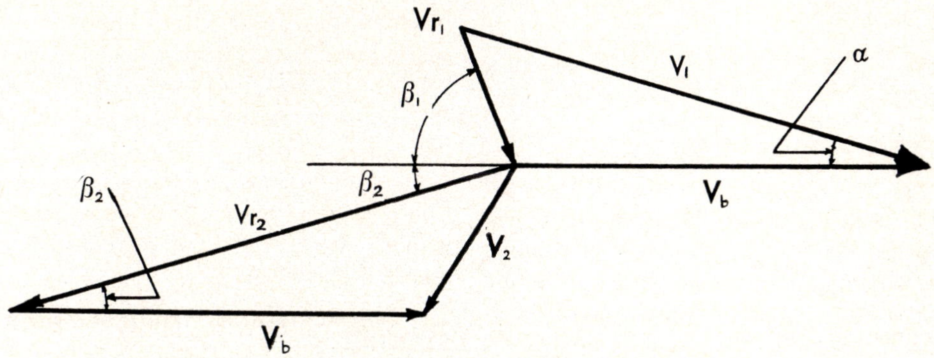

FIG. 4-20. Velocity Vector Diagram for Frictionless Reaction Stage.

Thus, for all practical purposes, the relative velocity developed in the moving blades may be taken as equal to the velocity developed in the nozzles. The velocity diagram can now be completed, using the above value for V_{r2} and the bucket exit angle, β_2. This angle is very nearly equal to the nozzle angle α in pure reaction stages, and is constructed equal to α in Figure 4-20.

Under the condition assumed above, that is, no friction losses, a negligible effect by V_{r1} on the magnitude of V_{r2}, and assuming $\beta_2=\alpha$, the two triangles shown in Fig. 4-20 are identical. If V_1 is changed so that $V_1 \cos \alpha = V_b$, then V_{r1} is vertical (its minimum magnitude) and correspondingly V_2 also becomes a minimum. Since V_2, the absolute exit velocity from the stage, represents lost kinetic energy to the stage, maximum work is obtained when

$$V_b = V_1 \cos \alpha. \tag{24}$$

When bucket friction is considered, Equation (23) must be modified to include its effect as follows

$$V_{r2} \approx C_b V_1.$$

Also β_2 may differ slightly from α. However, in a well designed reaction stage, Equation (24) still holds very nearly true. Figure 4-21 is a typical diagram of a stage with bucket friction and with $\beta_2 \neq \alpha$. In the figure V′ and V″ are the two axial flow velocity components of the steam entering and leaving the moving blades, respectively.

The stage work is obtained in the same manner as it was obtained in the impulse stage. Since the expression derived there (Equation 17) was not dependent on how the various velocities were produced, but only on their magnitude, the expression is the same in this case, that is

$$W_b = \frac{V_1^2 - V_{r1}^2 + V_{r2}^2 - V_2^2}{2g}. \tag{17}$$

Similarly, the expression for diagram efficiency is

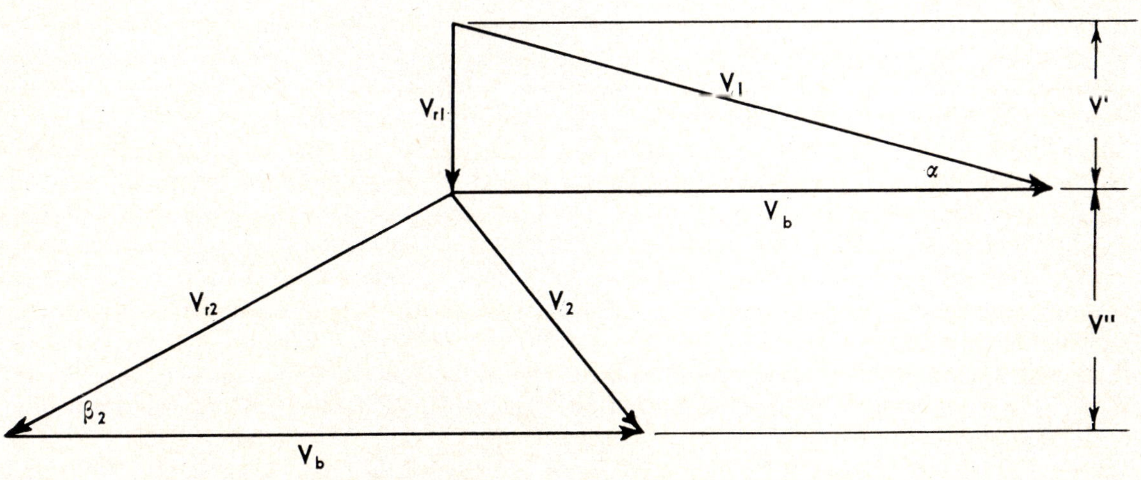

FIG. 4-21. Velocity Vector Diagram for Reaction Stage, (Blade Speed not Ideal).

the same,

$$\eta_b = \frac{V_1^2 - V_{r1}^2 + V_{r2}^2 - V_2^2}{V_1^2} \cdot \qquad (18)$$

418. REACTION STAGE LOSSES.

The pressure drop across the moving blades of the reaction stage gives rise to certain differences in construction between turbines containing nearly pure impulse stages and those containing stages with a high degree of reaction. These differences will be discussed in more detail in following articles, but it is necessary to briefly consider their salient features here.

Impulse buckets are carried on discs, secured to a shaft. The fixed blades, or nozzles are mounted in diaphragms, secured to the casing. Since there is little or no pressure drop across the buckets and therefore across the discs (discs often have holes in them to equalize pressure) there is little pressure on the disc area to produce end thrust. Also, there is little tendency for steam to flow around the edge of the disc instead of through the buckets; therefore generous clearance is provided between the disc circumference and the casing in way of the discs.

Reaction staging, on the other hand, is characterized by a pressure drop across the buckets. Drum construction is used in order to minimize the area exposed to the pressure change. That is, instead of mounting the blades on discs, the shaft is made in the form of a drum, on which the blades are directly mounted. Nevertheless, the pressure change across the blades acts upon the annular area of the blades and produces considerable end thrust. In addition, there is a strong tendency for steam to leak around the outer ends of the blades, necessitating small tip clearances between blade ends and the casing at that point.

With the above features in mind, the losses characteristic in a reaction stage (in addition to friction losses of the steam passing through the nozzles and moving blades) are:

(a) Windage loss—Friction of steam on non-working portions of the moving blading and drum sets up turbulence. (Smaller than in a corresponding impulse stage.)

(b) Tip leakage loss—Unused energy in that part of the steam which leaks past the blade tips without performing work. This occurs between the moving blade tips and the casing, and between the fixed blade tips and the drum. Since mechanical considerations limit the minimum clearance distance, the tip leakage area is a larger per cent of total flow area in short, or high pressure blading, than in long, or low pressure blading.

(c) Moisture losses—Moisture losses in reaction stages occur for the same reasons and have similar effects to those described for impulse stages in Article 414(e).

419. COMPARISON OF IMPULSE AND REACTION STAGING.

Because of the pressure change across the moving blading in a reaction turbine, it is not feasible to construct a partial admission arrangement, as is done in many high pressure impulse stages. The nozzle flow area must be kept to its proper value by using short nozzle blading (and bucket blading) and restricting the stage diameter. Therefore, the nozzle end losses and the suction losses of a partial admission stage are not present in a reaction stage.

For impulse and reaction stages of equal enthalpy drop, the velocities produced in the reaction nozzles are considerably smaller than those in the impulse nozzles. This is obviously true because the enthalpy drop across the reaction nozzles is about one-half of that across the impulse nozzles, the remainder taking place across the buckets of the reaction stage. Thus, even though the wheel speed of such a reaction stage would be higher than in the impulse stage, the velocity of the steam relative to the wheel is somewhat less. Refer to Figures 4-14 and 4-20. In actual practice, the wheel speed is usually designed to the same maximum practicable value for both; this condition, then, determines the nozzle velocities and therefore, the permissible enthalpy drop. This means that reaction stages tend to have less enthalpy drop than similar impulse stages, still further reducing local steam velocities. Since fluid friction losses vary approximately as the cube of the velocity of the fluid passing the surface, these considerations indicate a definite advantage of the reaction stage over the impulse stage with respect to fluid friction. However, smaller enthalpy drops per stage also indicate more stages for a given turbine, meaning larger and more expensive turbines.

In view of the last two paragraphs, tip leakage considerations favor the use of impulse blading in high pressure staging, where the leakage area is a fairly large proportion of the steam flow area; and fluid friction losses favor the use of reaction blading in low pressure staging where the effect of tip leakage loss is small.

The below comparison will show the effect of tip leakage area in typical H.P. and L.P. staging.

Example 43: Blading in a H.P. stage is 1″ long, mounted on a drum of diameter 20″ and having a tip clearance of .050″. In an L.P. stage, the blading is 8.5″ long, mounted on a drum of 24.5″ diameter, and having a tip clearance of .050″. Find the tip leakage area as a percentage of total flow area in each case.

Solution: For H.P. stage:

$$\text{Flow area} = \frac{\pi}{4}(22)^2 - \frac{\pi}{4}(20)^2 = \frac{\pi}{4}(484-400)$$

$$= 66.0 \text{ sq. in. (approximately)}$$

$$\text{Leakage area} = \pi(22)(.050)$$

$$= 3.46 \text{ sq. in. (approximately)}$$

$$\%\text{Leakage} = \underline{\underline{5.24\%}}$$

For L. P. stage:

$$\text{Flow area} = \frac{\pi}{4}(43.5)^2 - \frac{\pi}{4}(24.5)^2$$

$$= \frac{\pi}{4}(1890-600)$$

$$= 1015.0 \text{ sq. in. (approximately)}$$

$$\text{Leakage area} = \pi(43.5)(.050)$$

$$= 6.85 \text{ sq. in. (approximately)}$$

$$\%\text{ Leakage} = \underline{\underline{.68\%}}$$

420. THE VELOCITY COMPOUNDED IMPULSE STAGE.

It is necessary to consider another common type of stage, which is identified by any of the following names: *the two-row impulse stage, the velocity compounded stage,* or *the Curtiss stage.*

Consider the idealized zero angle stage in Figure 4-22. Steam issues from the nozzle with a velocity V_1. It approaches the first moving bucket with a relative velocity $V_1 - V_b$. It is reversed, becoming $V_b - V_1$, but its absolute velocity is $V_b + V_b - V_1$ or $2V_b - V_1$. It strikes a fixed bucket which again reverses its direction, but does not alter its magnitude. Its velocity is now $V_1 - 2V_b$. It approaches the second moving bucket, whose speed is V_b, with a relative velocity of $(V_1 - 2V_b) - V_b$, or $V_1 - 3V_b$. It is reversed again, becoming $3V_b - V_1$ relative to the blade and $V_b + 3V_b - V_1$, or $4V_b - V_1$, in absolute magnitude. To minimize this velocity, and thus

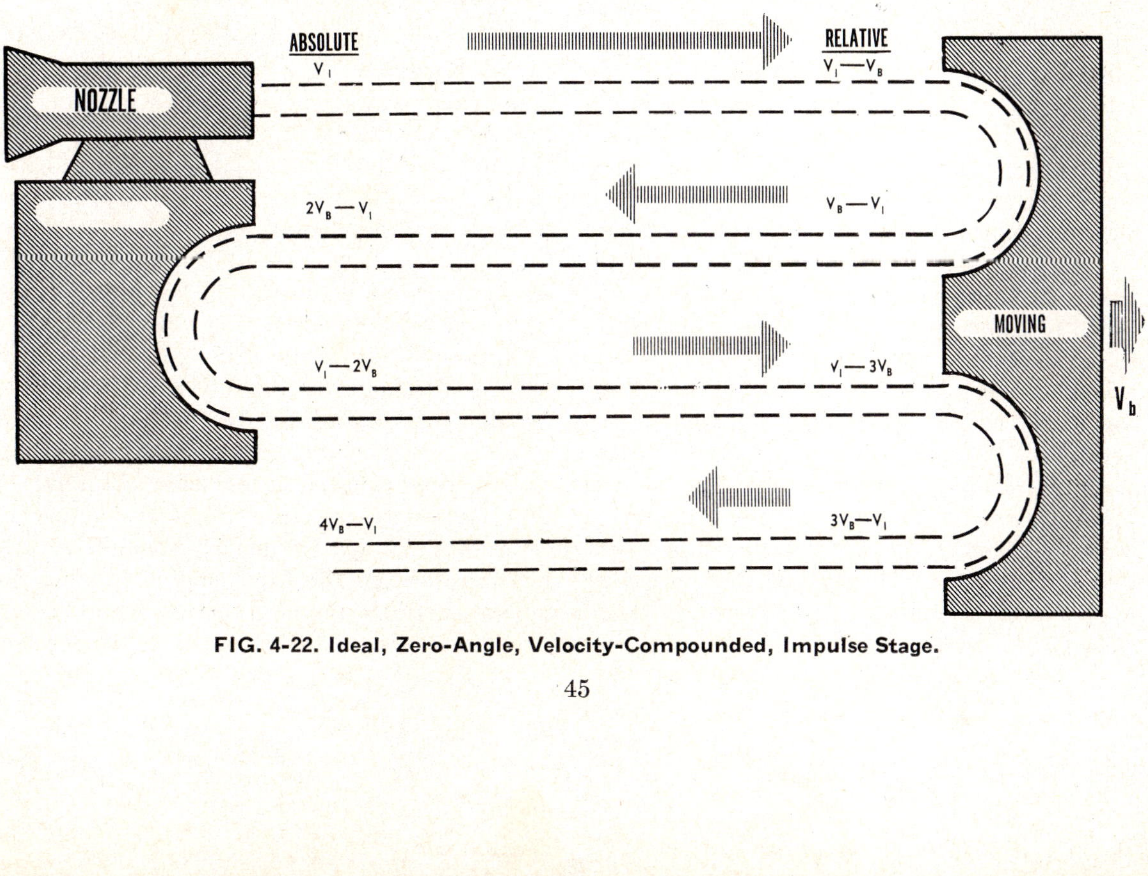

FIG. 4-22. Ideal, Zero-Angle, Velocity-Compounded, Impulse Stage.

the leaving kinetic energy, we set this velocity equal to zero.

$$4V_b - V_1 = 0$$

or

$$V_b = \tfrac{1}{4}V_1. \qquad (25)$$

Equation (25) shows that, for minimum leaving loss, the wheel speed is one-fourth the entering steam speed.

It may be shown that for an ideal stage with a finite nozzle angle, the blade speed-steam speed relation for minimum leaving loss is given by

$$V_b = \tfrac{1}{4}V_1 \cos \alpha, \qquad (26)$$

an expression similar to Equation (14).

Equation (26) shows that, for a given wheel speed, a velocity compounded stage can utilize a steam velocity of twice that of the single row impulse stage. This means an enthalpy drop four times as great as a single row impulse stage; therefore, such a stage would appear to be the energy equivalent of approximately four impulse stages. Unfortunately, the losses of such a stage are comparatively high, and its stage efficiency is less than either of the two types previously described. However, it is quite advantageous for use as the first stage in a turbine for the reasons discussed below.

It is highly desirable to decrease the pressure and temperature of the steam before its entry into the turbine casing proper. Turbine casings are made in the form of a cylinder, split and flanged into two halves along the axis of the cylinder. For very high pressures in the cylinder, the cylinder material must be stronger, heavier, and more expensive to manufacture. Also it is more difficult with high pressures, and at the high temperatures involved, to fasten the flanges so that leakage from the joint can be eliminated. In addition, the moving blading is subjected to high stresses due to force of the steam on the blades and to the large centrifugal forces. If high temperature steam were also present, conditions conducive to creep of the metal of the blading would be present. Instead, the high temperatures and pressures are admitted only to the steam chest, which is a cavity cast integrally in the upper half of the casing only. A large enthalpy drop is taken then across the nozzles at the exit from the steam chest. Thus, the steam entering the turbine shell proper is greatly reduced in pressure and temperature. This is realized by use of a velocity compounded wheel with partial (confined to the upper half) admission in the first stage. For example, in a typical modern destroyer turbine at 30 kts, the steam chest pressure and temperature is 1165 psia and 935°F respectively. At the exit of the first stage nozzles, the conditions are 338 psia and 694°F. Conditions conducive to metal creep may still exist at full power, but since full power is utilized only a small percentage of the time, and since creep is a function of time as well as temperature and stress, the conditions are accepted, but kept to values designed to give a reasonable operating life before blade renewal and other overhaul work.

Figure 4-23 is a velocity diagram of an idealized finite angle velocity compounded stage, operating with minimum leaving loss. It is realized that in actual practice the idealized conditions are not met, but they are fairly closely approximated.

It is quite possible to build three-row and even four-row velocity compounded stages, but high losses make such designs unpopular, and they are seldom seen on main propulsion turbines. However, in some designs of auxiliary turbines, the large enthalpy drop so obtainable with relatively low wheel speeds makes these designs useful in spite of inefficiency. Such arrangements will be noted later.

421. MULTI-STAGE ARRANGEMENTS.

After the above consideration of individual types of staging, it is now possible to better understand arrangements of turbine staging. From the statements previously made, it is obvious that a velocity compounded stage is a logical choice for the first stage. It is logical to follow this with impulse staging in the high pressure end to minimize tip leakage losses, and with the more efficient reaction blading in the low pressure end. This is generally modern practice.

Plate II is a sectional view of a modern high pressure turbine typifying the arrangement used by all manufacturers. Plates III and IV show typical arrangements of low pressure turbine construction by two large manufacturers. In Plate III, note the drum construction of the rotor. The moving blading is attached directly to the rotor and the fixed blading forming the nozzles is attached directly to the turbine casing. In Plate IV the rotor is forged in the form of

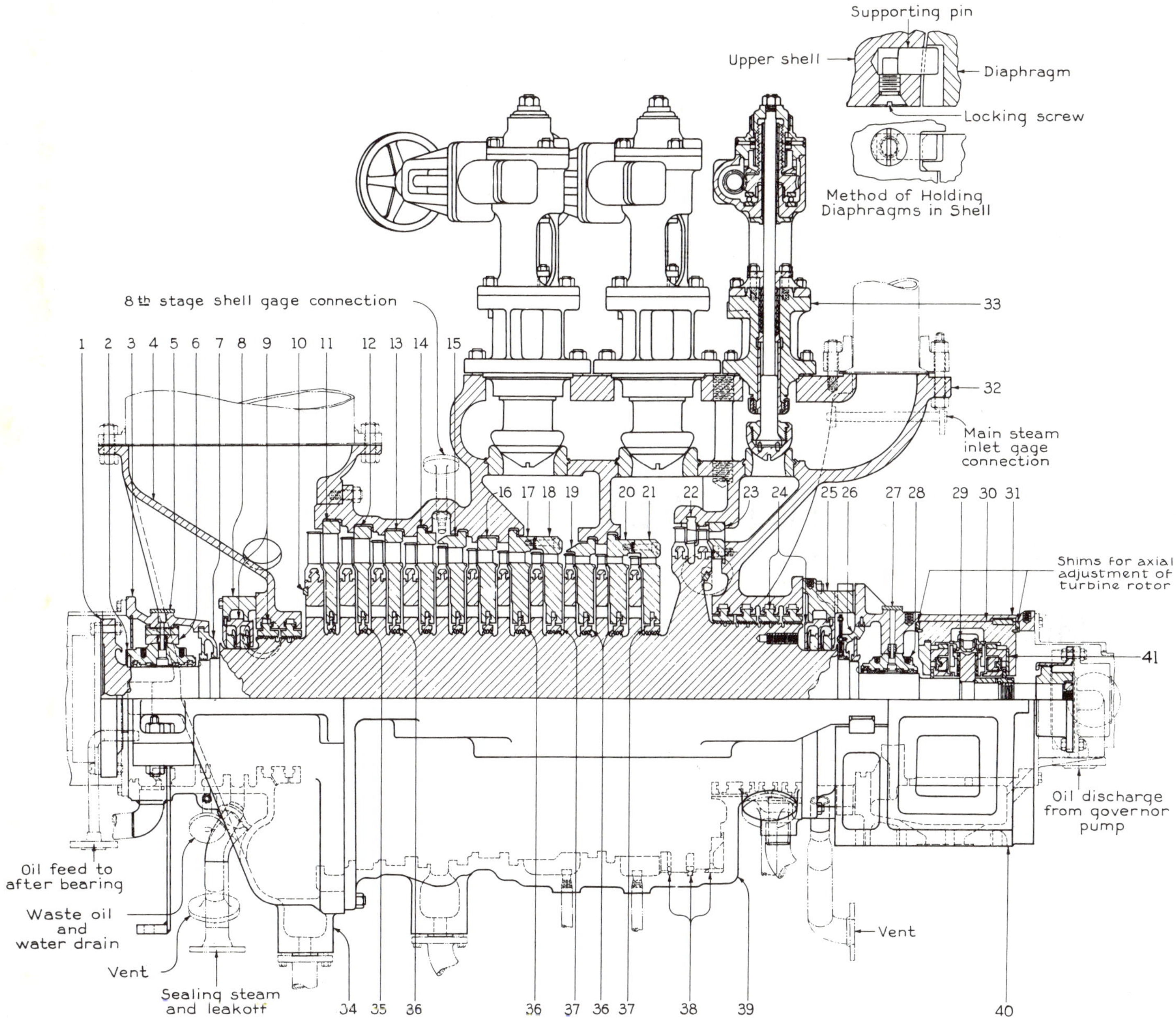

1 Rotor assembly
2 Bearing, after end
3 Bearing cap, after end
4 Exhaust casing, upper half
5 Cover over hole for bearing-wear gage
6 Bearing ring
7 Oil deflector
8 Packing casing, after end
9 Packing assembly, after end
10 Balance plug
11 Nozzle diaphragm, 12th stg
12 Nozzle diaphragm, 11th stg
13 Nozzle diaphragm, 10th stg
14 Nozzle diaphragm, 9th stg
15 Nozzle diaphragm, 8th stg
16 Nozzle diaphragm, 7th stg
17 Nozzle diaphragm, 6th stg
18 Nozzle diaphragm, 5th stg
19 Nozzle diaphragm, 4th stg
20 Nozzle diaphragm, 3rd stg
21 Nozzle diaphragm, 2nd stg
22 Intermediate bucket
23 Nozzle plate, 1st stg
24 Packing assembly, forward end
25 Packing casing
26 Oil deflector
27 Cover over hole for bearing-wear gage
28 Bearing, forward end
29 Thrust bearing
30 Thrust casing, inner
31 Bearing cap
32 Turbine casing, upper half
33 Handvalve assembly
34 Exhaust casing, lower half
35 Diaphragm packing, 11th and 12th stg
36 Diaphragm packing, 3rd, 4th, and 6th to 10th stg
37 Diaphragm packing, 2nd and 5th stg
38 Steam shield
39 Turbine casing, lower half
40 Bearing standard
41 Filler piece

PLATE II. Allis-Chalmers, Destroyer High Pressure Turbine.

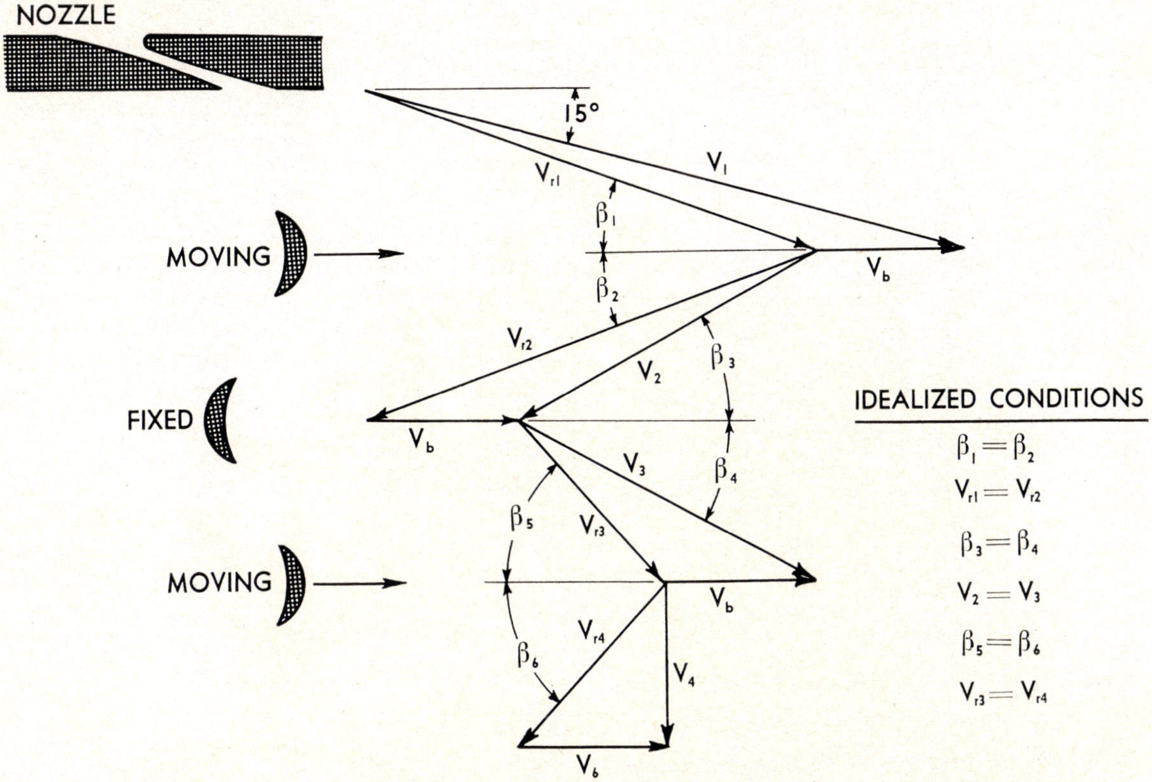

FIG. 4-23. Frictionless, Velocity Compounded, Impulse Stage.

shaft and wheels, with the moving blading attached to the wheels, or discs. The fixed blading forming the nozzles is attached to diaphragms, which in turn are seated in casing grooves. In both plates, the moving blades have a large degree of reaction, though not necessarily the same in both. Historically, the rotor and blade type construction has come to be known as reaction type construction and the disc and diaphragm construction has taken the title of impulse type. This is often confusing since in both types of construction, the moving blading in Plates III and IV is essentially of the reaction type.

422. TURBINE PERFORMANCE.

Calculations involving the turbine as a whole are based on the energy balance, obtained essentially by use of the general energy equation (Equation 3). It is rewritten here in a form which includes losses, L_m, which occur as a result of mechanical friction. The potential energy terms, Z/J, are considered as being negligibly small as is the kinetic energy term at the turbine inlet, $V_0^2/2g$. Also, in a well insulated turbine, heat exchange with the surroundings is insignificant. Equation (3) then takes the form

$$h_a = \frac{V_e^2}{2gJ} + h_e + \frac{wk}{J} + L_m$$

or

$$wk = (h_a - h_e) - \frac{V_e^2}{2gJ} - L_m \qquad (27)$$

where

h_a = enthalpy of steam at turbine inlet, BTU/lb

h_e = enthalpy of steam at turbine exit, BTU/lb

V_e = velocity of steam at turbine exit, ft/sec

L_m = mechanical losses, BTU/lb

wk = work produced at turbine shaft, BTU/lb.

The three quantities on the right hand side of Equation (27) are recognized as change of enthalpy in the working fluid, kinetic energy of the fluid leaving the turbine, and mechanical losses. These are examined individually.

(a) *Enthalpy Change.* If the turbine were ther-

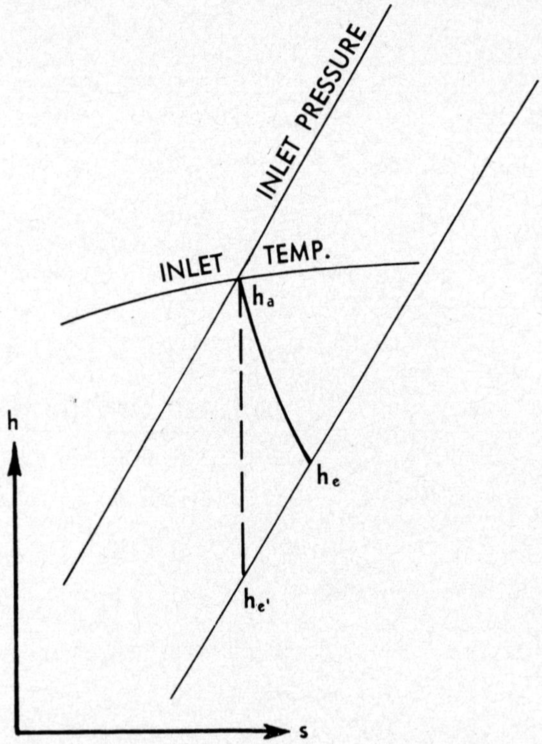

FIG. 4-24. h-s Diagram for Turbine (Simplified)

modynamically perfect, the enthalpy change could be represented as an isentropic line on $h-s$ coordinates, where the upper point is fixed by inlet steam pressure and temperature and the lower point by the exit pressure. This is shown by the line connecting h_a and h_e' in Figure 4-24. However the thermodynamic losses, caused by throttling action of steam leakage, nozzle and blade friction, and other factors previously discussed, result in a displacement of h_e' to the right as shown by the actual point, h_e, in the figure. The solid line connecting h_a and h_e is called the "state-line" or "condition-curve." Each point on it represents the actual state of the steam at the corresponding point in the turbine. The *line is nearly a straight line and little error results by ignoring its curvature in turbine analysis.* Thus it is very closely determined by the location of its lower end, called the "state line end point," abbreviated to S.L.E.P.

The position of the S.L.E.P. will vary with different loads, representing decreased efficiencies at loads away from the point of maximum turbine efficiency. In turbines having a velocity compounded inlet stage, there is characteristically a break in the line after the first stage, indicating the lower efficiency of this stage. Also, there is characteristically a rise in enthalpy in the cross-over pipe between the cruising and high pressure units, and between the high and low pressure units. This is due to reconversion to thermal energy of a part of the kinetic energy present in the steam when it leaves the blades of the stage preceding the cross over pipes. This is discussed more fully in (b) below. Some throttling is usually present between the inlet valve and the steam chest, giving a short, horizontal portion at the top of the curve. These features are indicated in Figure 4-25. Mollier charts, containing the state lines, will be found in most turbine instruction books aboard ship.

For analysis of the overall performance, sufficient information for our purposes is given by the location of the inlet condition point and the S.L.E.P.

Assuming the S.L.E.P. known at full power (marked h_e in Fig. 4-25) the *internal efficiency* of the turbine may be found. Considering the throttle as part of the turbine, the energy available to the turbine is h_a-h_e'. The energy delivered to the wheels is h_a-h_e. Therefore the internal efficiency is

$$\eta_i=\frac{h_a-h_e}{h_a-h_e'}\,. \tag{28}$$

(b) *Leaving Losses.* Leaving losses for individual stages, represented by the kinetic energy of the steam leaving the stages, have previously been discussed. These individual stage losses are not necessarily lost to the turbine as a whole. Normally, about 75% of the kinetic energy is effective energy "carry-over" to the next stage. Exceptions to this are the velocity-compounded stage which usually has a larger diameter than the subsequent stage, and a generous dead space between its exit and the next stage inlet. (See Plate II.) Even in this space the kinetic energy is converted back to thermal energy, so that the major loss is that associated with an entropy increase which restricts the subsequent availability of energy from the steam. These factors control in considering the leaving loss of the last stage of the high pressure turbine, though here the effect of entropy increase is even more marked. The major leaving loss occurs at the exit of the last stage of the low pressure turbine, for

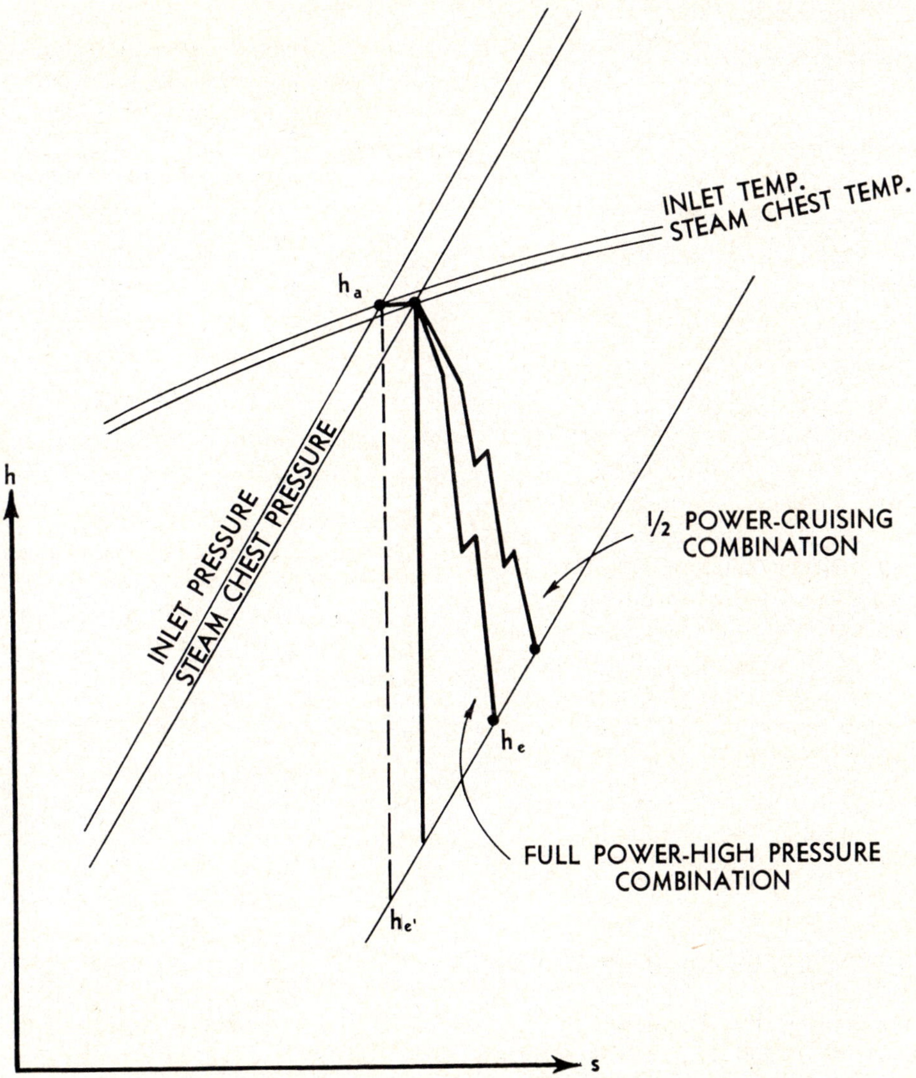

FIG. 4-25. h-s Diagram for Actual Turbine (Two Conditions).

here the kinetic energy is lost to the turbine. If the exit velocity, V_e, is known or computed from a velocity diagram, the loss is readily calculated. If the diagram is not known, the velocity can be determined approximately by consideration of the last stage steam flow area, the quantity of steam flowing, and its specific volume, from the continuity relation, as follows:

$$V_e = \frac{Qv}{A} \qquad (29)$$

where

V_e = exhaust velocity (normal to blade row), fps

Q = quantity of steam flow, lbs/sec

v = specific volume of steam at exit, ft^3/lb

A = annular area of last stage, ft^3.

Equation (29) is approximate because it gives only the axial flow component. However, the exhaust component is nearly axial at high powers when Q is large, so the equation is useful in this simplified form.

Example 44: The annular exhaust area of a L.P. turbine on a DE is 4.16 square feet. The S.L.E.P. at full power is 5″ Hg and 4.7% moisture. Steam flow is 144,600 lbs/hour. Find the leaving loss in BTU/lb.

Solution: Using Steam Tables:

$p = 5''$ Hg

$m = 4.7\%$

$v = 136.5$ ft³/lb

Also, $Q = \dfrac{144{,}600}{3600} = 40.1$ lbs/sec

From Eq. (29):

$$V_e = \frac{40.1 \times 136.5}{4.16} = 1315 \text{ ft/sec}$$

$$\text{Leaving loss} = \frac{V_e^2}{2gJ} = \frac{(1315)^2}{64 \times 778} = \underline{\underline{34.5}} \text{ BTU/lb}$$

(c) *Mechanical Losses.* The mechanical losses in the turbine are those associated with friction in the turbine bearings and shaft packing. Heat thus produced is carried away by circulating lube oil and is not available for conversion to work. This loss is small. Its exact determination can be made only by calculations based on test measurements, but a fair approximation is given by the following relation, adapted from Brown and Drewry.[1]

$$L_m = \frac{1.46(h_a - h_e)}{\sqrt{\text{H.P.(rated)}}} \text{ BTU/lb} \tag{30}$$

where

$h_a - h_e$ = actual enthalpy drop at rated power, BTU/lb

H.P. = horsepower at rated power.

Example 45: A 40,000 H.P. turbine at full power receives steam at 1050 psig and 940°F and discharges to a condenser at 5″ Hg pressure. The quality of steam at exit is 92.5%. Estimate the mechanical losses.

Solution:

$h_a = 1469$ (approx.; from steam tables for inlet condition)

$h_e = 1043$ (approx.; from steam tables for S.L.E.P.)

$$L_m = \frac{1.46}{\sqrt{40{,}000}}(1469 - 1043) = \underline{\underline{3.11}} \text{ BTU/lb}$$

[1] Brown, E. H., and Drewry, W. K., "Economy Characteristics of Stage Feed-water Heating by Extraction," *Trans. A.S.M.E.*, 1923.

This loss may be considered constant (on a per pound of steam flow basis) at all powers, but note that the formula is valid only when rated power values of enthalpy change and horsepower are used.

(d) *Summary.* In summary, the *mechanical efficiency* of the turbine may be defined as the energy at the turbine shaft divided by the energy available to the turbine:

$$\eta_e = \frac{(h_a - h_e) - \dfrac{V_{e2}}{2gJ} - L_m}{h_a - h_e'} \tag{31}$$

where the symbols have the meaning previously given. This is often called *engine efficiency.*

Example 46: The DE turbine in the Example 44 (for leaving loss calculation) delivers 20,000 SHP at full power. Its steam inlet conditions are 540 psig and 840°F. Find internal efficiency and engine efficiency.

Solution:

$h_a = 1432$ BTU/lb (from steam tables, for inlet conditions)

$h_e = 1071$ BTU/lb (from steam tables, for S.L.E.P.)

$h_e' = 977$ BTU/lb (from steam tables, for isentropic drop to exh. pressure from inlet conditions)

$$\eta_i = \frac{h_a - h_e}{h_a - h_e'} = \frac{1432 - 1071}{1432 - 977} = \underline{\underline{79.3\%}}$$

$$L_m = \frac{1.46(1432 - 1071)}{\sqrt{20{,}000}} = 3.7 \text{ BTU/lb}$$

$$\eta_e = \frac{(1432 - 1071) - 34.5 - 3.7}{(1432 - 977)} = \frac{323}{455} = \underline{\underline{70.9\%}}$$

423. TURBINE CONTROL.

The control of the quantity of steam permitted to flow through the turbine is the obvious means of controlling speed and power. The most apparent way to do this is by use of a simple throttle valve to restrict the flow. However, such a valve also decreases the pressure on its downstream side, and reduces the amount of energy per pound of steam available for conversion to work. It is recalled that the throttling process is one of constant enthalpy. It is represented in Figure 4-26

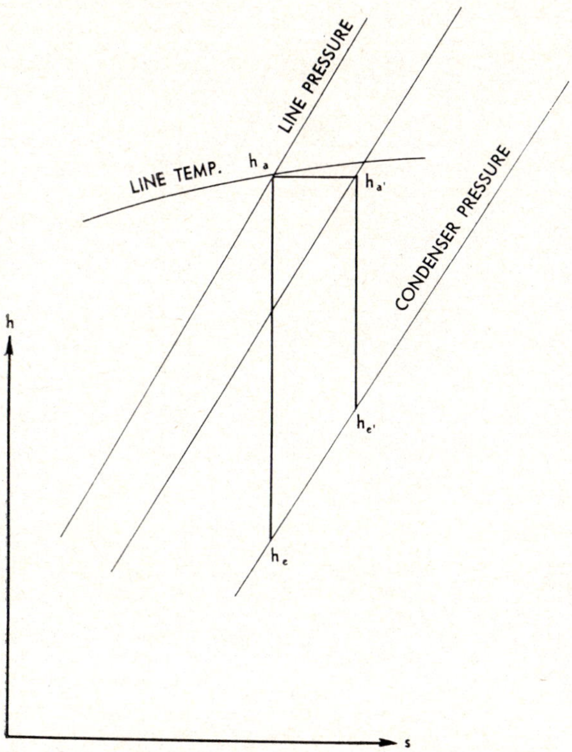

FIG. 4-26. h-s Diagram Indicating Loss in Available Energy Due to Throttling.

by the horizontal process line connecting h_a, representing throttle inlet conditions (at steam supply pressure and temperature) with h_a'. Note that the energy available to the turbine per pound of steam is decreased by the amount $h_e' - h_e$. Therefore, pure throttling is unsatisfactory as a means of control.

If steam can be supplied to the first stage nozzles at line pressure, and the nozzle area can be varied to control flow, the above is practically eliminated. This can be accomplished by arranging several valves to open nozzles, or groups of nozzles, in the first stage in sequence. Schematically, such an arrangement is shown in Figure 4-27. The forward end of the turbine is shown with a portion of the steam chest removed to reveal three separate chambers containing nozzle blocks. The three nozzle control valves open sequentially to admit steam to the nozzles. One valve is opened fully before the next commences to open. The valves are all controlled by a single handwheel and operated by a cam and spring arrangement. In some modern installations, the control handwheel operates a power unit which in turn controls the valves. When any of the valves are partially open, some throttling still occurs. From two to five nozzle control valves may be found on modern installations.

An alternative arrangement is shown in Figure 4-28. In this arrangement the nozzle control valves are individually operated by hand, and are either wide open or closed. Speed combinations requiring 4, 6, 7, 9, 11, 12, or 14 nozzles can be set with the three nozzle control valves. Speeds intermediate between these combinations are obtained by necessary adjustment of the main throttle valve.

The problem of efficient turbine control over a wide speed range is not completely solved by the above scheme. Variation in blade speed-steam speed ratios occurs at different wheel speeds. Recall that the enthalpy drop across the turbine depends on the initial pressure and temperature of the entering steam and the final pressure of the leaving steam. For an isentropic turbine with nozzle valve control, this total drop would be constant at all loads, except for minor variations in initial conditions caused by some throttling, and in the exit conditions caused by slight changes in the condenser pressure with load. The portion of the enthalpy drop taken across the first stage varies considerably with the load, but the remaining drop is divided about equally among the remaining stages, and each stage enthalpy drop varies only slightly with the load. Since the nozzle velocity in each stage is proportional to the square root of the enthalpy drop, this indicates that the nozzle velocity in each

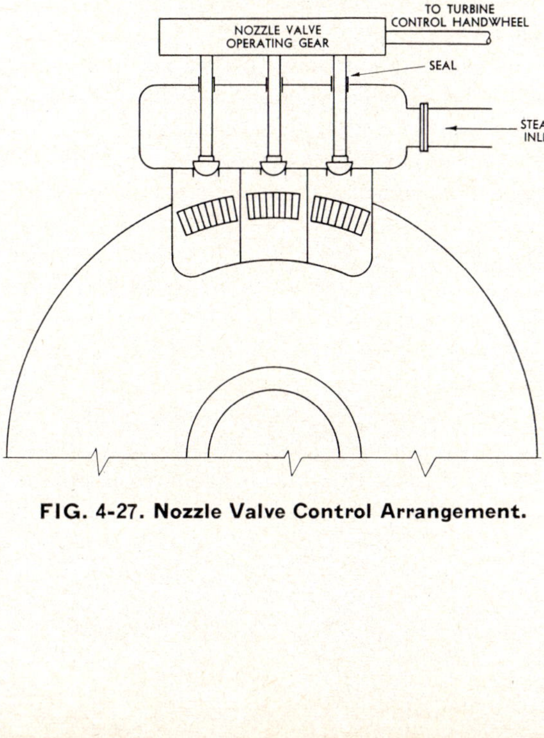

FIG. 4-27. Nozzle Valve Control Arrangement.

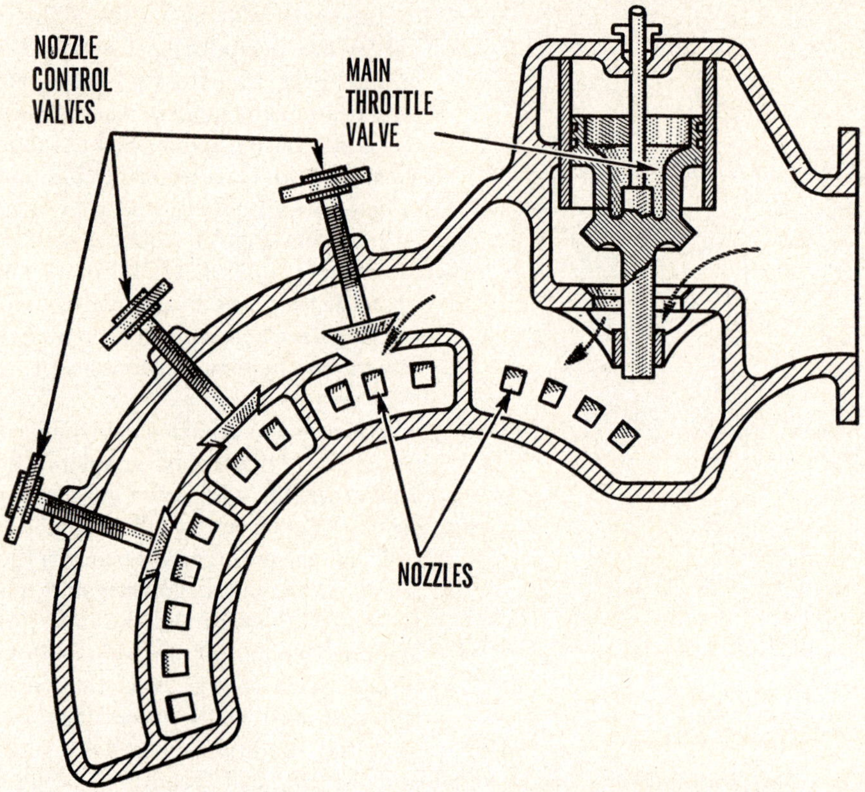

FIG. 4-28. Nozzle Valve, Manual Control Arrangement

stage varies only slightly at varying loads. On the other hand the wheel speed varies over a wide range. For a turbine designed with the ideal velocity ratio at high powers, it appears that smaller stage enthalpy drops and consequent lower nozzle velocities would be desirable at low powers. With a given total enthalpy drop, the only way to obtain smaller stage drops is to add more stages.

The use of the cruising turbine permits variation in the number of stages. The cruising turbine, placed ahead of the high pressure turbine, is added to the steam flow path at lower powers and bypassed at higher powers. In addition, on many naval turbines at very high powers the first few stages of the high pressure turbine are bypassed. The combination of these two schemes permits maximum conversion of thermal energy to work in each used stage over a wide speed range. A typical bypassing arrangement is shown in Figure 4-29. Efficiency considerations aside, bypassing also permits steam to enter the turbine at a point where the total nozzle area is greater than in the first stage, and so permits greater flow rates and consequently a higher power from a given turbine. The bypassed stages do no effective work. The valve areas are so arranged as to cause a small pressure drop across these stages, so as to provide a continuous flow of steam to prevent a dangerous build up of temperature due to windage in "dead," or nonflowing steam.

One disadvantage of bypassing in the high pressure turbine is the exposure of the turbine casing to essentially line pressure and temperature and the turbine blading to the high temperatures. The avoidance of this condition dictated use of a velocity compounded first stage wheel as previously discussed. With steam conditions (600 psig, 825°F) in common use in recent years, the blading could not be *continuously* exposed to the temperature, but exposure for the relatively small percentage of total turbine life spent at high powers was accepted. With more recent steam conditions (1200 psig, 950°F) even short periods of exposure are unacceptable and the velocity compounded wheel is not by-passed.

Another scheme for varying the number of stages is embodied in the CVA-60 turbine, shown

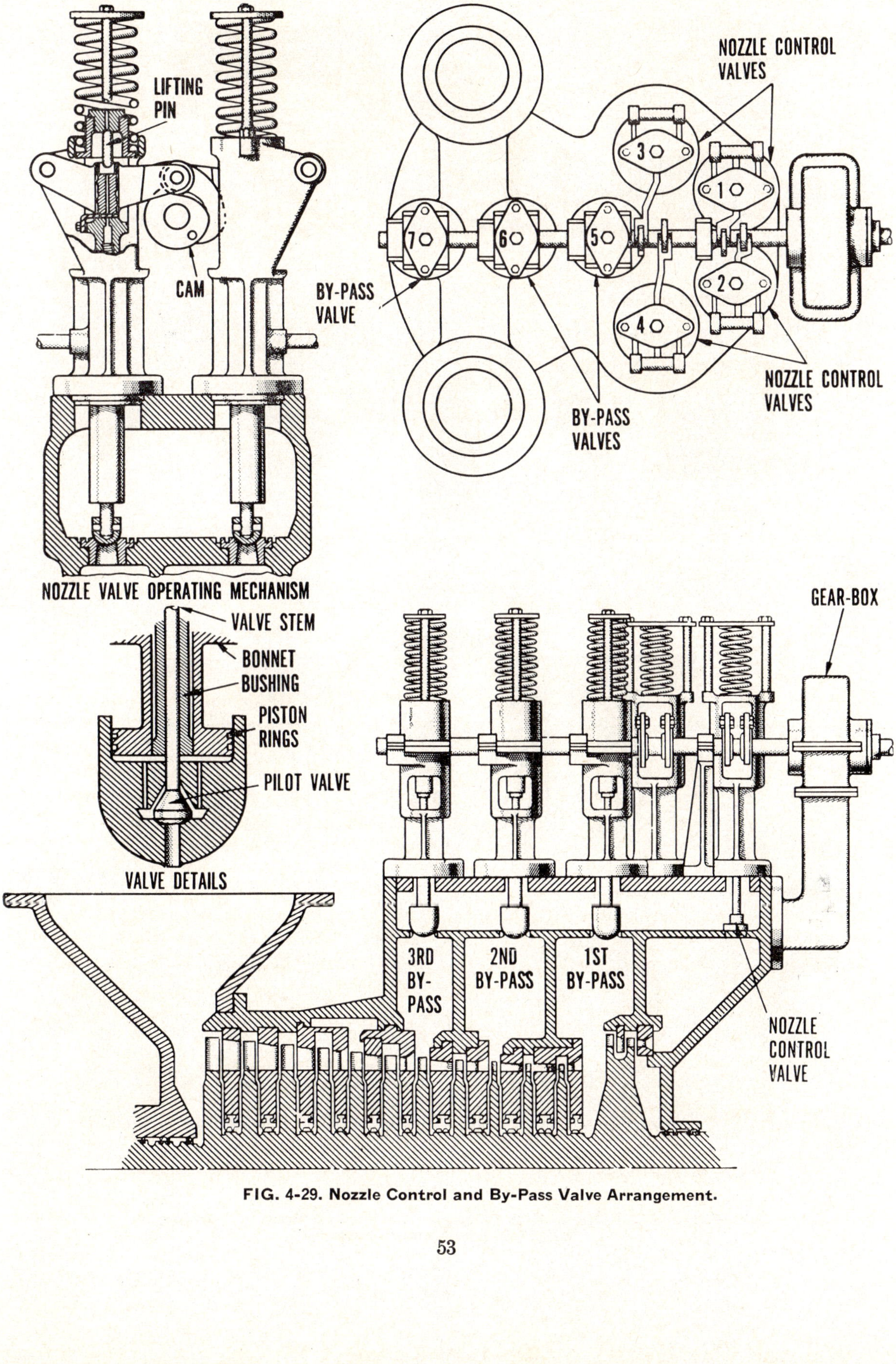

FIG. 4-29. Nozzle Control and By-Pass Valve Arrangement.

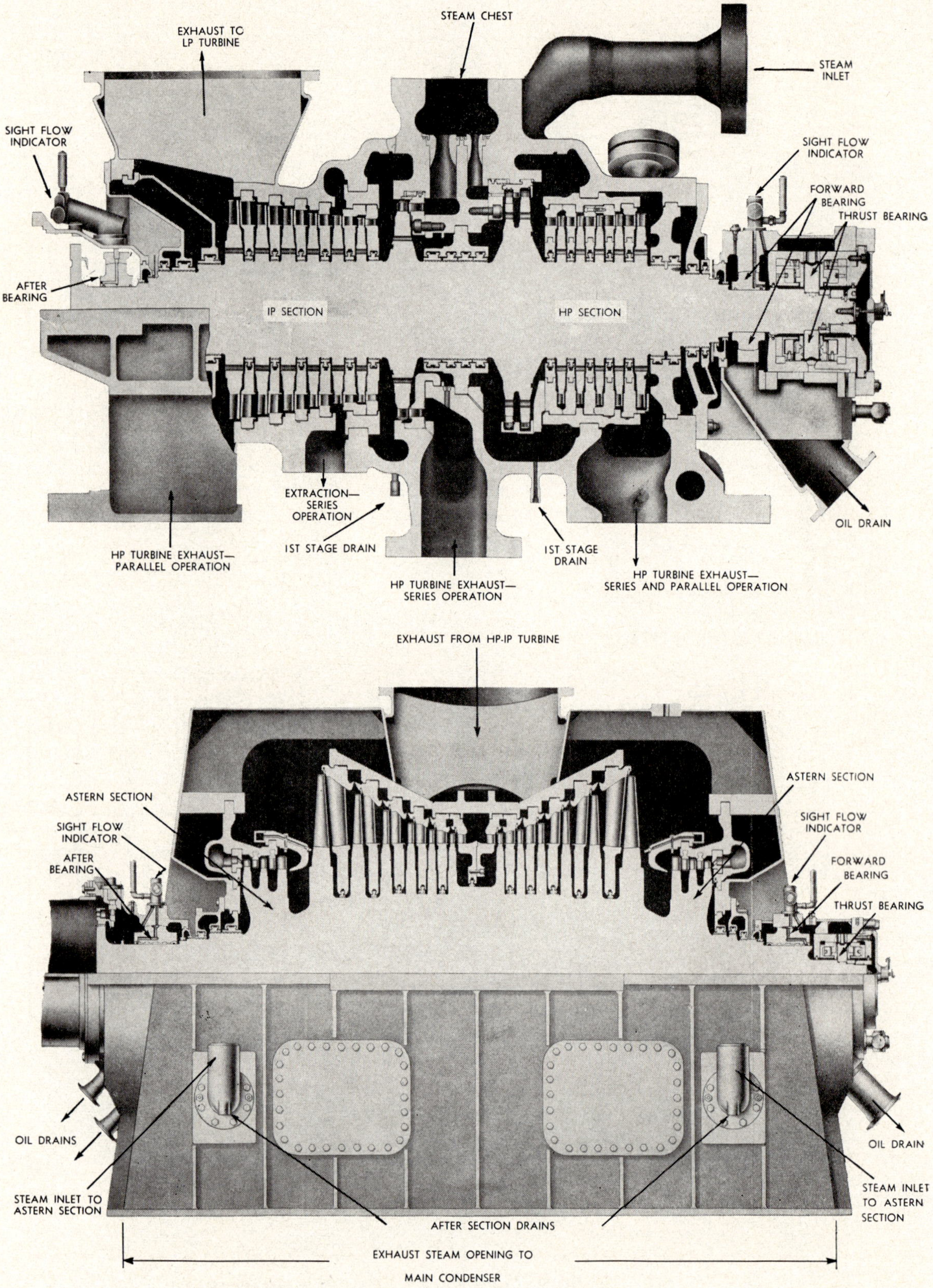

FIG. 4-30. (a) CVA-60 High Pressure—Intermediate Pressure Turbine. (b)CVA-60 Low Pressure Turbine.

in Figure 4-30. In this turbine, line steam is admitted in the center of the High Pressure-Intermediate Pressure unit. At low powers, only the nozzle control valves to the HP section are open. Steam flows through the velocity compounded wheel and the five remaining stages of the HP section, then through a transfer line to the seven stages of the IP turbine, from which it passes to a conventional low pressure turbine. At high powers, the nozzle control valves to both sections are opened. The entering steam divides and a portion passes through the HP and IP sections simultaneously. Valves in the transfer line are arranged to join the exhausts from these two turbines to pass together into the low pressure unit. A large enthalpy drop is taken in the first stage IP (note the larger wheel diameter giving greater blade speed) although it is not velocity compounded. The turbine also has an extraction opening to provide steam to the D.F.T. for feed heating, a feature seldom found on naval turbines at the present time.

Good turbine control, then, embodies a steam admission arrangement to minimize throttling losses, and a scheme for varying the number of stages. Such complex controls are seldom found on small auxiliary turbines, which are usually controlled by a simple throttle valve; the higher percentage losses being acceptable because of their relatively low powers. Ship service generator turbines, being larger, have nozzle valve control, but no bypass arrangement.

424. NAVAL TURBINE TYPES.

Previous discussion has made reference to the classifications "impulse," "reaction," "High Pressure," "Low Pressure," and others. Many descriptive terms are used in categorizing turbines. They are classified generally under the following headings:

(a) *Type of staging.* The terms "impulse" and "reaction" have been previously used to describe the thermodynamical events taking place in the stages. They are also common terms applied to a description of the turbine as regards its construction. As mentioned before, turbines made up of several impulse stages are characterized by the bucket-wheel, nozzle diaphragm type of construction as shown in Plate II. When no velocity compounded stage is present, such a unit is called *a pressure compounded impulse* turbine. When a velocity compounded stage is present, the unit is termed a *pressure-velocity* compounded unit.

Turbines comprising reaction stages are typified by a drum rotor construction, usually having the fixed blades secured directly to the casing, with the moving blading fixed directly to the drum rotor. In the past, numerous combinations of reaction and impulse construction have been used in the same unit. Presently the rotor construction is seldom found on H.P. naval turbines, being confined almost exclusively to the L.P. unit.

It is again emphasized that the impulse type construction often contains blading designed for a certain degree of reaction. This is especially true in those L.P. units which use the impulse type construction.

Summarizing, present design almost invariably includes the impulse construction in the H.P. (and cruising) units, and may utilize either reaction or impulse construction in the L.P. units.

(b) *Flow path.* Turbines may be described by the flow path of the steam as it passes through them. If the complete expansion takes place by flowing through a single turbine from one end to the other (as it does in the ship's service generator turbine) it is called a *single flow* turbine. If it takes place in two separate turbines mounted on the same shaft, it is called a *tandem flow*, or tandem compound flow turbine. If the complete expansion takes place by passing in series through two separate turbines, on separate shafts but geared to a common output, it is called *cross-compound flow.* If the steam enters at the center of the turbine and divides to flow toward each end, it is called *parallel flow* or *divided flow.* Present naval practice in high powered units consists of a cross-compounded turbine unit with a divided flow L.P. turbine. In units of lower power, a single flow L.P. turbine may be used. The CVA-60 unit described in Article 423 is an example of a divided flow H.P. unit, though this is termed a series-parallel unit to indicate its operation in both modes.

Auxiliary turbines are classified in two other descriptive ways, (a) by direction of flow with respect to the turbine wheel and shaft (radial, axial or tangential), and (b) by repetition of flow. These will be discussed later.

(c) *Astern turbines.* It is desirable here to mention provisions for astern propulsion to complete the major operational characteristics of propulsion turbines. Astern operation is provided for by the inclusion of astern stages in the L.P. turbine. These stages generally consist of one or two velocity compounded stages, sometimes in combination with an impulse stage or stages. These elements are located at each end of divided flow L.P. turbines or at the exhaust end of single flow L.P. turbines. Steam admission is controlled by an astern throttle valve. Since these are few-stage turbines, they are characterized by high steam exit velocity, high leaving losses, and generally poor efficiency. To minimize overheating caused by astern operation due to the "compressor" effect of the ahead blading working on the astern steam exhaust, a baffle is placed between the last stage of the ahead blading and the astern element. See Plates III and IV. This helps to route the astern exhaust flow directly to the condenser. Nevertheless, overheating of the ahead stages occurs during extended periods of astern operation because of friction and windage losses due to dead steam surrounding the ahead elements. No practical provision has been developed to prevent this, and extended time of operation at full astern power is limited for this reason.

425. TURBINE MECHANICAL FEATURES.

Mechanical features of turbines discussed herein are (a) the means for supporting the rotor, (b) the means for maintaining it in the correct axial position, (c) the means by which steam leakage between stages and leakage of steam out of or air into the turbine are minimized. Many other features such as the valve gear operation, removal and replacement of blading and diaphragms, and rotor balance are important to the operation and maintenance. Detailed description of all mechanical features are found in Instruction Books for specific turbines, available in each ship.

(a) *Rotor Support.* The turbine shaft pierces the turbine casing at both ends. Each end passes through a sealing arrangement as it leaves the casing to be discussed subsequently, and then through *bearings* which carry the weight of the rotor. The portion of the shaft passing through the bearing is called the *journal,* and must be finished with an exceedingly smooth surface. The journal rests in a sleeve type bearing (as opposed to a ball or roller type). Two important types are shown in Figures 4-31 and 4-32. Note that both are provided with a babbitt lining which is cast

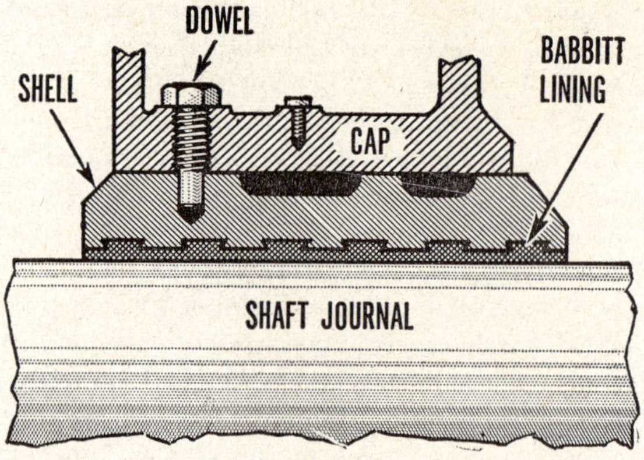

FIG. 4-31. Cylindrical Journal Bearing.

into a steel shell. The shell is, in turn, mounted in a housing rigidly attached to the turbine casing and supported on the turbine foundation. The housing is flanged so that its upper portion, or cap, can be removed to take out the bearing. Bearings are made in two halves, so that the upper half can be lifted off when the cap is removed and by jacking the rotor weight off the lower half, the latter can be rolled out around the shaft. The upper half is doweled to the cap to prevent rota-

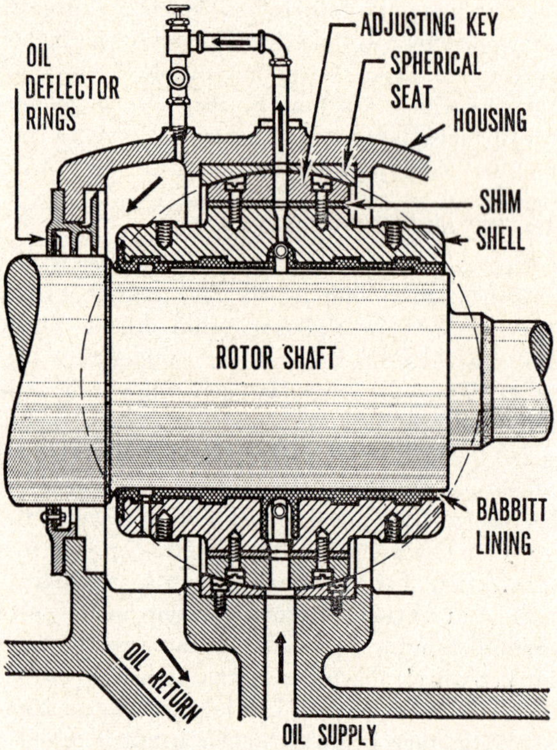

FIG. 4-32. Spherically-seated Journal Bearing.

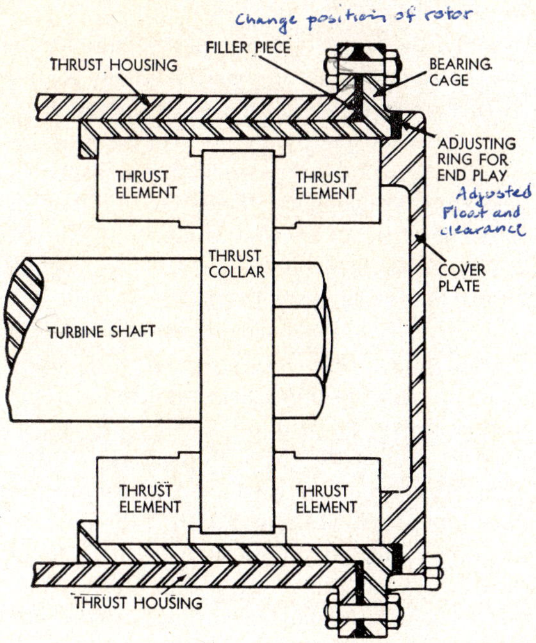

FIG. 4-33. Schematic of Turbine Thrust Bearing.

tion of the bearing with the shaft. The spherically seated type in Figure 4-32 is most commonly found on present installations. The spherical seating feature allows the bearing to automatically seek exact alignment with the shaft. Note the provision for positive lubrication, and the oil deflector rings which prevent oil loss from the lubrication system and its contamination by steam present on the turbine casing side of the bearing.

In addition to carrying the weight of the rotor, the bearing provides a means of adjustment for its exact radial position. This is important since radial clearances are critical in the turbine labyrinth packing and, in many cases, between the blade ends and the casing. Initial radial adjustment and subsequent compensation for wear can be made by the selection of proper shim thicknesses between the shell and the housing, or seat. See Figure 4-32.

Another type journal bearing in recent use is the pivoted shoe type, which will be discussed in Article 430, in the section on lubrication.

Bearing loads of journal bearings are measured by dividing the total weight carried by the bearing by its projected bearing area, the length times the diameter. Bearings are made large enough to limit these unit pressures to about 200 psi.

(b) *Rotor Axial Position.* Since the forces acting on the blades cannot be maintained exactly in the direction of blade motion at all speeds, some force in the axial direction is always present on the turbine rotor. It is necessary to absorb this force while maintaining the close axial clearances required between the moving blades and the fixed blades, or nozzles. The turbine thrust bearing performs this function. It is usually located at the extreme forward end of the turbine shaft. It is diagrammatically represented in Figure 4-33. The thrust collar is made of smooth surfaced steel and the thrust elements are babbitt lined. The thrust elements are of the pivoted shoe type and their action will be fully described in Article 429. Every turbine is provided with a device to indicate the axial position. To set this correctly and to adjust it after thrust bearing wear, the filler piece and adjusting ring shown in Figure 4-33 are provided. Note that the adjusting ring provides for adjusting the bearing clearance

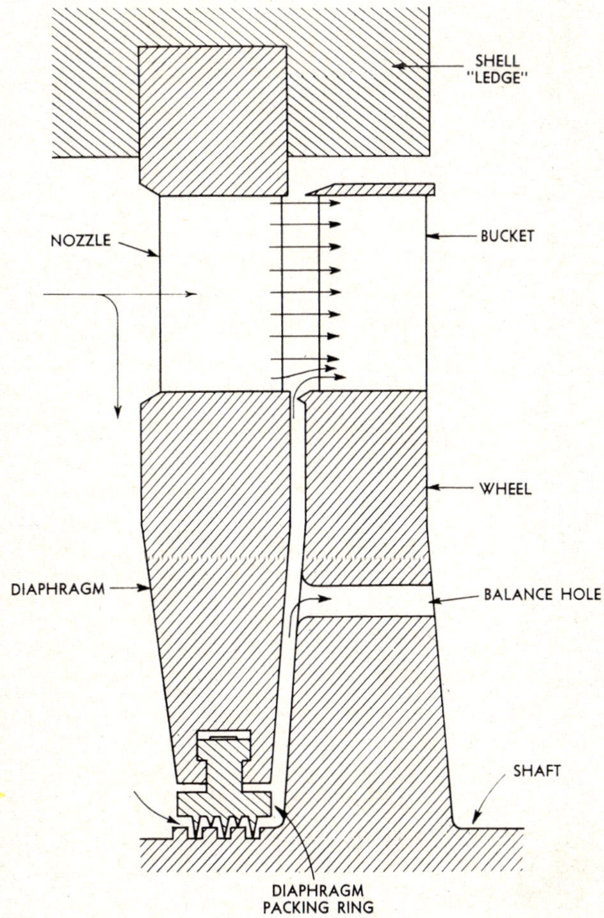

FIG. 4-34. Steam Leakage in Impulse Stage.

(clearance between thrust collar and thrust elements) and the filler piece provides for adjusting the axial position of the bearing and rotor, thus providing for correct axial clearances within the turbine.

(c) *Seals.* The necessity for sealing between stages has been previously discussed. The leakage path in the impulse stage is between the rotating shaft and the fixed diaphragm, as indicated in Figure 4-34. In the reaction stage, the leakage path is around the ends of the fixed end moving blades, as shown in Figure 4-35. In both cases the small annular passages between the sharp projections and the shaft, or casing, act as flow-restricting orifices. These are effective to the degree of reducing to about 4-5% the amount of steam that would pass through a minimum friction passage (or nozzle) with the same cross section flow area and the same pressure drop. The effectiveness increases with the number of "teeth" in the packing, and longer labyrinths will be found between stages where the designed pressure drop is higher. The cross section area is kept small by limiting the clearance between the packing projections and the shaft to about 0.010″.

Another vital sealing point is the juncture of the shaft and turbine casing. The seal here serves two functions: (a) It minimizes steam leakage from the casing when the casing (shell) pressure is higher than atmospheric; (b) it prevents the leakage of atmospheric air into the turbine (and thus into the condenser) when the shell pressure is less than atmospheric. Both possibilities may exist at one time or another at all seals, except at the low pressure exhaust end, where the shell pressure is always lower than atmospheric.

Labyrinth type seals, called labyrinth packing, are used alone or in conjunction with carbon packing rings at points where high temperatures and pressures may be present. The carbon packing, when used with labyrinth packing, is always on the outboard end of the latter, where pressures and temperatures are low. Figure 4-36 illustrates an arrangement. At points such as the L.P. turbine exhaust end, carbon packing alone may be used on some installations, though often labyrinth packing is found here also.

In Figure 4-36, the leak-off points will be noted. When the shell pressure is above atmospheric, these bleed off leakage steam to prevent its loss. Steam throttled from the auxiliary exhaust line is supplied under low pressure, $\frac{3}{4}$ to $1\frac{1}{2}$ psig, to the gland seal connection. This steam divides, a part leaking back along the labyrinth

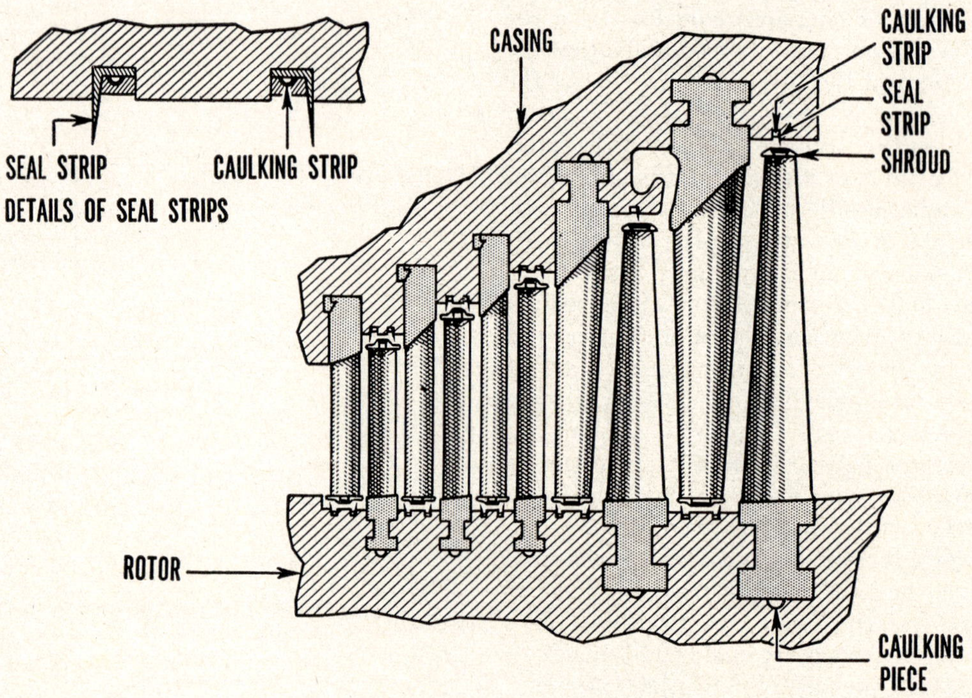

FIG. 4-35. Typical Reaction Blading.

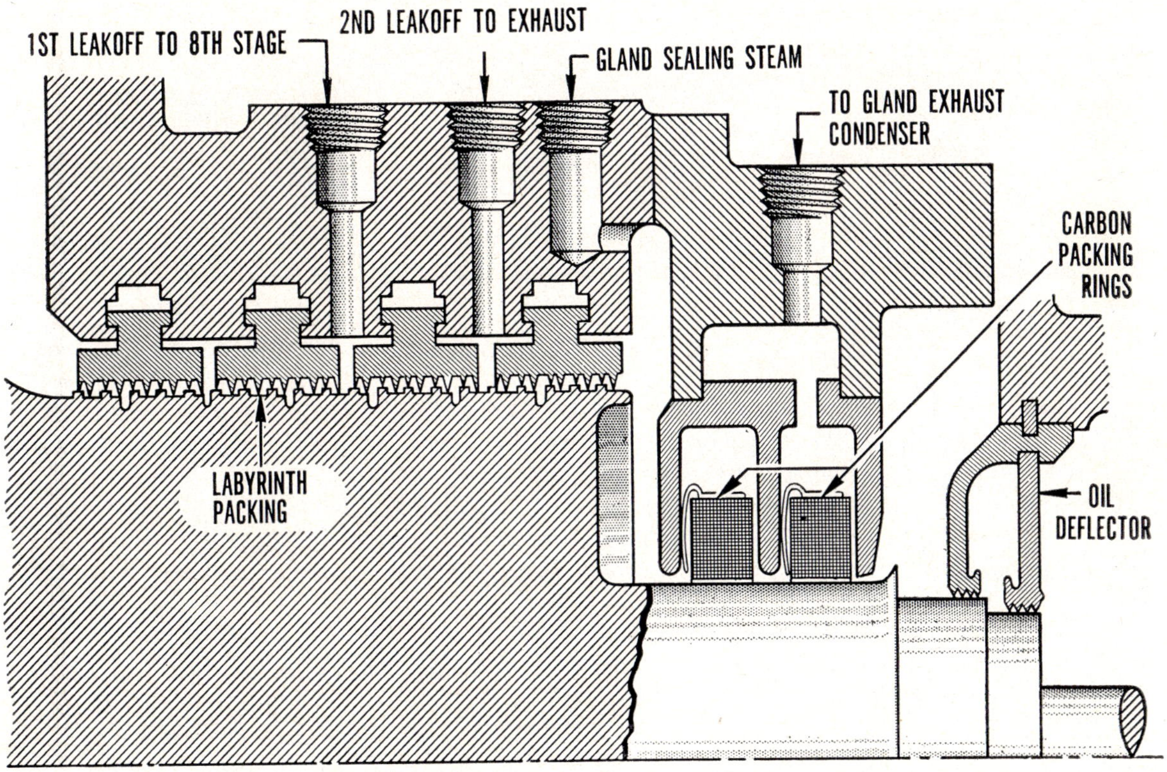

FIG. 4-36. Typical High Pressure Turbine Gland.

packing to the exhaust connection, a part through the carbon packing (or additional labyrinth packing in many cases) to the shaft end. This is a small quantity, and condenses here and drips off, the remainder passing into the atmosphere. This steam is from a system so valved that it supplies steam only when the shell pressure is lower than atmospheric, its flow effectively preventing the leakage of air through the packing. The detailed arrangement of the gland seal system differs considerably on various installations. The carbon packing rings are made up in segments, held tightly around the shaft and against vertical sealing faces by springs.

LUBRICATION

426. LUBRICATION PRINCIPLES AND SYSTEMS.

It is difficult to imagine any physical process in which energy is not wasted, or converted to a non-usable form. Even the kinetic energy of the rotation of the planets around the sun is gradually being dissipated. In all machinery plants, loss of energy in the process of mechanical friction is one of the major design and operating difficulties. Reduction of this mechanical friction, and of its consequent waste of energy is the major reason for lubrication. In addition to reducing loss of energy proper lubrication prolongs the life of machinery, and in many cases, extends its operational characteristics into higher speed and power regions. It is important to examine the characteristics of friction before studying the necessary steps for its reduction.

427. FRICTION.

When two solids are in contact under even a small amount of pressure, forces which tend to prevent their relative motion may arise from three sources: *adhesion, shearing,* and *plowing.* In Figure 4-37(a), the adhesive action is shown. The two pieces of metal in all three figures represent finely machined surfaces, with their surface profiles enormously exaggerated in the picture. The force necessary to move the one piece relative to

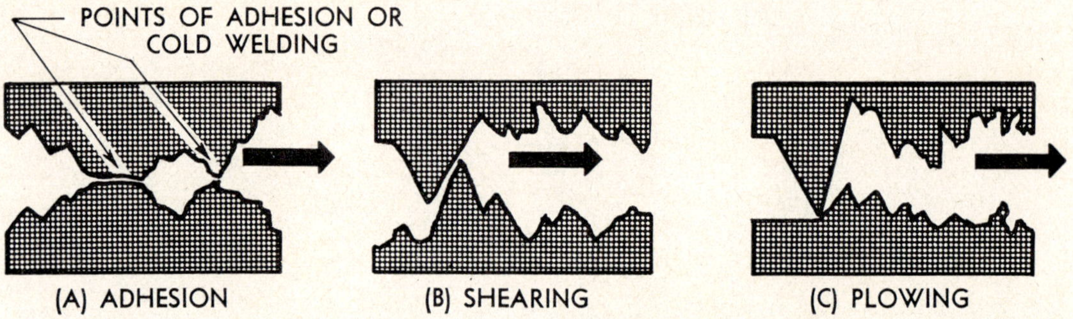

FIG. 4-37. Three Types of Friction.

the other must be sufficient to actually shear the metal at the welded points. Electrical conductivity tests have shown that the area of the welded surfaces is proportional to the load normal to the interface, so that the contact stress (normal load divided by contact area) remains practically constant *regardless of the total area of the rubbing surfaces*. This explains why total contact area has negligible effect on the friction force. (It will be recalled that a rectangular block will commence sliding down an incline at practically the same angle, irrespective of whether it rests on either of its sides or its end.) Furthermore since the shearing force is proportional to the total welded area, and since the total welded area is proportional to the normal load, the shearing force is also proportional to the normal load. This proportionality constant is, of course, the coefficient of friction. Carefully controlled experiments and microscopic examination have shown that practically perfect welds actually take place, even between dissimilar materials. This process is believed to continue even when relative motion exists between the contact faces. Adhesion is thought to be the primary cause of friction between well machined surfaces.

Shearing is shown in Figure 4-37(b). It is brought about by the shearing off of the mutually interfering projections. This action is believed to occur particularly with rough surfaces, and especially in "new" contact areas, being significantly present in new bearings and disappearing somewhat after the bearings are well "run in." The action doubtless is accompanied by adhesion, since the fracture energies of the projections cause high, localized temperatures.

The third major type is plowing, and is present when the two surfaces have different hardnesses. (See Figure 4-37(c).) Plowing may plastically deform the softer material without fracture.

The presence of foreign solid material between the two areas can cause any or all of the above actions.

It is noted that in rolling friction, the latter two types of friction are probably largely eliminated, but adhesion forces still exist.

428. LUBRICATION.

It is the function of lubrication to separate the two metallic surfaces, so that they do not touch, and thus to substitute smaller forces of fluid shear for the types of forces described above. Three mechanisms of lubrication are well understood, *hydrostatic*, *hydrodynamic*, and *boundary lubrication*.

Hydrostatic lubrication depends on oil pressure supplied from an outside pump to keep the surfaces apart. If the surfaces shown in Figure 4-38 are rectangular, and high pressure oil is forced into the supply orifice, the surfaces will separate so that the peripheral area around the step in the lower plate will be large enough to accommodate the supply flow. Tremendous loads can be accommodated in this manner. The moving part of the Hale telescope, weighing approximately 1,000,000 lbs, is supported on three such bearings or pads, with a combined recess area of 84 square inches, and it is operated by a 1/12 H.P. clock

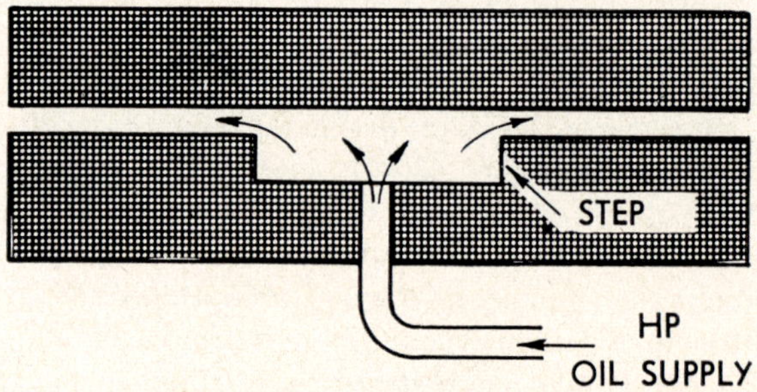

FIG. 4-38. Hydrostatic Lubrication.

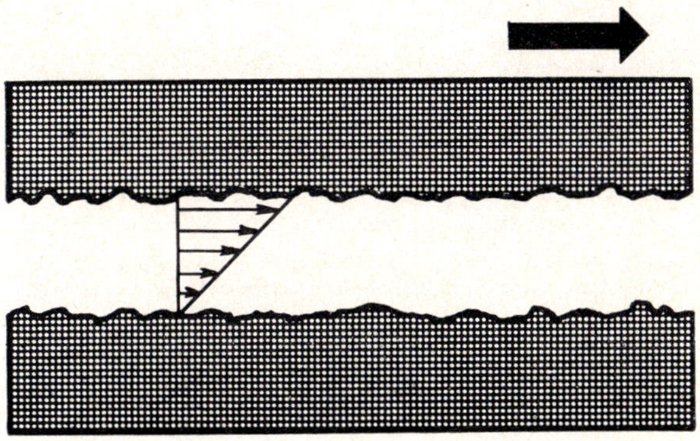

FIG. 4-39. Oil Film Velocity Distribution.

motor! When several such bearings support a weight, it is a virtual necessity that a separate pump be supplied for each bearing so that maintenance of the flow to each bearing can be insured. Tilt of the load to one bearing would allow all the flow of a single pump to pass through the others. Therefore, this system has not been adopted for naval machinery use, the additional necessary pumps complicating the reliability requirement.

Hydrodynamic lubrication also depends on oil pressure separation of the bearing surfaces, but here the pressure is developed by the motion, or pumping action of the bearing itself. When two surfaces move past each other, as in Figure 4-39, oil clings to both surfaces. The oil at the upper surface has the velocity of that surface; the oil at the lower surface is at rest. The velocity of the molecules of oil between the surfaces varies from one to the other. It will simplify the subsequent discussion to assume a linear velocity distribution as shown in the figure. Since flow is set up, there is a pressure build up. The discussion at Article 429 will consider this in detail. This is the type lubrication found in nearly all large shipboard bearings.

Boundary, or thin-film lubrication results from an extremely thin layer of oil, in the order of one ten-millionth to a maximum of one forty-thousandth of an inch in thickness, which is "bonded" to the metal by adsorption or, in some cases, by chemical reaction. Furthermore, the molecules of oil in contact with the surfaces are capable of resisting displacement to a great degree. The molecules (long hydro-carbon chains) are believed to be bonded at one end to the surface and to orient themselves approximately perpendicularly. "Oiliness" is the term which describes the ability of a lubricant to form such a boundary film. It is found to a high degree in animal and vegetable oils and the chemical family of fatty acids, but to an extremely limited degree in natural mineral oils. However, small amounts of additives in mineral oils are sufficient to impart to it the property of oiliness. Boundary lubrication is schematically illustrated in Figure 4-40. The film gradually squeezes out under load and with time. This film is the only lubrication

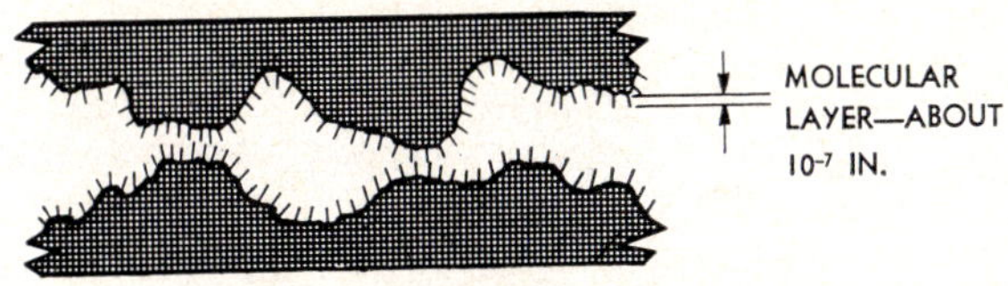

FIG. 4-40. Boundary Lubrication.

available to the friction surfaces of shaft journals when starting up. For this reason, it is important to keep it established by periodic "turning over" or operation of machinery. This type lubrication is important also in ball and roller bearings, and is the principal type of lubrication in gear meshes.

429. HYDRODYNAMIC FILM FORMATION—THE THRUST BEARING.

It will be useful to examine first the formation of the hydrodynamic film between two plane bearing surfaces.

Imagine that Figure 4-41 represents the cross-section of two plane surfaces A and B, both of which are infinitely long in a direction perpendicular to the plane of the paper. A is fixed, while B is moving at constant speed. Assume, in the first case, that the two surfaces are separated by a continuous film of oil, and that there is no force tending to cause one surface to move nearer the other. Now, the oil will cling to both surfaces, and hence the particles of oil adhering to A will be stationary, while those adhering to B will have the same velocity as B. Particles lying between A and B will have a velocity proportional to their distance from A. Hence, if bb_1 represents the

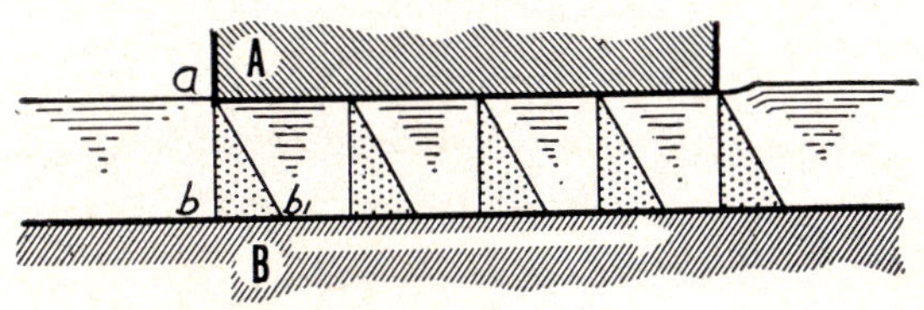

FIG. 4-41. Velocity Distribution Due to Motion Only.

velocity of B, the triangles abb_1 are velocity triangles, and the dotted area represents the volume of oil passing any section per unit depth perpendicular to the plane of the paper. Hence the same volume of oil flows past each section, and the arrangement simply constitutes a pump.

Assume, now, that both surfaces A and B are at rest, but that a force is applied which will gradually cause the two surfaces to approach each other and to squeeze out the oil film. If the dashed line represents the initial conditions in Figure 4-42 and the full lines the conditions at some other time, it is clear the areas a_1 and a_2 of the bulges will be equal respectively to $x_1\Delta t$ and

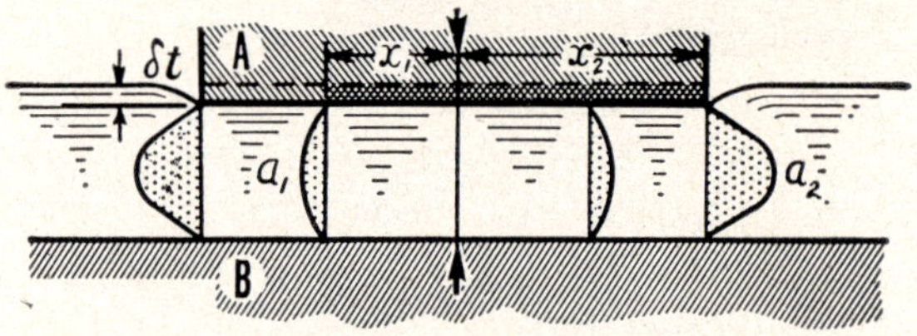

FIG. 4-42. Velocity Distribution Due to Pressure Only.

$x_2\Delta t$. It will be understood that the curved bulging lines represent successions of oil molecules which were originally represented by the straight lines. The existence of the bulging lines is evidence of an internal pressure in the oil, which is momentarily balancing the external forces applied to A and B.

Now consider that the conditions assumed in the foregoing considerations exist simultaneously, i.e., that B is moving and that a force is applied to A and B tending to bring them together as in Figure 4-43. It is to be further assumed that the surfaces are so constrained as to remain parallel. From what has been said before, it will be apparent at once that the dotted areas in Figure 4-43 represent the volume of oil passing any section of unit depth in unit time. But these areas are not equal, and it would seem that more oil is passing out from between A and

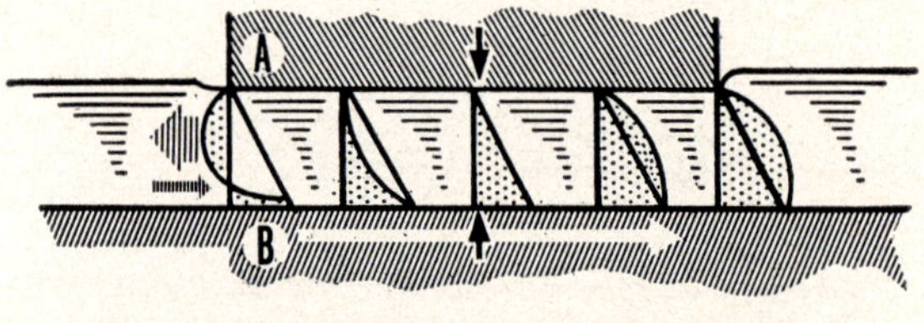

FIG. 4-43. Velocity Distribution Due to Combined Motion and Pressure—Surfaces Parallel.

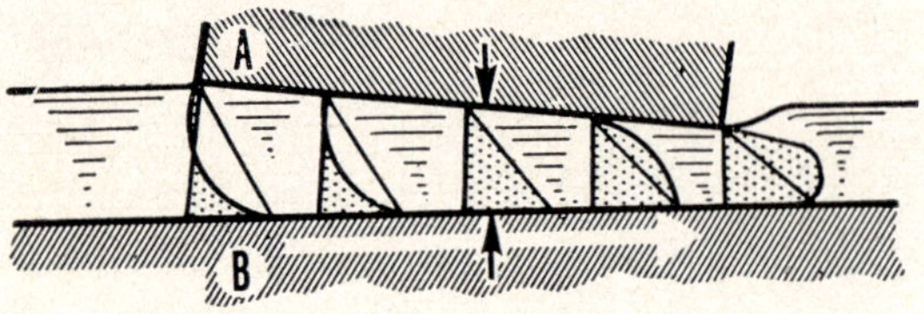

FIG. 4-44. Velocity Distribution Due to Combined Motion and Pressure—Equilibrium Condition.

B than is entering, and therefore that in a short time no oil will separate the surfaces A and B, and they will be in metallic contact.

The conditions assumed are those which exist in that type of thrust block where one or more collars, turned on the shaft, abutt against fixed shoes. The shoes are lined with white metal, and the oil is fed into the space between the shoes and collars at various points. In actual practice the performance of this type of thrust bearing is notoriously bad at high speeds, and the maximum allowable pressure is only about 40 to 50 lb. per sq. in. of effective area.

Imagine, now, that the block A is freed from constraint and allowed to tilt at any angle to the surface B as shown in Figure 4-44. Then if the angle is such that the dotted areas are equal at all points, a constant quantity of oil will flow through the bearing; and since the oil is dragged in due to its adherence to the surface B and its internal viscosity, the surfaces A and B will always be separated by a film of oil.

The principle of the pivoted-shoe thrust bearing is that of the wedge-shaped film described above (Fig. 4-44). When the film is supplied constantly with fresh oil, there is a complete separation of the surfaces, and hence no wear.

The pivoted-shoe thrust bearing (Kingsbury thrust bearing) makes possible the automatic formation of wedge-shaped oil films under a thrust load. This result is secured by dividing the stationary bearing element into segments (called shoes), usually three or six in number (Figure 4-45). These segments are so supported and pivoted that they are free to tilt slightly. The rotating bearing element (or collar) is attached to, and turns with, the turbine or propeller shaft. The entire assembly is submerged in oil. When the shaft rotates, some oil is dragged in (between the collar and each shoe) at the leading edge of the shoe. The thrust on the shaft or collar naturally has a tendency to squeeze the oil out again. It can leave at three sides of the shoe. Hence the trailing

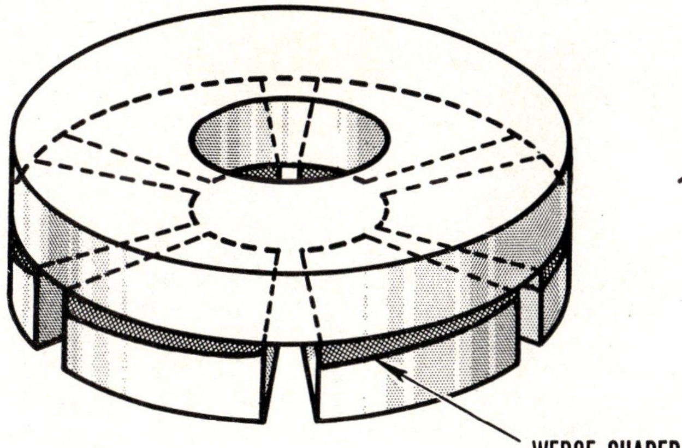

FIG. 4-45. Thrust Collar and Shoes.

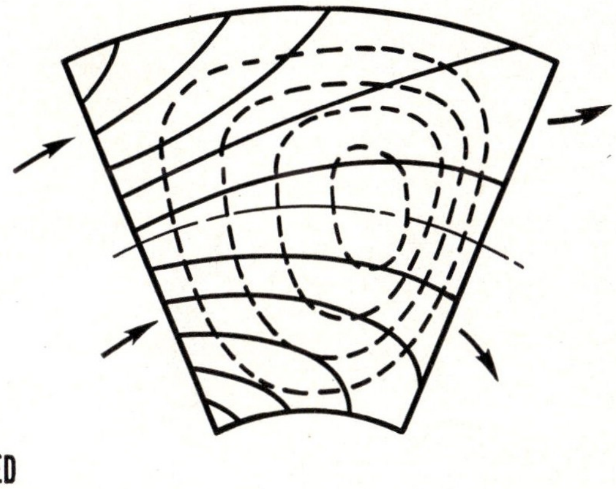

FIG. 4-46. Flow Lines and Pressure Distribution on Shoes.

edge need present a much narrower passage for the oil than the leading edge. The wedge-shaped film is established. It should be obvious from the above discussion that the oil film automatically assumes whatever taper is required by the speed, load and oil viscosity. The bearing pressure of a typical pivoted shoe bearing is about 250 lb. per sq. in. of effective area. The flow is as indicated in Figure 4-46, which also shows the lines of equal pressure, and indicates the region of peak pressure.

The essential elements of a Kingsbury thrust bearing are:

(1) The thrust collar made of steel and keyed to the turbine or propeller shaft; the collar transmitting the thrust from the shaft to the oil film and the shoes;

(2) The tilting segments, called shoes, which maintain the proper oil wedge between the collar and the shoes and transmit the thrust to the leveling plates;

(3) The upper leveling plates upon which the shoes rest and the lower leveling plates upon which the upper leveling plates rest; the function of these leveling plates is to equalize the thrust load among the shoes; and

(4) The base ring, upon which the lower leveling plates rest. The base ring is the support which holds the leveling plates in place and transmits the thrust on them to the ship's structure. Figures 4-47 and 4-48 are developed and exploded views, respectively, of a Kingsbury thrust bearing.

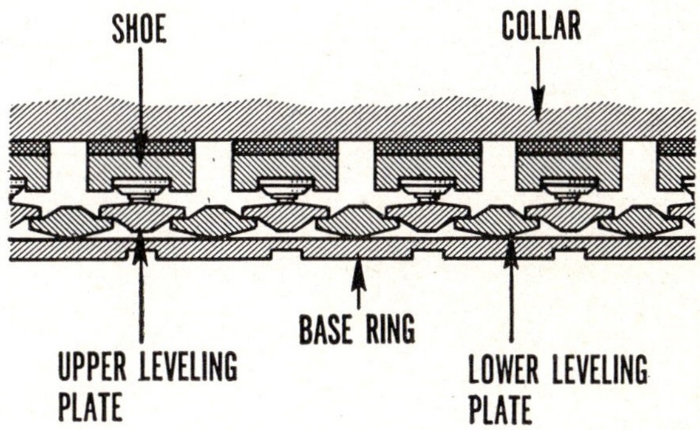

FIG. 4-47. Schematic Cross Section of Leveling Plates, Shoes and Collar.

430. HYDRODYNAMIC FILM FORMATION—THE JOURNAL BEARING.

The formation of the wedge shaped film in a journal bearing is very similar to that in the thrust bearing. Figure 4-49 shows the progressive formation as the journal accelerates from zero to normal speed. In (a) the journal is at rest. In (b) it is rotating slowly, and due to high friction is "climbing up" the left side of the bearing, stopping at a position where the upward component of friction force balances the weight. (The position in the figure is exaggerated for clarity.) Meanwhile the journal is dragging or pumping oil into region *A*; increased pressure here tends to move the journal back to the right, resulting in the position shown in (c). But the journal continues to pump oil in to region *A* and the shaft seeks its final position at running speed to the right of the center of the bearing. Figure (d)

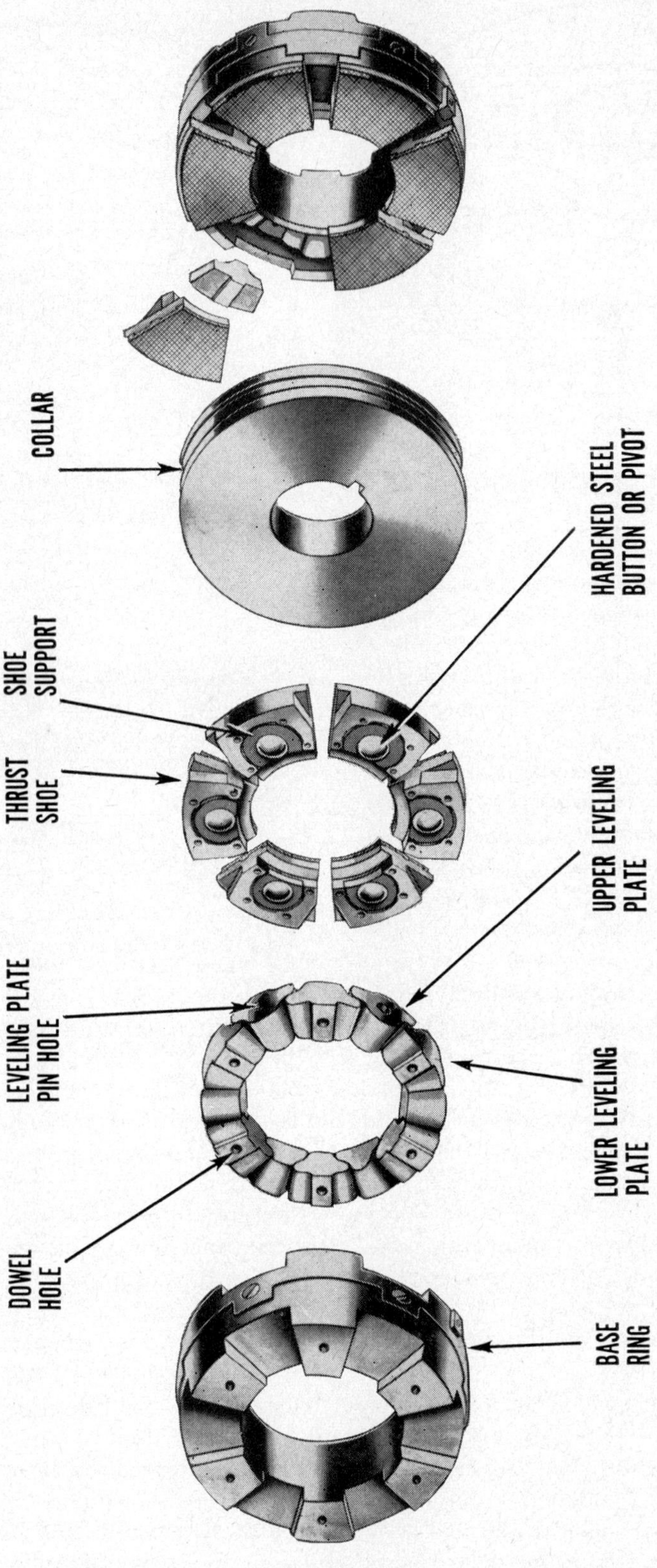

FIG. 4-48. Exploded View of Pivoted Shoe Thrust Bearing.

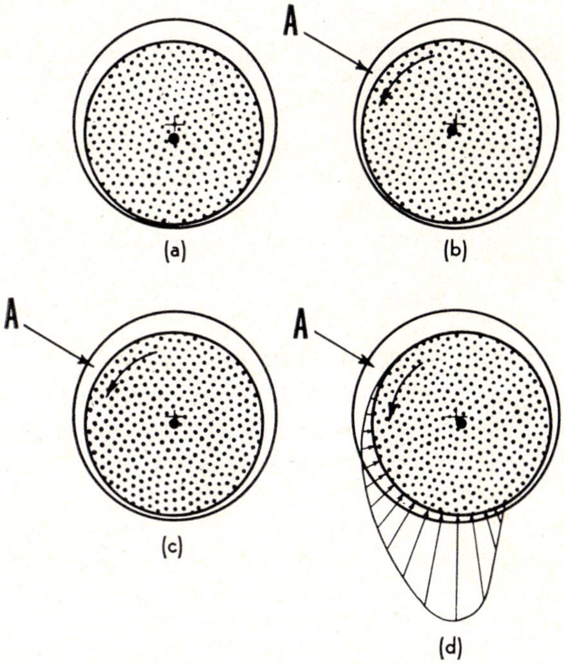

FIG. 4-49. Formation of Oil Film in Journal Bearing.

shows this together with the pressure distribution in region *A* which combines in a resultant force *R*. This force holds the shaft in equilibrium up the "incline" of the right side of the bearing.

Hunsaker and Rightmire (see reference at end of chapter) have shown that the resultant force due to the film pressure is always at right angles to the line of centers. The journal, therefore, always positions itself in the bearing so that the line between the center of the journal and the center of the bearing is at right angles to the direction of the load. The magnitude of the load determines the degree of eccentricity of the journal center from the bearing center, larger loads causing a larger eccentricity. The above statements hold true for any shaft carrying a load which is constant in magnitude and direction. They do not hold true in cases where the magnitude varies, as in the crankshaft of an internal combustion engine.

In recent years, the pivoted-shoe bearing has been adapted to journal bearings. If the shoes are properly pivoted, they will have a freedom of motion which permits them to automatically align themselves to the shaft, and yet hold the shaft very close to its proper axial position. The motion of the center of the journal in Figure 4-49 may amount to several thousandths of an inch. A pivoted shoe bearing, depicted schematically in Figure 4-50, is capable of restricting this motion to a much smaller amount.

A phenomenon occurring in high speed, lightly loaded journal bearings is "oil whirl." It results from an inability of the forces present (see Figure

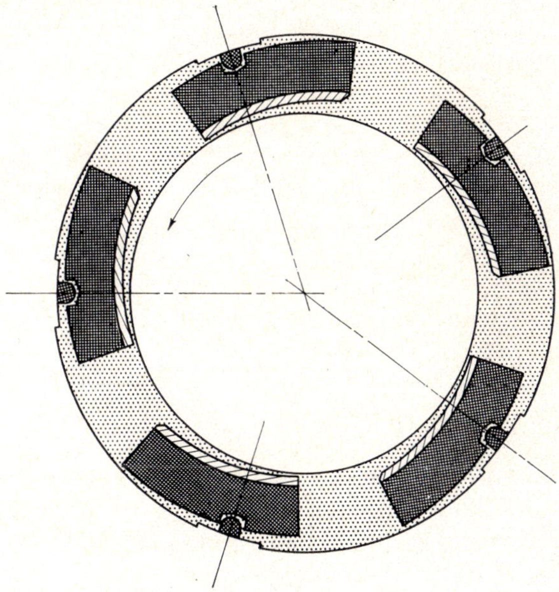

FIG. 4-50. Schematic of Pivoted Shoe Journal Bearing.

4-49) to reach equilibrium, with the consequence that the shaft whirls rapidly around in the bore of the bearing, setting up intense vibration. The shaft journal and bearing may be destroyed under these conditions. The pivoted-shoe journal bearing has been very successful in the remedy of this condition.

431. LUBRICATING OIL PROPERTIES.

It is evident that certain fluid properties are associated with proper lubrication. "Oiliness" has been discussed. Viscosity is another highly important property of a lubricant. Viscosity may be defined as the resistance of a fluid to shearing forces. High viscosity fluids will support heavier loads, but have a greater resistance to shear, or a greater internal friction. High viscosity fluids require larger bearing clearances. On the other hand, low viscosity fluids require less bearing clearance and have a smaller degree of internal friction. The selection, therefore, is a compromise, considering load, clearance, and operating temperature. The latter is important because viscosity decreases with an increase in temperature. Too high an operating temperature may result in an unacceptably low load carrying ability. Lubricants are now treated with additives which tend to suppress the change of viscosity with temperature.

As a practical matter, a good lubricating oil for turbines and reduction gears must have other characteristics. It must resist oxidation, which produces organic acids which may react chemically with the bearing materials to corrode the bearing surfaces. Oxidation is promoted by high temperatures and the presence of metal oxides or other impurities which may catalyze the reaction. The presence of water in the lubricating oil promotes rust, or oxidation of metal surfaces, and thus indirectly promotes oxidation of the oil itself. Additives acting as anti-oxidants and rust inhibitors aid in this problem. Since it is difficult to prevent the entry of any water into the lubrication system, emulsification of water and oil may occur, the resulting fluid having poor lubrication qualities, and promoting corrosion. Thus a good oil must have a high *demulsibility*, or ability to separate rapidly from an emulsion.

In summary, the requirements of a good turbine lubricant are:

(1) That it will readily form and maintain a boundary film. (Adequate oiliness)

(2) That it will form a hydrodynamic film with low friction losses. (Proper viscosity)

(3) That it will be resistant to chemical and physical changes. (Slow to oxidize and good demulsibility)

432. LUBRICATING OIL AS A COOLANT.

The foregoing discussion has been concerned with the use of lubrication to reduce friction and accompanying power losses. Friction cannot be eliminated. It has been shown that fluid friction replaces dry friction, but power is still lost through the generation of heat in the bearings. Lightly loaded or slow moving bearings can dissipate the heat generated by radiation and conduction, but energy in the order of a few hundred horsepower may be lost in the bearings of large turbines at high powers. Thus, an additional important function of lubricating oil is to serve as a coolant. This is accomplished by circulation and cooling in a heat exchanger. The circulation has a two-fold effect—it disposes of the heat generated in the bearings while simultaneously preventing the oil from reaching a temperature range where a change in its physical or chemical properties would endanger its lubricating properties. The system described in Article 433 is the major lubricating oil system in the propulsion plant. In addition to it, it is customary for each auxiliary component to have a self-contained lubrication system. These will be mentioned as the components are studied.

433. THE MAIN LUBRICATING OIL SYSTEM.

The main lubricting oil system, then, serves (a) to provide a continuous supply of oil to turbine and reduction gear bearings at the temperature for proper lubrication of the bearings; (b) to reject heat generated in the bearings; (c) to remove water and solid contaminants from the oil.

In order to accomplish these functions the lubrication system consists of the following essential parts: (a) Pumps for delivering oil to the various parts of the system; (b) Oil drain tanks or sump tanks in which the oil is collected after it leaves the lubrication points; (c) Oil coolers which keep the oil at the desired temperature; (d) Oil settling tanks in which water and other impurities settle out, and in which used oil is stored; (e) Oil strainers and filters which remove larger particles of foreign matter from the oil; (f) Oil purifiers, which remove water and other impurities that do not settle out in the settling tank; (g) Oil heaters which heat the oil prior to its entry to the purifier to facilitate purification, and which may be used to heat the oil when starting up the plant under low temperature conditions; (h) Gauges, thermometers, relief valves, and warming devices necessary to operate and control the system.

The following arrangement is one found on a large, modern aircraft carrier. It is described in some detail as representing a typical arrangement. The system described services one main propulsion plant.

Three pumps are provided for the system: (a) an attached pump which is geared to and driven by the main reduction gear, (b) a turbine driven pump, which has a governor control operated by oil pressure from the most remote bearing, and (c) an electric motor driven pump with a pressure switch control. Under normal, ahead operation, the attached pump delivers pressure to the system, varying with speed but

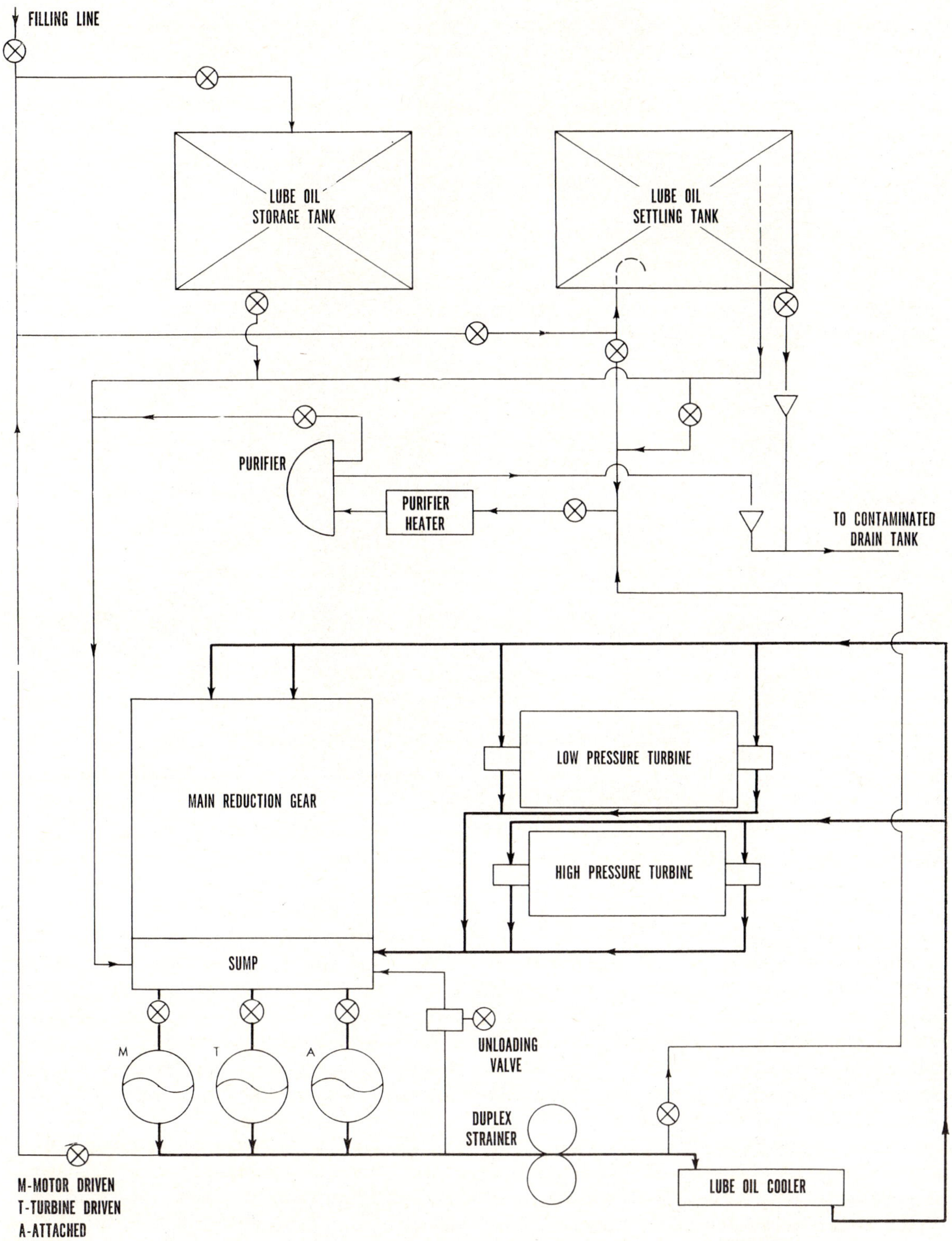

FIG. 4-51. Schematic Arrangement of Typical Main Lubricating Oil System.

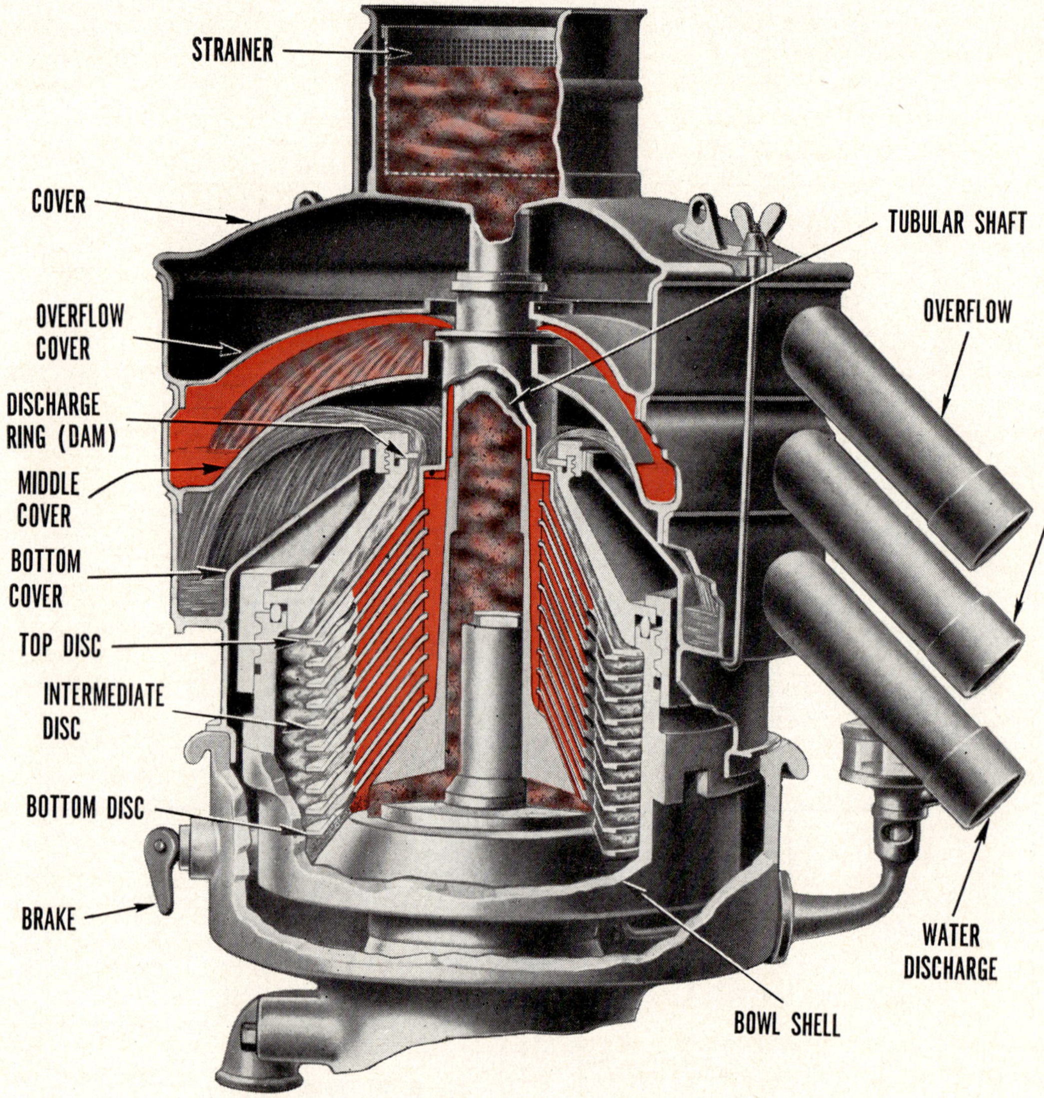

FIG. 4-52. Cross Section of Typical Disc Type Purifier Bowl.

limited to a maximum of 25 psig by an automatic unloading valve. The turbine driven pump is set to idle at 60 rpm in a stand-by condition while the attached pump is operating normally. At low speed, stop, or astern operation, the turbine governor is actuated when the pressure at the most remote bearing drops to 7 psig, and the pump turbine automatically speeds up and increases pressure to 25 psig. It cuts out automatically when the ship speeds up to the point where the attached pump can provide the pressure. The motor driven pump is normally secured, but starts up automatically if bearing oil pressure should fall to 5 psig. It must be secured manually. This pump can also be used as a lube oil transfer pump, and to supply lubrication when the engine is jacked over in port.

The sump is simply a drain tank located underneath the reduction gear, attached to the lower section of the reduction gear casing. Oil from the reduction gears, reduction gear bearings, and turbine bearings drains to the sump by gravity. The pumps take suction directly on the sump.

The operating pump delivers oil to the strainer, a duplex basket type so arranged that the oil can be routed through either basket while the other can be removed for cleaning. The baskets are fitted with removable magnetic assemblies to attract magnetic particles.

The oil flows from the strainer to the cooler, where it passes around a bundle of tubes through which salt water flows. By throttling the flow of salt water through the cooler the exit oil temperature is controlled. The oil then passes to the turbine and reduction gear bearing where it completes the cycle. Figure 4-51 is a simplified schematic diagram of the system.

A portion of the oil may be routed to the L.O. Settling Tank, where heating coils promote separation of water and solid contaminants; or a portion may be continuously fed to the purifier where mechanical separation occurs. When underway, the purifiers are operating about half the time. About once a month, when the plant is secured, all of the oil is pumped from the sump to the settling tank. The sump is cleaned out after which the oil is returned through the purifier.

Figure 4-52 illustrates the construction of a disc type purifier. Oil enters the top through a strainer and passes down the tubular shaft. It then passes into the stack of discs revolving at high speed. Water and heavier than oil particles are forced outward by centrifugal force. The particles strike the surfaces of the conical discs and cling, or are forced outward and downward. The water, heavier than the oil, is forced to the outside. Continuous flow forces the water upward around the outer periphery of the discs to the discharge opening. Similarly, the clean oil rises along the central portion of the disc stack into a separate discharge. Flow into the purifier is controlled by a regulating valve (not shown). Periodically, it is necessary to remove the discs and clean the sediment from their surfaces. The construction of the unit allows this to be done readily. The discs separate the oil into thin layers and permit more rapid separation of solid particles than would be possible with one large rotating mass.

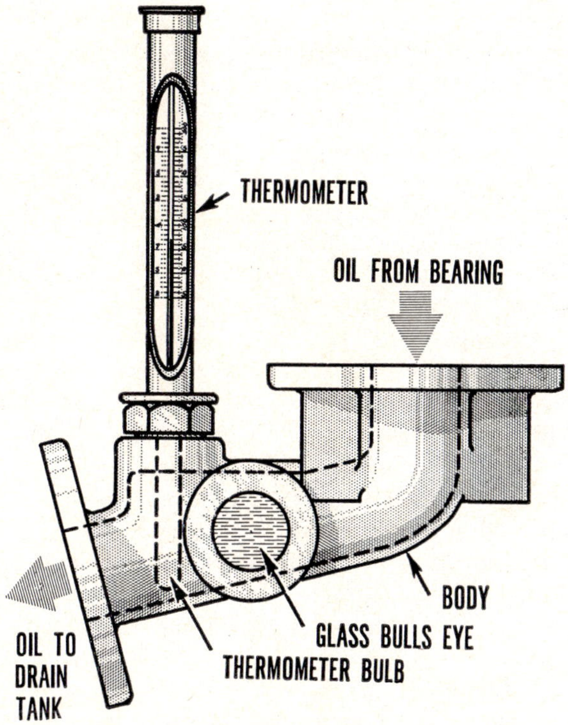

FIG. 4-53. Combined Sight Flow and Thermometer Fitting.

Turbine bearings normally should have an oil inlet temperature of 120°–130°F. The outlet temperature should be about 140°–160° but never greater than 180°F. The maximum temperature rise allowable in the bearing is 50°F, even though such a rise may cause a final temperature of lower than 180°F. Bearings are fitted with thermometers for continuous check of temperature (some later ones also have thermocouples installed) and with sight flow gauges to insure continuous supply. Figure 4-53 is a drawing of a fitting provided for the above purposes.

GEARING

434. REDUCTION GEARS.

From the material on turbines it is recalled that higher rotor speeds lead to a larger energy conversion per turbine stage and therefore more powerful turbines for a given size. However, the mechanical energy developed by the turbine must be converted to thrust delivered by the propellers; and propellers work efficiently at relatively much lower speeds. (It is interesting to note that the last direct turbine drive, ocean going vessel was built in 1914.) Some means of coupling the higher turbine speed with the desired lower propeller speed must be provided. Historically, this has been done in two principal ways: by electric drive and by mechanical reduction gears. Electric drive found favor in many large naval installations as the turbine began to supplant the reciprocating engine.

In electric drive, the turbine is used to propel a high speed generator which in turn furnishes power to a lower speed motor. By varying motor pole combinations for various speed steps, the speed range of the turbine can be kept within relatively narrow limits, an advantage in turbine design. Another advantage to certain types of ships is that full astern power can be applied to the shaft by reversing electrical connections while maintaining constant turbine direction. Still another advantage is the possibility of connecting several generators, each driven by a relatively small unit, to one large motor driving a single shaft. These advantages of electric drive for high power installations are offset by the larger size and complexity of the electric systems as compared to a mechanical reduction gear and some decrease in efficiency of power transmission as opposed to direct mechanical means.

The development of facilities and techniques for construction of large mechanical gears in this country has given their use a decided advantage, particularly in large, high powered, combatant ships. Such gears are cut under the most exacting conditions of temperature control by machines that obtain tolerances of one forty-thousandth of an inch while working on gears of diameters in the order of ten feet. This accuracy has resulted in a very high degree of operational reliability and a high efficiency of power transmission. For double reduction gears, the efficiency varies somewhat with speed and load, but remains in the range of 96.5% to 98%.

435. REDUCTION GEAR PRINCIPLES.

When two gears are in mesh, the ratio of the angular speed of the follower to the angular speed of the driver is called the *train value*, the symbol for which is r. The linear velocity of the pitch circles is the same for both gears and it is related to the angular velocity as follows:

$$V = R\omega = 2\pi RN = \pi DN$$

where

V = linear velocity, fps
ω = angular velocity, radians per second
R = radius of pitch circle, ft
N = angular velocity of gear, revolutions per second
D = pitch diameter, ft.

Then, for two gears, a and b, in mesh,

$$V_a = V_b = R_a\omega_a = R_b\omega_b = 2\pi R_a N_a = 2\pi R_b N_b$$

whence

$$r = \frac{\omega_b}{\omega_a} = \frac{N_b}{N_a} = \frac{R_a}{R_b} = \frac{D_a}{D_b}. \quad (32)$$

But the number of teeth, T, on each gear is proportional to its radius; therefore

$$r = \frac{T_a}{T_b}. \quad (33)$$

Modern HP turbine speeds are in the order of 6500 RPM; LP turbine speeds are about 5000 RPM. Propeller speeds are in the range of 200 RPM for large ships, and around 350 to 400 RPM for smaller ships. Assuming a 300 RPM propeller, a 6500 RPM HP turbine, and a reasonable turbine gear diameter of about 12 inches, the diameter of the large gear attached to the shaft, for a single reduction, would be

$$D_b = \frac{D_a N_a}{N_b} = \frac{12'' \times 6500}{300} = 260''$$

or over 21 feet.

In order to reduce the size and weight of the large gear, a double reduction gear is universally used. Figure 4-54 shows a common type. Figure 4-55 shows an alternative arrangement. Both of these are known as *locked train, double reduction gears*. The term "locked train" will be discussed later. Note that each turbine is connected to a first reduction pinion. These are often called simply HP pinions or LP pinions, to denote the turbine which drives them. Each pinion is meshed with two gears called the first reduction gears. They are mounted on quill shafts, each of which carries a smaller gear termed the second reduction pinion. These pinions drive the second reduction gear, commonly referred to as the "bull" gear. The arrangement of dividing the load to the four pinions permits lower tooth loads for a given power.

Examination of Figure 4-56 will show that the force exerted on one gear by another is not tangential to the pitch circles at the point of contact, but acts along a line of action which lies at an angle with the tangent, called *the pressure angle.* The product of this force, the radius of the gear, and the cosine of the pressure angle, is the torque transmitted by the gear. This force must be opposed by an equal and opposite force acting at the center of the gear. The opposition force is transmitted to the bearing. Thus, bearings which

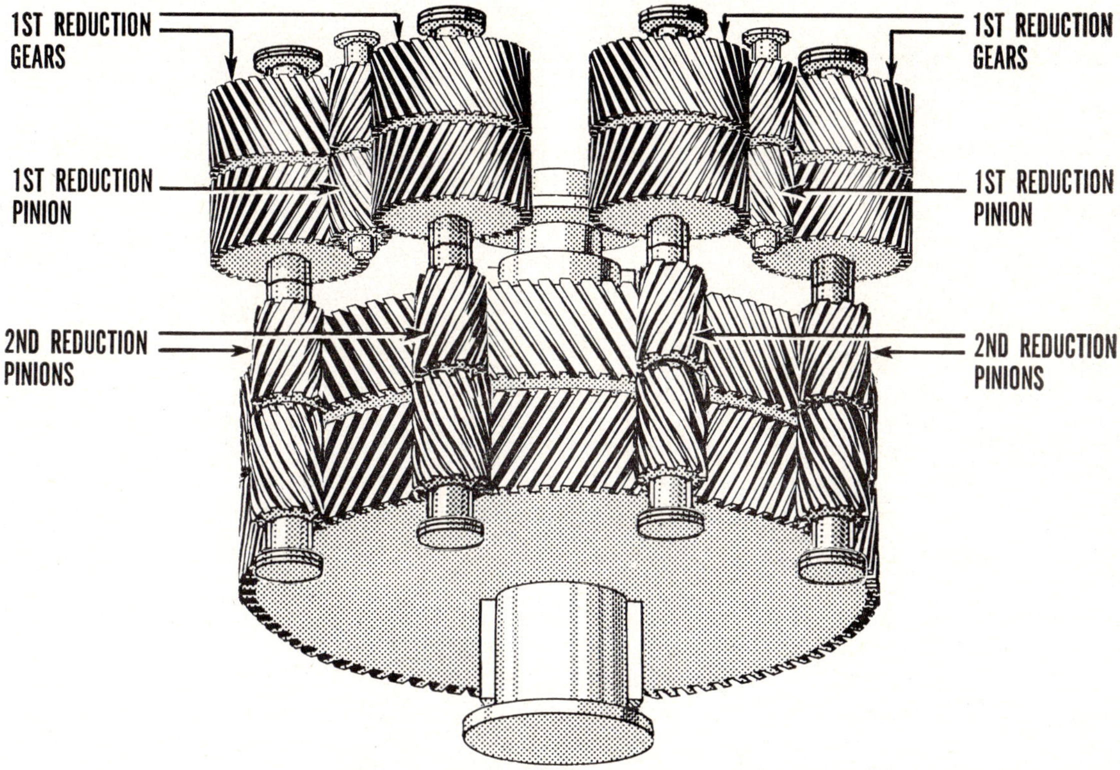

FIG. 4-54. Locked Train Double Reduction Gear.

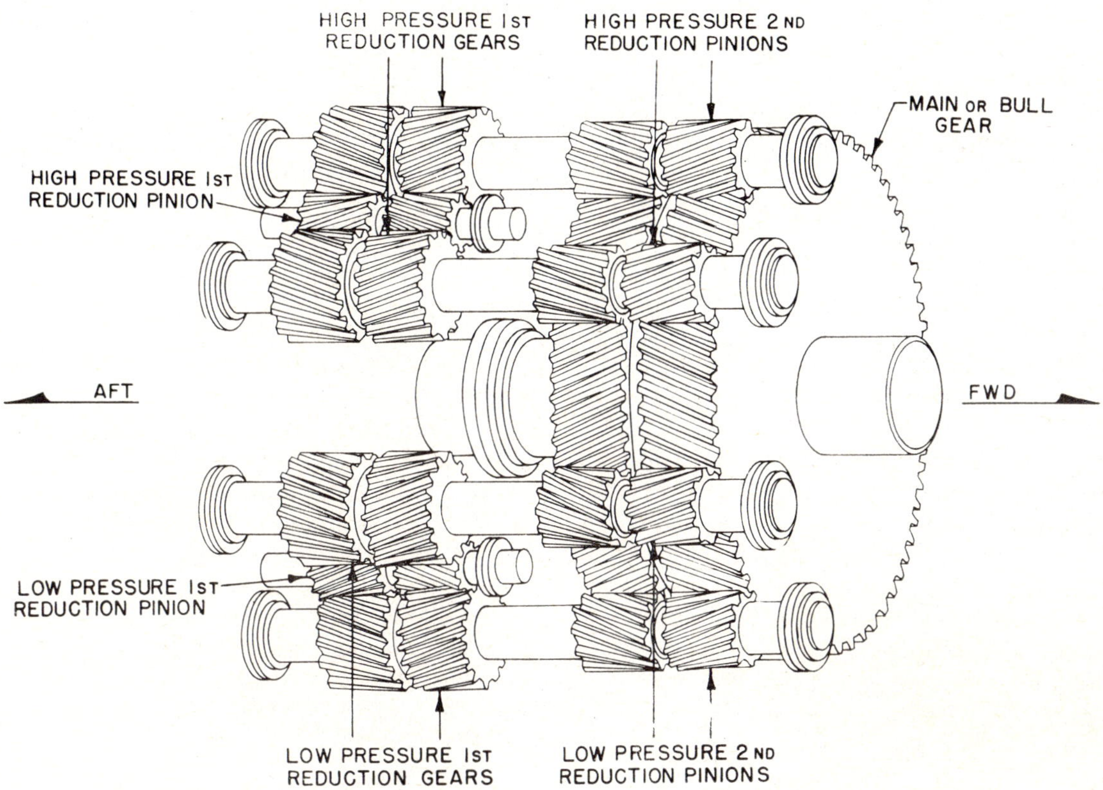

FIG. 4-55. Destroyer Escort Double Reduction Gear.

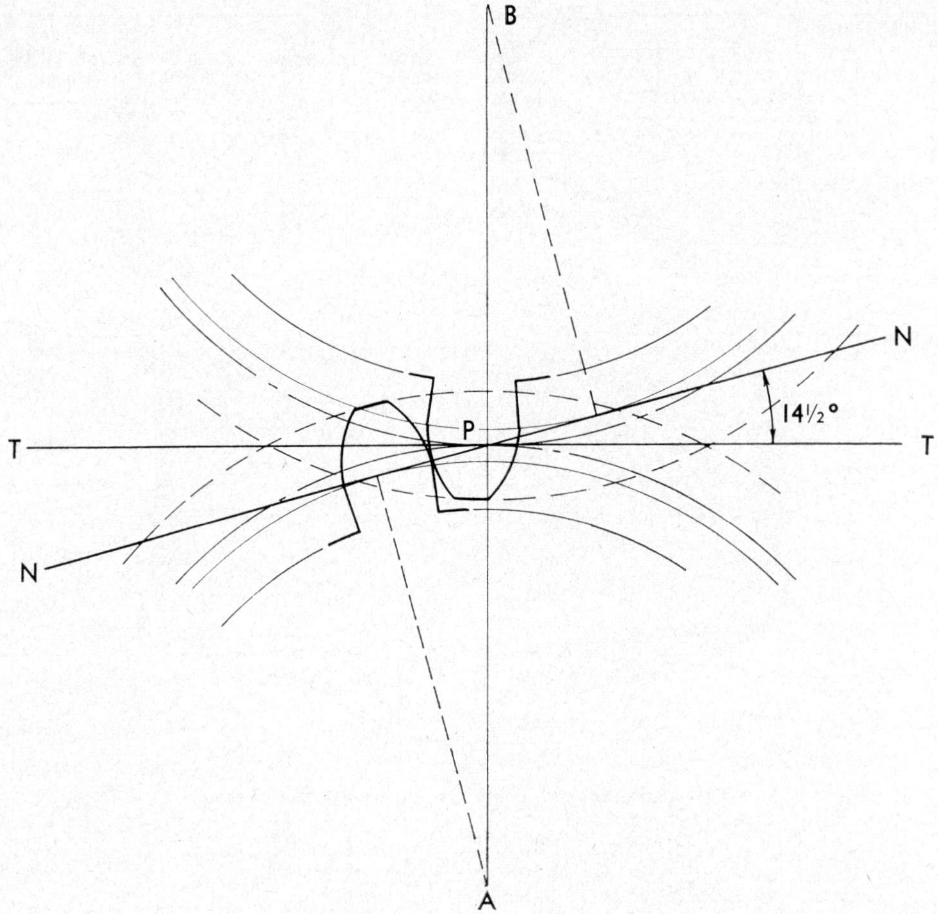

FIG. 4-56. Line of Action of Force in a Gear Mesh.

support shafts carrying gears must carry not only the load imposed by the weight of the gear and its shaft, but also an additional load which is a function of the torque transmitted. In general, these two loads act in different directions, and their resultant is the vector sum of the two loads.

In Figure 4-57(a), the small gear, A, is driving gear B. It exerts a force F_G along the line of action, which is opposed by the reaction force F_R, acting at the center of gear A. Figure 4-57(b) shows F_R added to the weight of gear A and its shaft to determine the magnitude and direction of the resultant force acting on the bearings. The two halves of the bearing shell are normally located so that their division is perpendicular to the resulting force on the bearing at full load. The load on the bull gear bearings is the vector sum of the reaction forces caused by all four meshing pinions and its weight. When one gear is meshed simultaneously with two other gears diametrically opposite each other, the reaction forces cancel out. This is the case with both the HP and LP turbine pinions in the locked train gear.

436. REDUCTION GEAR DESCRIPTION.

Referring to Figures 4-54 and 4-55, it is noted that on each gear the teeth are cut in a double helix. The opposing helices act to balance end thrust on the gears which results from a driving force on a single helix, and to keep meshed gears rigidly positioned with respect to each other axially. However, there is a requirement for axial flexibility between the first and second reductions. The main shaft, connected to the bull gear has a slight axial freedom, caused by the variable oil clearance in the main Kingsbury thrust bearing. This causes the bull gear to shift axially a few thousandths of an inch under various thrust conditions. This motion is transmitted

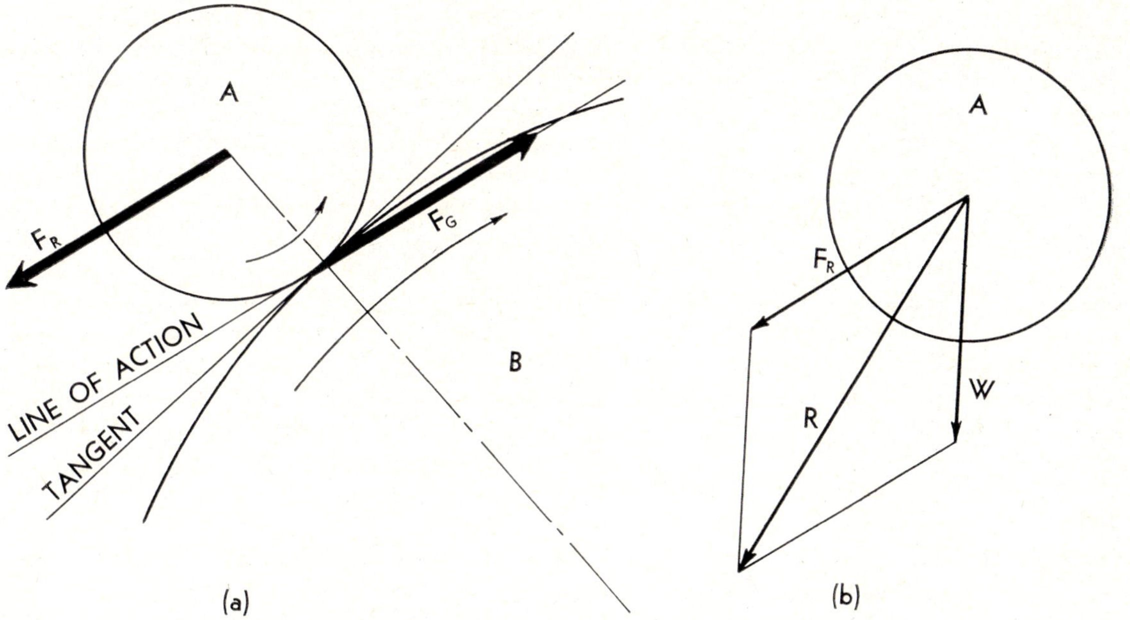

FIG. 4-57 (a) Force Couple in a Gear Mesh. (b) Resultant of Forces on Gear.

to the second reduction pinions by the double helical tooth mesh. The motion is absorbed in the quill shaft arrangement which joins the second reduction pinions to the first reduction gears and it is not transmitted to the other gears.

The quill shaft assembly also provides some flexibility to compensate for slight radial misalignment. The arrangement is shown in detail in Figure 4-58. Note the central shaft which passes through the hollow gear and pinion. On each end is a spline, or gear type coupling, by which it is attached to the gear and pinion. This coupling permits axial motion of the gear and pinion relative to each other. The length of the shaft provides considerably more torsional flexibility than a short shaft, connecting the inner faces of the two gears, would provide.

The term *locked train* is used to indicate that the power is distributed equally between the two quill shafts on each side of the assembly. Con-

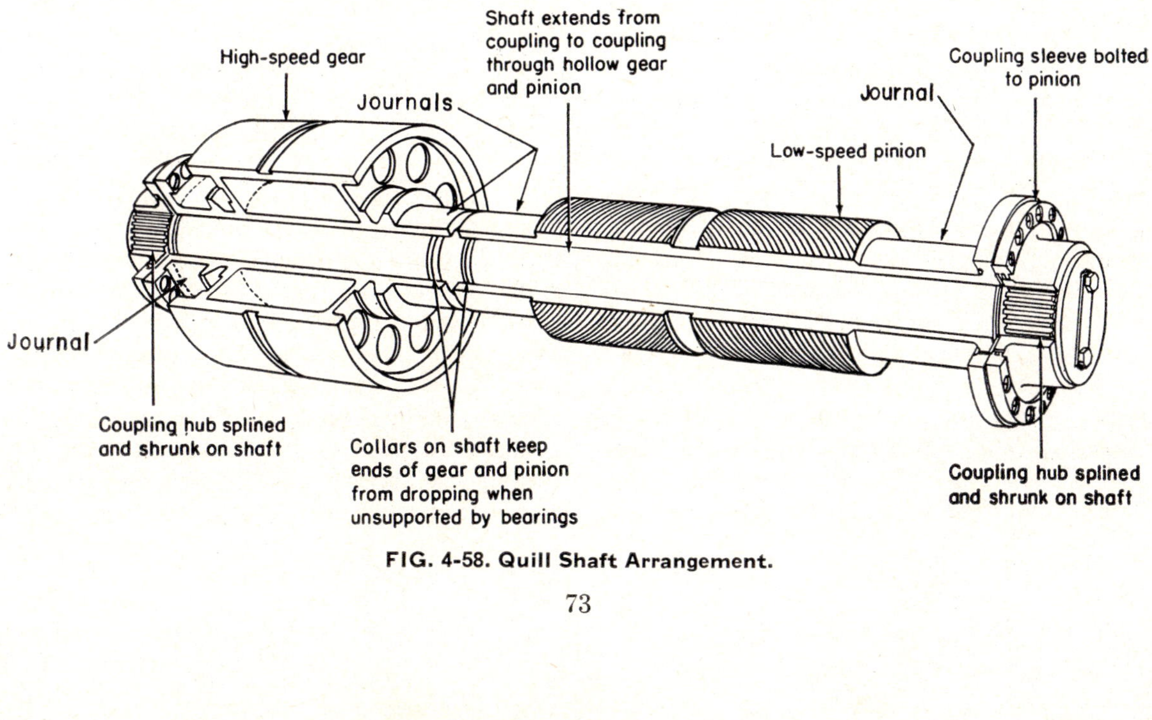

FIG. 4-58. Quill Shaft Arrangement.

sider the HP turbine side of the gear unit. In order for the gear to function properly it is necessary that the HP pinion deliver its torque equally to the two first reduction gears. It is also necessary that the second reduction pinions deliver equal torques to the bull gear. To insure this, it is necessary that the rotational position of the second reduction pinions be exactly adjusted so that each delivers the same tooth pressure to the bull gear. This is made possible in assembly of the gear by the vernier action provided by the relationship between the number of teeth on the various gears and on the quill shaft couplings, and by the fact that the circumferential pitch of the teeth in the first reduction, second reduction, and quill shaft couplings are different. Thus it is possible by proper assembly to set the two second reduction pinions in an angular relationship to each other so that each meshes with the bull gear with exactly the same tooth pressure. Automatic equalization of power between the two quill shafts is not assured without such adjustment. The gear train is said to be "locked" when it is adjusted properly.

The table below gives data for the CVA-60 reduction gear. Horsepower and RPM are given for full power.

	First Reduction				Second Reduction	
	High Pressure		Low Pressure		Pinions	Gear
	Pinion	Gears	Pinion	Gears		
HP	35,000		35,000			
RPM	6543	1450	4034	1450	1450	170
Pitch Dia.	10.25″	46.25″	13.75″	38.25″	18.575″	158.395″
No. Teeth	41	185	55	153	55	469
Circular Pitch	←0.7854″→				←1.0610″→	

The gears are mounted in a casing assembly consisting of a weldment of heavy steel plate. Figure 4-59 is a typical casing arrangement. The lower casing supports the bull gear and serves as a sub-foundation for the rest of the assembly. It is bolted directly to the top of the gear foundation built into the ship. The upper casing carries the quill shafts and supports the covers which protect the gear from the entry of foreign matter, and it supports oil lines and spray nozzles arranged to supply a continuous flow of oil to the gear meshes. The whole casing also serves as a support for accessories such as the shaft turning gear, the revolution counter drive, and the attached lube oil pump drive. The lube oil system sump tank, to which all the lube oil from the turbines and reduction gear is led, is built into the foundation immediately below the lower casing. The gear cover is provided with fittings for lube oil sight flow indicators and thermometers to the bearings (not shown in the figure). In addition several hinged inspection plates are provided for periodic inspection of spray nozzle oil flow to the gear meshes. These covers are kept padlocked to prevent inadvertent admission of foreign matter to the gears.

In some installations, where the bull gear may extend below the surface of the sump oil level, a crescent shaped sheet metal pan is fitted around the lower portion of the gear. This prevents the gear from turning directly in the body of oil and thus churning and overheating the oil, which may cause foaming. This fitting is often called the "oil excluding pan."

437. SHAFT TURNING GEAR.

It is often necessary to turn over the main shaft and the turbines without the use of steam to the turbines. Some necessary occasions are: (1) Immediately after securing the turbine, the turbine rotor should be turned continually for a period to assure even cooling of the rotor to prevent unequal thermal stresses from developing; (2) In port, the shaft is turned over at regular intervals to insure the maintenance of the oil film in the turbine and gear bearings; (3) When starting up a cold plant, the turbine is turned over before

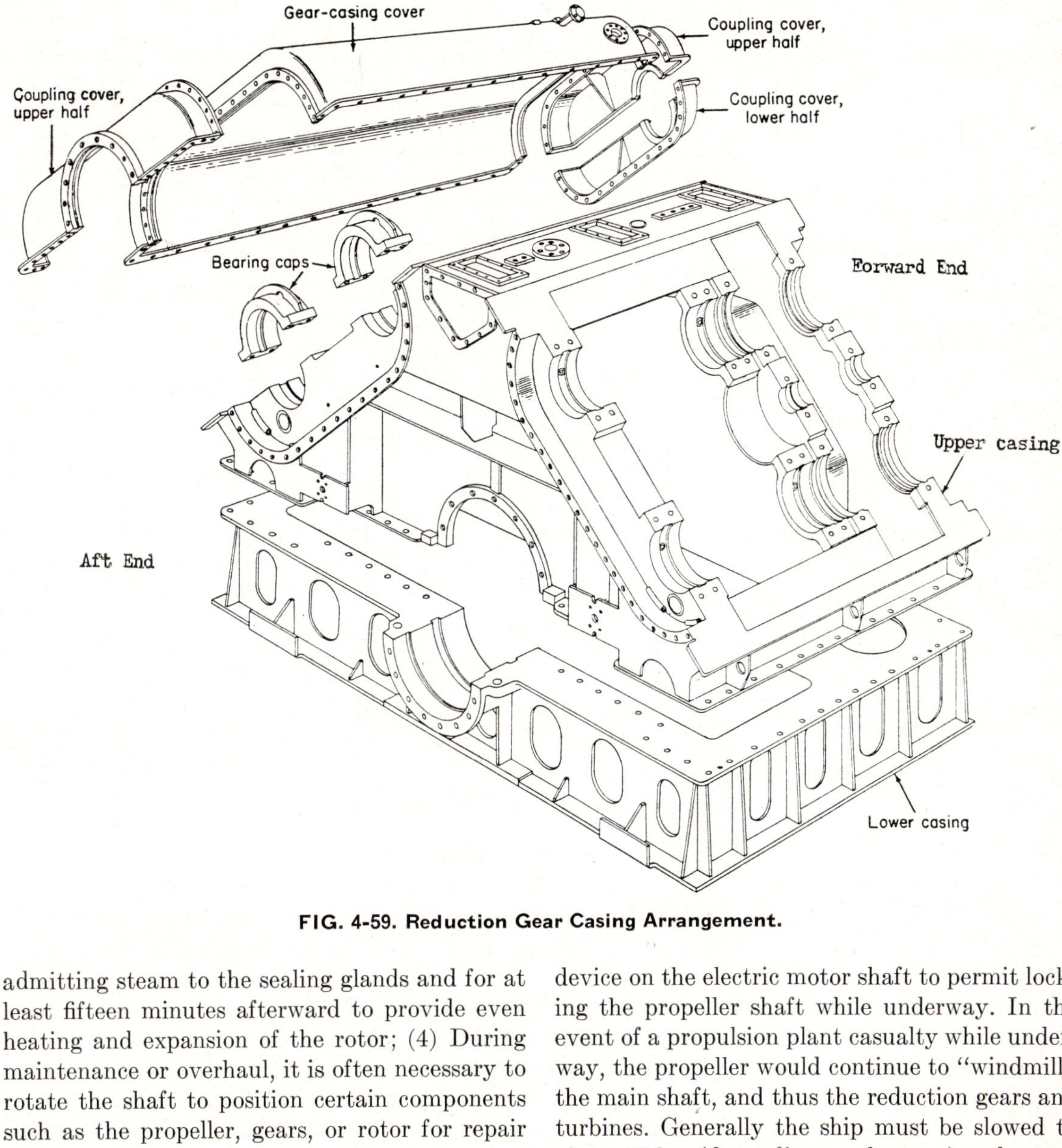

FIG. 4-59. Reduction Gear Casing Arrangement.

admitting steam to the sealing glands and for at least fifteen minutes afterward to provide even heating and expansion of the rotor; (4) During maintenance or overhaul, it is often necessary to rotate the shaft to position certain components such as the propeller, gears, or rotor for repair work.

An electric motor powered gear train is provided for the above purpose. The gear train is normally arranged to engage the HP pinion shaft inside the reduction gear casing. Figure 4-60 shows an isometric sketch of the arrangement. A typical installation is driven by a 5HP, 1100 RPM electric motor, and the gear train has a reduction ratio of 283 to 1, so that the HP turbine rotor turns over nearly 4 times a minute.

Another important function of the shaft turning gear is permitted by the addition of a locking device on the electric motor shaft to permit locking the propeller shaft while underway. In the event of a propulsion plant casualty while underway, the propeller would continue to "windmill" the main shaft, and thus the reduction gears and turbines. Generally the ship must be slowed to about 15 kts (depending on the type) and astern steam admitted to the damaged plant to stop it from spinning. Then the jacking gear is engaged and the locking device set. The ship may then resume full power on the remaining shafts. The locking device varies on different installations, but may be in the form of a friction brake or a locked gear arrangement. During the locking operation, the RPM of the other shafts and the astern chest pressure used to stop the disabled shaft are recorded. These conditions are then duplicated in the unlocking operation.

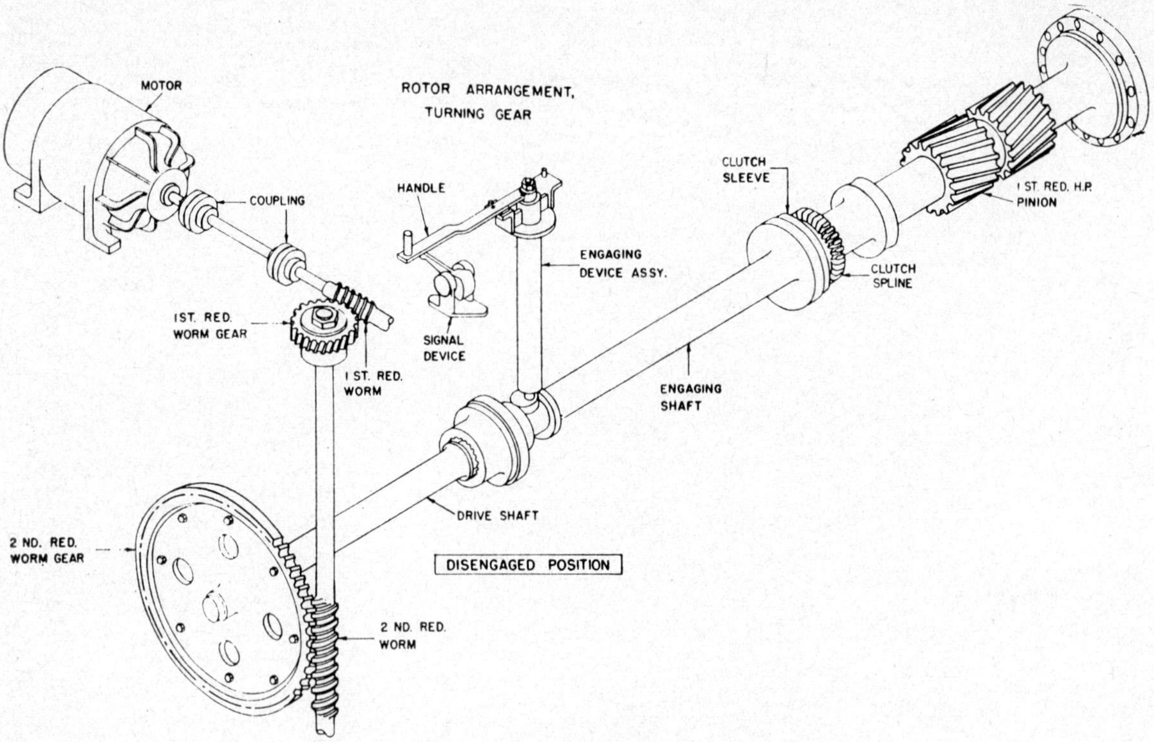

FIG. 4-60. Shaft Turning Gear, Isometric Sketch.

SHAFTING AND PROPELLERS

438. PROPULSION SHAFTING—PRINCIPLES.

Propulsion shafting serves the dual function of transmitting the torque of the propulsion unit to the propeller and transmitting the thrust developed by the propeller to the thrust bearing. The main thrust bearing is often located at the reduction gear in which case the entire shaft is subjected to both the torque and thrust load. In many installations, the main thrust bearing is farther aft, in which case the portion of the shaft forward of the bearing transmits only torque, the remainder of the shaft transmitting both torque and thrust. Another load which the shaft must carry is a varying torque produced by the uneven conversion of torque to thrust by any single propeller blade in its various positions during one revolution. A number of causes result in this effect. The principal one is the unequal distribution of flow into the propeller disc caused by the shape of the hull and the position of its fittings in the proximity of the propeller. The net effect produces a varying torque on the shaft, and when its frequency approaches the natural torsional frequency of the shaft-propeller-gear system, the shaft is loaded with peak torques higher than the average torque delivered to the propeller. Specifications limit the magnitude of this varying torque to 10% of the full load torque for turbine installations. For similar reasons, a longitudinal vibration may be set up which could become dangerous at the natural longitudinal frequency of the system. To limit this, the system is designed so that its natural longitudinal frequency does not occur at any point between 50% and 115% of full power. Another load carried by the shaft is its own weight. The last section, or tail shaft, must also support the weight of the propeller and the inertial forces caused by its lateral motion when the ship is pitching, rolling, or yawing. Other loads are caused by bending of the shafting due to working of the hull in a seaway. These loads may be summarized as static and dynamic loads.

A. Static Loads.
   1. Torque
   2. Thrust
   3. Weight of shaft
   4. Weight of propeller (tail shaft only)

B. Dynamic Loads.
   1. Torque variations
   2. Thrust variations
   3. Bending of shaft
   4. Propeller inertia (tail shaft only)

The static thrust load causes a compressive stress (when the ship is going ahead) which is the total thrust divided by the cross-sectional area of the shaft, or

$$S_c = \frac{P}{A}$$

where

S_c = compressive stress, psi
P = thrust, lbs
A = cross sectional area of shaft, in^2.

The stress caused by torque is a shear stress and varies over the cross section of the shaft, being zero at the center and reaching its maximum at the outer fibers of the shaft. Its value is

$$S_s = \frac{Tr}{I_p} = \frac{Tr}{\frac{\pi}{4}(R_1^4 - R_2^4)} \qquad (34)$$

where

S_s = shear stress, psi
T = torque, in-lbs
r = distance from center to radial section investigated, inches
I_p = polar moment of inertia of shaft cross section, in^4
R_1 = outer radius of the shaft, inches
R_2 = inner radius of the shaft, inches.

The combined stress caused by these two stresses, increased by the dynamic longitudinal and torsional stresses, plus the stresses arising from any bending loads, must be kept well below the endurance limit of the shaft material, to prevent fatigue failure.

It is customary to make the shaft hollow, since the outer portions of the shaft carry the major portion of the torsional load. The inside diameter is usually about 60% of the outside diameter. For a given cross section area, this increases the torque which a shaft will carry at the limiting value of shear stress. For instance, a twelve inch solid shaft can carry a torque of T_1 at a stress value of S_s. A hollow shaft, with $D_2/D_1 = 0.6$, will have the same area when

$$\frac{\pi}{4}(D_1)^2 - \frac{\pi}{4}(.6D_1)^2 = \frac{\pi}{4}(12)^2$$

or

$$D_1 = 15''$$

and

$$D_2 = .6 \times 15'' = 9''.$$

For equal values of S_s, from Equation (34)

$$\frac{T_1 \times 6}{\frac{\pi}{4}(6)^4} = \frac{T_2 \times 7.5}{\frac{\pi}{4}(7.5^4 - 4.5^4)}$$

or

$$\frac{T_2}{T_1} = \frac{6}{7.5} \times \frac{(7.5^4 - 4.5^4)}{6^4}$$
$$= 1.70.$$

Thus the hollow shaft can carry 70% more torque, or power at a given speed, than the solid shaft of the same area.

439. PROPULSION SHAFTING ARRANGEMENT.

The whole length of propulsion shafting is divided into two main parts, (a) the inboard, or *line* shafting, and (b) the outboard, or *wet* shafting. Each of these parts is made up of individual sections of shafting for ease of manufacturing and assembly.

The line shafting comprises all of the sections between the reduction gear and the section which pierces the hull of the ship. The *stub* shaft is the short section built integrally with the reduction gear. Either the stub shaft, or one of the sections of the line shafting, is fitted with a thrust collar which transmits the shaft thrust to the main thrust bearing.

The line shafting is made of forged steel, with the center bored out to the correct inside diameter. The ends of each section are upset to form a flange. The flanges are machined so that their faces are perpendicular to the center line and the flanges are provided with bolt holes. After the

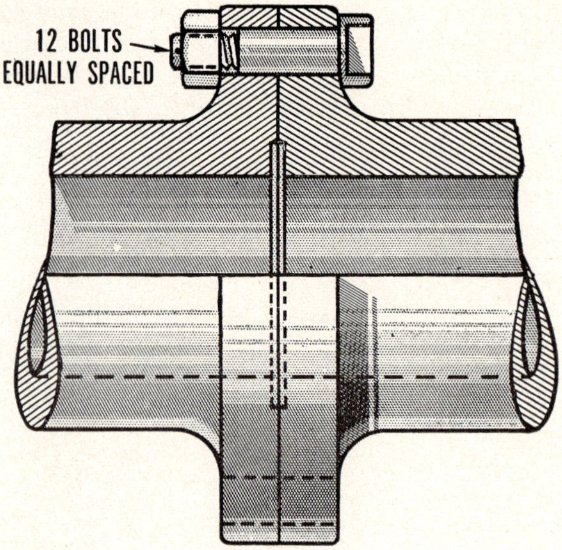

FIG. 4-61. Propulsion Shaft Flanged Coupling.

shafting has been put into place and aligned, the bolt holes in adjacent flanges are reamed and the coupling is made by the fitting of body bound bolts. Figure 4-61 shows a typical flanged coupling.

The line shafting is supported by bearings mounted in pedestals, or foundations, attached rigidly to the hull structure. These are called line shaft bearings or more simply, *spring bearings*. The bearings carry the weight of the shaft and fix its alignment. Spherically seated bearings are often used. Figure 4-62 shows a typical arrangement. The bearings are self-lubricated. An oil sump is provided in the lower part of the bearing assembly. The oil rings, carried around by the shaft dip into the sump and carry oil to the top of the shaft.

Where the line shafting passes through watertight bulkheads, a stuffing box is fitted to preserve the watertight integrity of the compartments.

The line shafting terminates at a coupling just forward of the stern tube. The stern tube is built into the structure of the ship. It contains bearings at each end to support a shaft section, being fitted with a stuffing box on the forward end to provide a watertight seal around the shaft. The section of shafting contained in the stern tube is a part of the wet shaft assembly. This shaft section must be installed, and removed for repairs

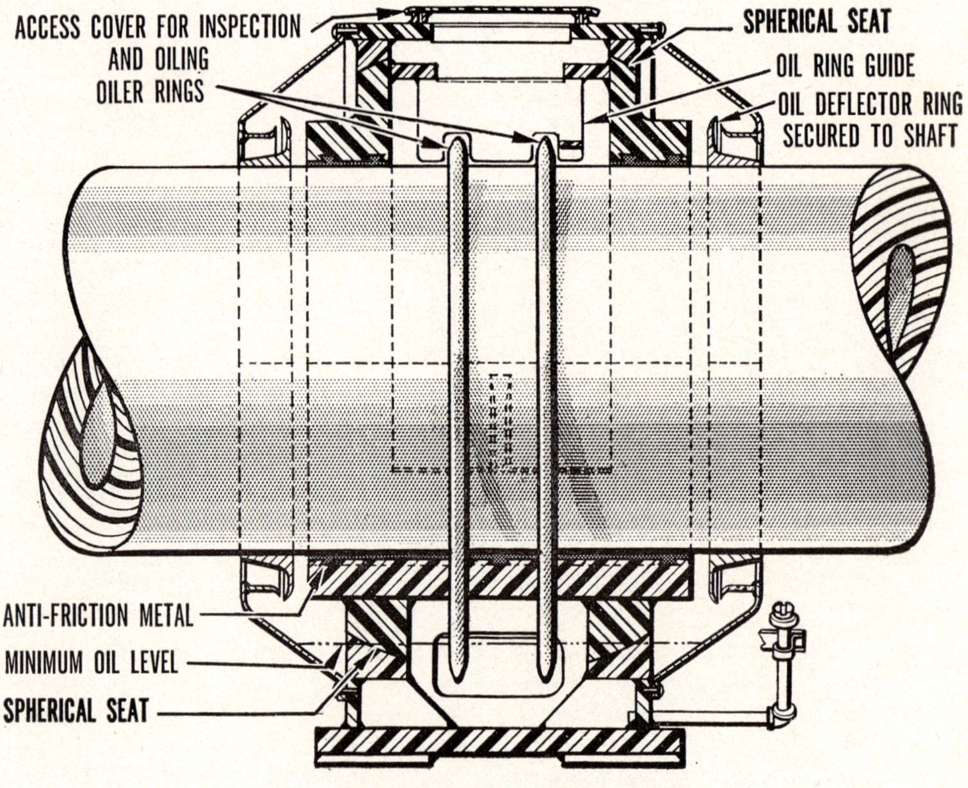

FIG. 4-62. Line Shaft Pedestal Bearing.

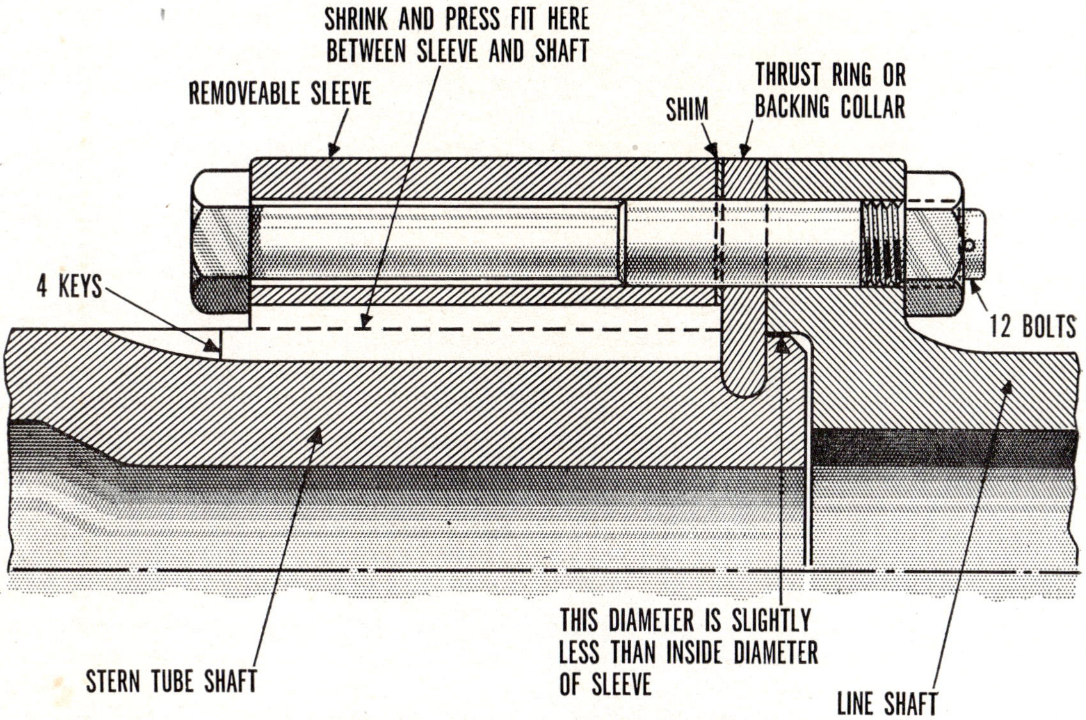

FIG. 4-63. Removable Shaft Coupling.

if necessary, in drydock. It is usually removed by withdrawing outboard. Since its end must pass through the stern tube, it must be fitted with a removable flange. Figure 4-63 shows how such a flange is fitted. The four keys carry the torque, the thrust ring transmits astern thrust, and the shaft ends butt (not shown in figure) to carry the ahead thrust.

The stern tube stuffing box is attached to the forward end of the stern tube. Figure 4-64 shows the cross section of a typical arrangement. The stuffing box is packed with rings of square sectioned flax or metallic foil wrapped packing, which is held in place and compressed to the proper degree around the shaft by adjusting the gland nuts which position the gland. To prevent overheating of the packing, a small amount of water is allowed to leak through the inboard end when underway. Tightening the nuts slightly stops this leakage when in port. The flushing connection is provided to force salt water from the fire main through the stern tube, opposite to the normal direction of flow, to flush out silt and mud which can accumulate from shallow water operation. It is good practice to use this connection at all times in shallow water operation to prevent entry of foreign matter.

The stern tube carries a bushing, or hollow bronze cylinder, at each end to carry the stern tube bearings. These bearings, being immersed in water, are designed to be water lubricated. Details are discussed in the following paragraph.

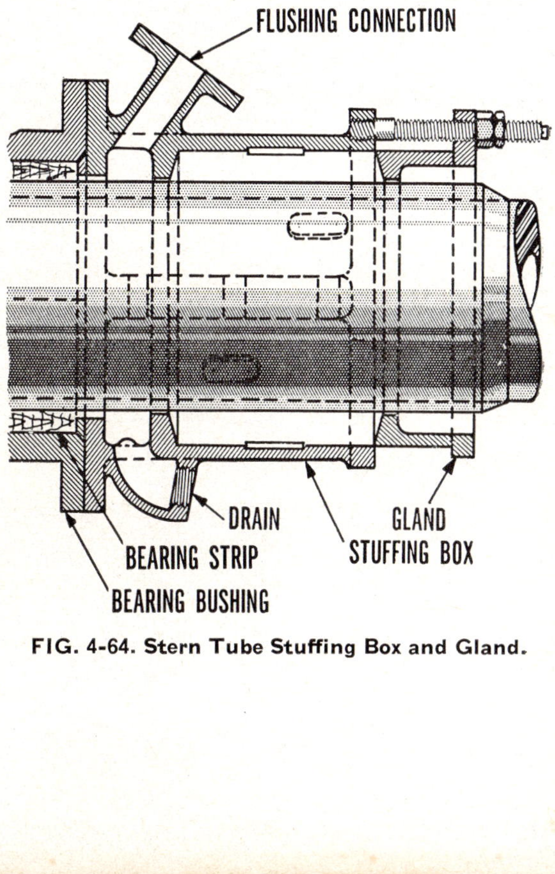

FIG. 4-64. Stern Tube Stuffing Box and Gland.

When the propeller is not mounted immediately aft of the stern tube (as in destroyers, cruisers, and the two outboard shafts of aircraft carriers) the wet shafting is supported by bearings mounted in *struts* which are rigidly attached to the hull structure. These bearings are also water lubricated bearings. The struts carry split bronze bushings to support the bearing material. One end of the bushing is flanged and drilled for bolting to the end of the strut. The bearings are made of strips of bearing material of one of the three following types: (1) Lignum vite (a hard tropical wood); (2) synthetic rubber bonded to brass strips, or (3) laminated plastic consisting of layers of cotton bonded with phenolic resin. The synthetic rubber now is almost universally used in larger installations. The strip construction provides for a free circulation of sea water through the bearing. The strips fit into longitudinal grooves nachined in the bushing halves. Figure 4-65 shows underwater bearing details.

The machined bearing surfaces on the shaft must be made of a non-corrosive material. Consequently, bronze sleeves are mounted on the shaft with a shrink fit in way of the bearings. The surface of the shaft between sleeves is covered with a protective coating consisting of synthetic rubber, neoprene, or other plastic material bonded to the surface. Machined portions of outboard flanged couplings are protected by enclosing the entire coupling with a cover fabricated of light steel plate and fitting snugly around the shaft on each side of the coupling. The coupling cover is filled with tallow after installing and effectively protects the flanges from sea water. Another type of coupling frequently found on wet shafts is the *muff* coupling, shown in Figure 4-66. This coupling is designed to keep the machined surfaces dry by the tight fit of the tapered shaft in each end. The feather keys are the primary means of transmitting torque. The cross keys wedge the taper shaft into the coupling when driven up, assist in torque transmission, and transmit astern thrust. The coupling cover is provided with threaded holes at its center through which red lead is "gunned" after assembly to prevent water seepage. The holes are then closed with screw plugs. This type coupling, on many existing installations, is not being used on new construction at present, since the flange coupling and coupling cover is cheaper and simpler to construct and maintain.

The portion of the shaft carrying the propeller is called the *tail shaft*. The first section of the wet shaft is called the *stern tube shaft*. In some cases, the entire wet shaft is enclosed in a free flooding

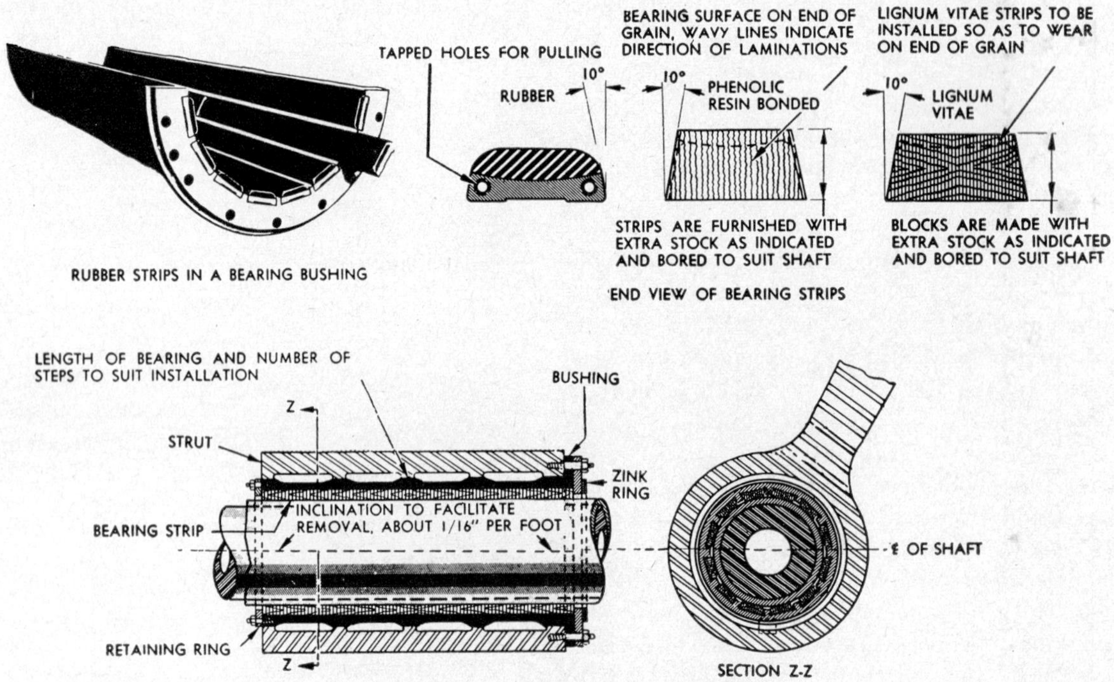

FIG. 4-65. Underwater Bearing Details.

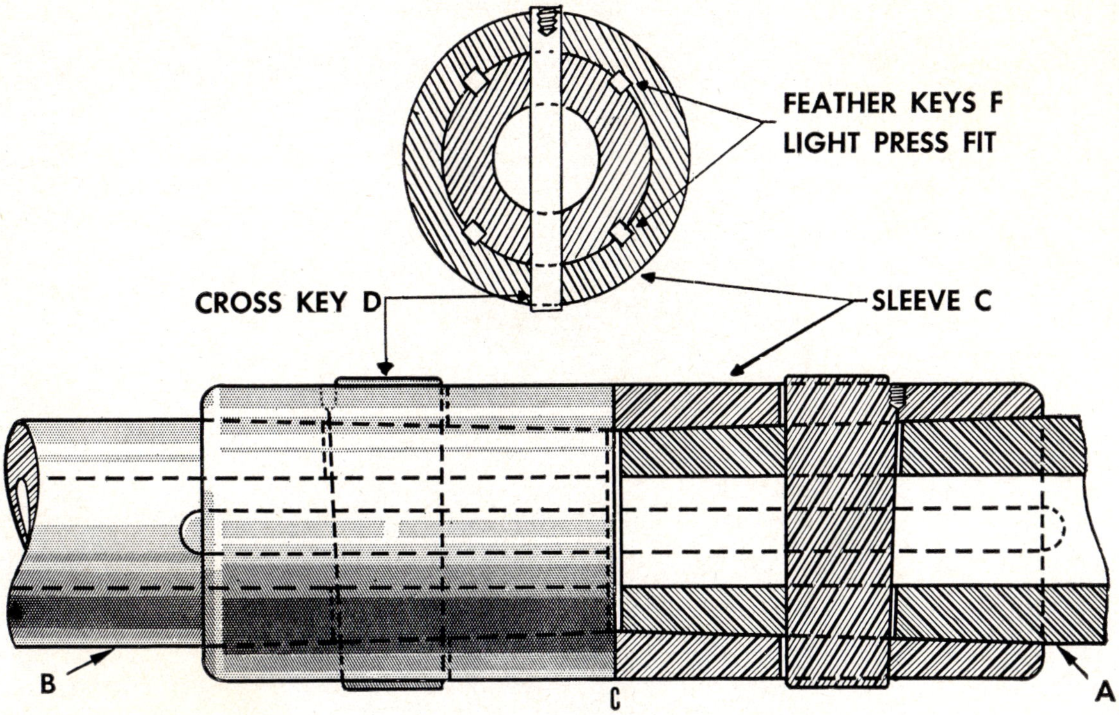

FIG. 4-66. Muff Coupling.

skeg, which is a structural portion of the hull. The skeg structure in these cases forms the support for the wet shaft bearings. This arrangement is found on large carriers and battleships on the two inboard shafts. See Figure 4-67.

Plate V shows a typical destroyer shafting arrangement.

440. PROPELLERS.

The basis of the design of a screw propeller is the helicoidal surface. Such a surface is generated by a straight line revolving about an axis at constant angular speed while advancing along it at constant linear speed. Figure 4-68 shows such a surface. The astern face of the propeller, or the driving *face*, is based on a surface of this type. The forward area of the blade, or the *back*, deviates from this surface due to the necessity for finite blade thickness and the desirability of giving certain shape to the blade section.

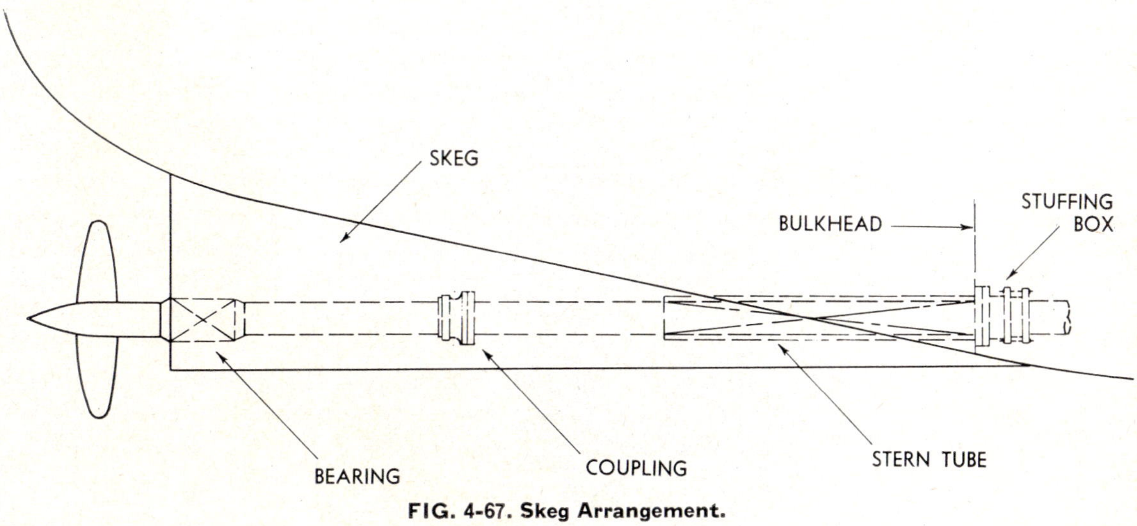

FIG. 4-67. Skeg Arrangement.

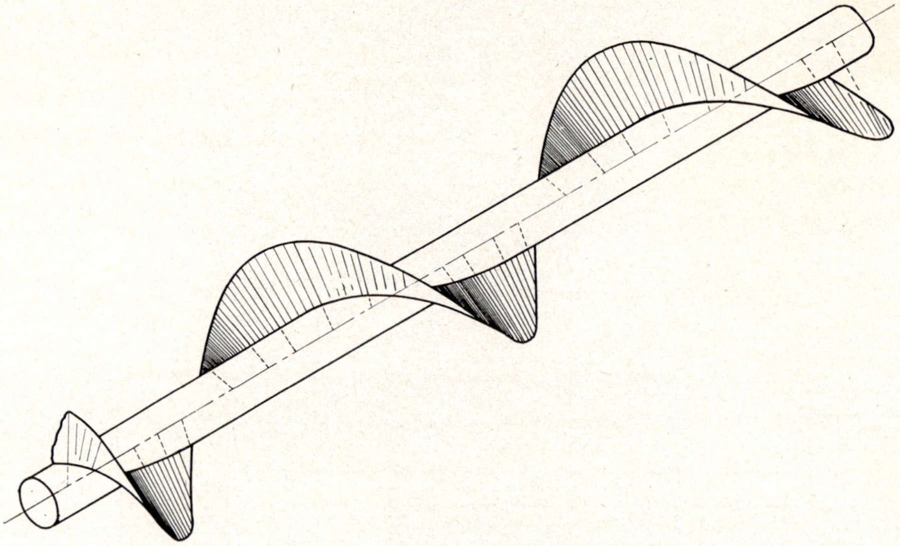

FIG. 4-68. Helicoidal Surface.

The pitch of the helicoidal surface, or of the propeller utilizing such a surface, is the distance along the axis the straight line generatrix travels while making one revolution. Note that all points of the generatrix travel the same distance during one revolution, so that the pitch of all parts of the surface from axis to tip is the same. A propeller which has a face which is exactly a helicoidal surface is called a constant pitch propeller.

It has been found that the efficiency of the propeller in converting torque to thrust can often be increased by changing the pitch of the blade slightly from hub to tip. A propeller with such blades is called a *variable pitch* propeller. The nominal pitch used in describing a variable pitch propeller is the mean effective pitch of the propeller, which is usually the pitch at about the 0.7 radius circle.

The action of the propeller blade on the water may be compared to the action of an airplane wing. Water divides as it passes over the blade section, the flow creating an increase in pressure on the face of the blade and a decrease in pressure on the back of the blade. This pressure difference results in "lift" or thrust forward along the shaft axis and a "drag" which opposes rotation of the propeller. As in an airplane wing, a high lift-drag ratio is most desirable. Under conditions of heavy loading, or thrust, the pressure on the back of the blade becomes low enough to equal the vapor pressure of the water, and bubbles of water vapor begin to form. This is called *cavitation* and results in immediate reduction in efficiency. The formation and collapse of cavitation bubbles has a highly erosive effect on the blade surface and rapidly causes pitting, which alters the blade contour and eventually reduces efficiency at all speeds.

High propeller efficiency is usually obtained with large diameters and slow rotational speeds. The actual diameter is made as large as possible considering the draft of the ship and clearance between propeller tips and adjacent hull structure. In destroyer construction, the propeller circle usually extends below the keel line so that larger diameter propellers can be used. Docking and channel draft considerations on larger ships prevent this. Adequate clearance between the blade tips and adjacent hull structure must be provided to prevent excessive hull vibration and to prevent harmful effect of the hull on the flow in the vicinity of the propeller.

Having the largest feasible diameter, the rotational speed and pitch of the propeller determine its power characteristics. The number of blades is generally chosen to minimize the vibration difficulties in the stern hull structure and the shaft system; other factors being equal, propellers with the smaller number of blades have a slight advantage in efficiency.

Various materials have been used for propellers, but a manganese bronze composition is the

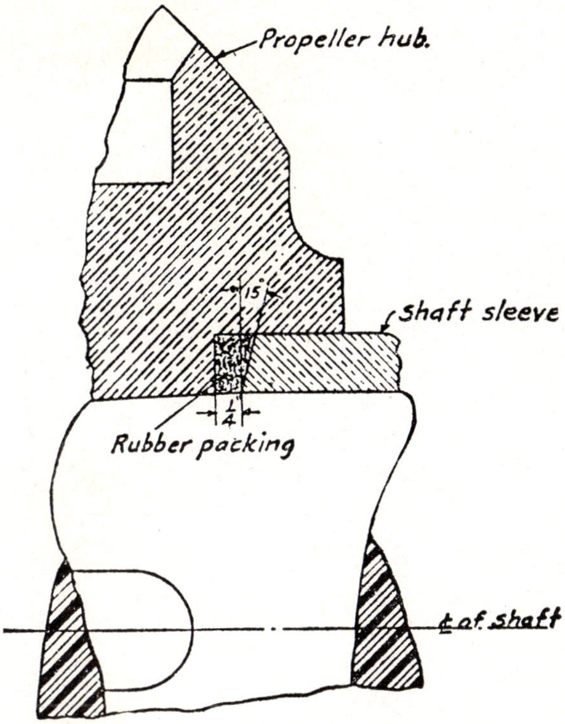

FIG. 4-69. Details of Propeller Mounting.

most common, particularly on large combatant types. Because the electrolytic action set up between the propeller, the wet shaft bearing sleeves, and the adjacent steel structure tends to accelerate corrosion of the steel hull and rudder portions near the propeller, sacrificial metal plates of pure zinc are bolted to the strut, rudder and hull structure in the vicinity of the propellers. Being higher in the electromotive series, the zinc corrodes selectively in place of the steel. This type protection is called passive protection. Active protection, consisting of impressing an electrical potential between the dissimilar metals and designed to counteract the electrolytic potential, has been under experimental development with some promise of success.

The propeller hub has a tapered bore, designed to fit on the tapered end of the shaft. A large longitudinal shaft key transmits the shaft torque to the propeller. The end of the shaft is threaded and a large nut holds the propeller in place. The after end of the propeller is fitted with a cap, called a hub cap or dunce cap, to cover the nut. The cap is filled with tallow to prevent entry of sea water. The forward end of the hub is counter bored to form a stuffing box. A packing ring holds rubber packing in place firmly around the shaft to prevent entry of water to the machined shaft taper. See Figure 4-69.

441. CONTROLLABLE PITCH PROPELLERS.

The fixed pitch propeller operates most efficiently at a fixed relationship among the variables: propeller rpm, ship speed, and shaft torque. The speed-torque characteristics of the engine driving the propeller are thus influential in its operating efficiency. Fortunately, the characteristics of the turbine are well matched to those of the propeller. However, the turbine torque increases with speed. In applications requiring high torques at lower speeds, the diesel engine has the attractive characteristic that its torque is practically independent of speed. Such applications are those involving moving heavy external loads at low speeds such as minesweeping, towing, retracting landing craft and ships from the beach, etc. Under these conditions, the controllable pitch propeller is of special interest, since its characteristics can be continuously matched to operating conditions to produce the maximum thrust.

Also, a controllable pitch propeller whose pitch can be completely reversed is capable of delivering astern power without shaft rotation reversal. This is an advantage with diesel and gas turbine drive installations.

The blades of the propeller are arranged to rotate in the hub, being positioned by a hydraulic piston and cylinder arrangement operating a mechanical linkage, all of which is enclosed in the propeller hub. The valve which controls the hydraulic device may be operated by (1) a hydraulic line running through the propeller shaft to the control station, (2) a pneumatic line running through the shaft or (3) an electric motor in the hub with its control wires passing through the shaft bore.

Problems connected with the controllable pitch propeller include (1) the difficulty of sealing the internal hub mechanism against the entrance of sea water, (2) the maintenance of oil pressure in the hub for actuating the blades (which must be supplied from outside the shaft system), and (3) the difficulty of obtaining reliable operation that does not require frequent drydocking for maintenance.

A few installations have mechanical rather

than hydraulic actuating devices within the hub. These may be controlled by an electric motor in the hub, or by a mechanical control shaft passing through the bore of the main shaft, and being actuated by brakes and gears in the engine room.

The controllable pitch propeller will find increasing use in installations having special requirements as its development improves through operating experience.

REFERENCES

Stodola, A. and Loewenstein, L. C.—*Steam and Gas Turbines,* 1945, Peter Smith Co.

Keenan, J. H.—*Thermodynamics,* 1941, John Wiley and Sons, Inc.

Seward, H. L.—*Marine Engineering,* 1942, Society of Naval Architects and Marine Engineers.

Rossell, H. E. and Chapman, L. B.—*Principles of Naval Architecture,* 1942, Society of Naval Architects and Marine Engineers.

Labberton, J. M.—*Marine Engineering,* 1943, McGraw-Hill Book Co., Inc.

Salisbury, J. K.—*Steam Turbines and Their Cycles,* 1950, John Wiley and Sons, Inc.

Church, E. F.—*Steam Turbines,* 1935, McGraw-Hill Book Co., Inc.

Fuller, D. D.—*Theory and Practice of Lubrication for Engineers,* 1956, John Wiley and Sons, Inc.

Hampton, R. E.—*Introduction to the Basic Mechanisms,* 1956, U. S. Naval Institute.

General Specifications for Ships of the U. S. Navy, Bureau of Ships, Navy Department, Washington, D. C.

Bureau of Ships Manual, Bureau of Ships, Navy Department, Washington, D. C.

Hunsaker, J. C. and Rightmire, B. G.—*Engineering Applications of Fluid Mechanics,* 1947, McGraw-Hill Book Co., Inc.

CHAPTER V
THE CONDENSATE SYSTEM

500. GENERAL.

The main functions of the condensate system are to condense the steam exhausting from the turbine and to provide a vacuum into which the turbine can exhaust in order to increase thermodynamic efficiency. In addition to the main functions, it normally provides several convenient features, such as: (1) the removal of a considerable amount of dissolved oxygen from the working fluid; (2) the conservation of heat in the plant by transferring it to the working fluid in heat exchangers to be discussed; (3) the provision of a convenient point in the cycle for introducing additional working fluid; (4) the provision of a convenient point in the cycle for collecting fresh water discharged from other shipboard systems apart from the main steam cycle.

The equipment in the condensate system includes heat exchangers, condensate pumps, air pumps (ejectors), and the necessary piping to properly connect the components. The exchange of heat in the condensate system is of such importance that some heat transfer relationships will be discussed prior to describing the system.

501. ENERGY BALANCE IN HEAT EXCHANGE.

When two or more fluids pass into a device designed to transfer heat from one fluid to the other, and to discharge the fluids separately, the final heat effects can be evaluated by application of the Law of Conservation of Energy. In engineering applications, the mechanical effects, such as change in kinetic energy, flow work done in the fluid excursion, and change in potential energy, are usually negligible compared to the changes in thermal energy and can be omitted. With this assumption, the total thermal energy entering the device per unit of time (conveniently expressed as BTU/min, BTU/sec or BTU/hr) can be equated to the total thermal energy leaving, expressed in the same units. A further simplifying assumption is usually justifiable: that the transfer of heat from the unit to the surrounding atmosphere is negligible. In fact, it is the practice to insulate the units when necessary to keep this transfer small.

For each fluid, the energy entering per unit time is the product of its flow rate, conveniently expressed in pounds per unit time, and its thermal energy, expressed in BTU per pound. For liquids, the specific heat times the temperature above 32°F will determine the thermal energy content. For gases, the enthalpy is a measure of the energy per pound, being obtained either from appropriate tables, or computed by multiplying the constant pressure specific heat times the temperature above 32°F. Figure 5-1 and the equations below illustrate the relations.

Enthalpy/lb of steam $= h_1$ (from Steam Tables for p_1 and T_1)

Energy/lb of air $= c_p(T_a - 32)$ (although c_p for air varies somewhat with both temperature and pressure, the use of $c_p = 0.242$ BTU/lb-°F will be of satisfactory accuracy for ranges of temperature and pressure found in this book).

Energy/lb of water $= c_w(T - 32)$ where

$$c_w = 1.0 \text{ for fresh water}$$

$$c_{sw} = 0.94 \text{ for sea water.}$$

Then, summing up, and equating energy entering to energy leaving:

$$w_s h_1 + w_a c_p(T_a - 32) + w_w c_{sw}(T_w - 32) = w_s c_w(T_2 - 32) + w_a c_p(T_a' - 32) + w_w c_{sw}(T_w' - 32)$$

or

$$w_s h_1 - w_s c_w(T - 32) + w_a c_p(T_a - T_a') + w_w c_{sw}(T_w - T_w') = 0.$$

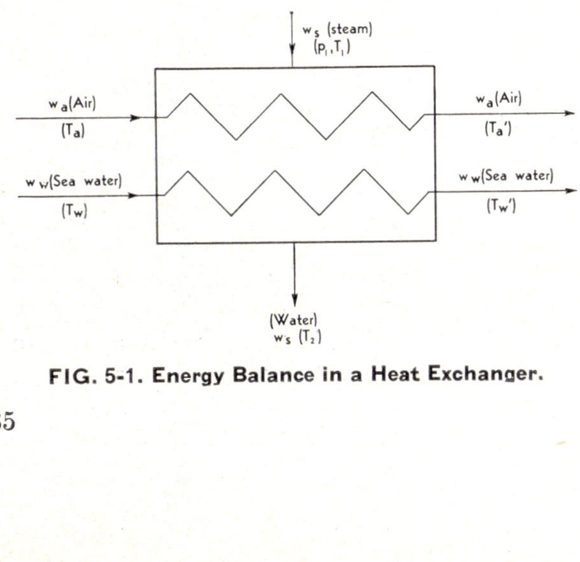

FIG. 5-1. Energy Balance in a Heat Exchanger.

When two liquids, or a liquid and vapor are homogeneously mixed, as happens in a heater where steam condenses and combines with the water it heats, the same general technique applies. Also, where the water is known to be saturated, its value of enthalpy from steam tables may be used in place of the expression, $c_w(T-32)$, used in the preceding.

502. HEAT TRANSMISSION.

When a material is interposed between two constant temperature sources, the rate of flow of heat from the high temperature source to the low temperature source becomes constant after equilibrium is established. The "potential" tending to cause heat flow is the temperature difference between the boundaries of the material. The "resistance" to flow is dependent on the character of the material itself, and the quantity of the substance through which the heat must travel. It is directly proportional to the length of the path of travel, and inversely proportional to the cross section area of the path. In heat transmission, it is often convenient to express "resistance" as a reciprocal quantity, *conductivity*. The conductivity is then, according to above statements

$$\text{conductivity} = \frac{kA}{L} \text{ BTU/hr-°F}$$

where

k = thermal conductivity, BTU/hr-°F-ft, a property of the conducting material

A = cross section of conductor, ft²

L = length of heat flow path, ft.

If the rate of heat flow is imagined as a heat "current," the three terms "potential," "resistance," and "current" immediately suggest the familiar Ohm's Law:

$$\text{Current} = \frac{\text{Potential}}{\text{Resistance}} \cdot$$

Then, the rate of heat transfer, q, is given by

$$q = \frac{kA}{L}(\Delta T) \qquad (1)$$

where

ΔT = temperature difference between the two surfaces of the material.

In actual practice, the temperature sources on either side of the conducting medium are fluids of some kind. The heat transmission from one fluid to the other is greatly altered by the thin films of nearly stagnant fluid which cling to the surfaces under nearly all conditions of flow. The magnitude of resistance offered by the fluid films is generally quite large compared to the resistance of the material itself. Since the film thickness is indeterminate by measurement, the conductivity characteristics, called *film coefficients*, are determined empirically on the basis of a large mass of experimental data. The film coefficients vary with the thermal conductivity of the fluid, its viscosity, density, and specific heat. When the fluid velocity is great enough to cause turbulent flow, the resistance of the film is decreased as velocity is increased. In the laminar flow region, the velocity has little effect. The film coefficient is designated $\mathbf{h}$ in heavy type to prevent confusion with h used for specific enthalpy. Its units are BTU/hr-ft²-°F. The effect of the film formation is indicated in Figure 5-2. For plane surfaces, where the area of the films and the conductor are equal, the overall thermal conductivity, U, is found by adding the resistances of the layers between the fluid bulk temperatures, T_1 and T_2, as follows

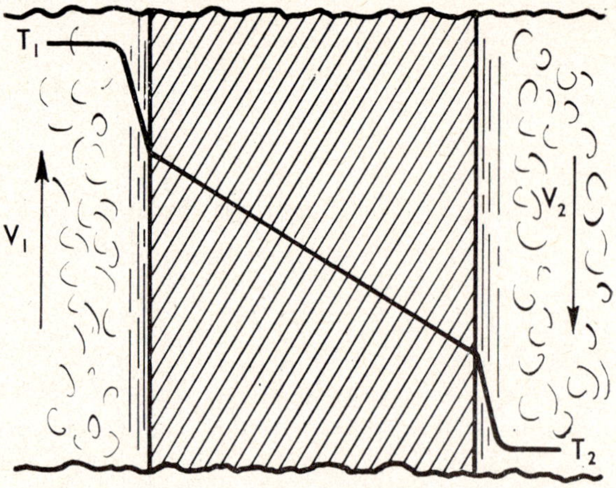

FIG. 5-2. Temperature Gradients through Fluid Boundaries.

$$\text{resistance of film 1} = \frac{1}{\mathbf{h}_1}$$

$$\text{resistance of conductor} = \frac{L}{k}$$

$$\text{resistance of film 2} = \frac{1}{\mathbf{h}_2} \cdot$$

Then

$$\text{overall resistance} = \frac{1}{h_1} + \frac{L}{k} + \frac{1}{h_2} \qquad (2)$$

and, overall conductivity

$$U = \frac{1}{\frac{1}{h_1} + \frac{L}{k} + \frac{1}{h_2}}. \qquad (3)$$

Equation (1) is then modified to

$$q = UA\Delta T. \qquad (4)$$

Representative values of film coefficients along with some thermal conductivities are given in Table 5-1.

Table 5-1

Representative Values of Film Coefficients and Thermal Conductivities

Fluid or Material	Film Coefficient-h (BTU/hr-ft²-°F)	Thermal Conductivity-k (BTU/hr-ft-°F)
Air	8	0.013
Water	1000	0.350
Oil	50	0.100
Steam (condensing on outer tube surfaces)	2000	—
Steel	—	27.0
Brass	—	60.0

Example 51: Find U for a brass plate 0.13″ thick with water on one side and steam on the other, using values from Table 5-1. Compare this with U for the brass plate if fluid films are neglected.

Solution:

$$U = \frac{1}{\frac{1}{1000} + \frac{0.13}{12 \times 60} + \frac{1}{2000}}$$

$$= \frac{1}{.00100 + .000181 + .00050}$$

$$= \frac{1}{.00168} = 595 \text{ BTU/hr-ft}^2\text{-°F};$$

with no films

$$U = \frac{60 \times 12}{0.13} = 5540 \text{ BTU/hr-ft}^2\text{-°F}.$$

When one of the fluids is confined in a pipe or tube, and the other surrounds it, the areas associated with the films and the conductor are different. In this case the overall conductivity, UA_1, is found by rewriting equations (2) and (3) to include the proper values of areas, as follows:

$$UA_1 = \frac{1}{\frac{1}{h_1A_1} + \frac{L}{kA_m} + \frac{1}{h_2A_2}} \qquad (5)$$

where

h_1 = film coefficient outside pipe
A_1 = area of outer surface
A_m = mean area of conducting substance
h_2 = film coefficient inside pipe
A_2 = area of inner surface.

As noted above, the film areas can be taken as the inner and outer areas of the pipe or tube. For accurate computation of A_m, the following relation is used

$$A_m = \frac{A_1 - A_2}{\ln \frac{A_1}{A_2}}. \qquad (6)$$

However, less than a $1\frac{1}{2}\%$ error is involved by using the arithmetical average of A_1 and A_2 when $A_2 \geq .7\, A_1$. Equation (6) is most useful when computing overall transfer through thick layers of insulation. For heat exhangers, the arithmetical average is generally satisfactory.

Equations (4) and (5) now give a satisfactory means of computing areas for heat exchangers *if the bulk temperatures of the fluids remain constant at T_1 and T_2 in the unit.* This is not generally the case.

503. TYPE OF FLOW AND TEMPERATURE DIFFERENCE.

Consider the situation which exists when a hot and a cold fluid flow in a parallel direction, one inside a pipe and one outside. The hot fluid will transfer its heat to the cold fluid. As the fluids progress along the pipe the difference in their temperatures will become smaller than the initial value. The situation is depicted in the graph in Figure 5-3, where ΔT_1 is larger than ΔT_2. It is necessary, then, to determine a mean value for the temperature difference, ΔT_m, in order to utilize equation (4). It is shown in texts on heat transmission that the proper value of ΔT_m is a

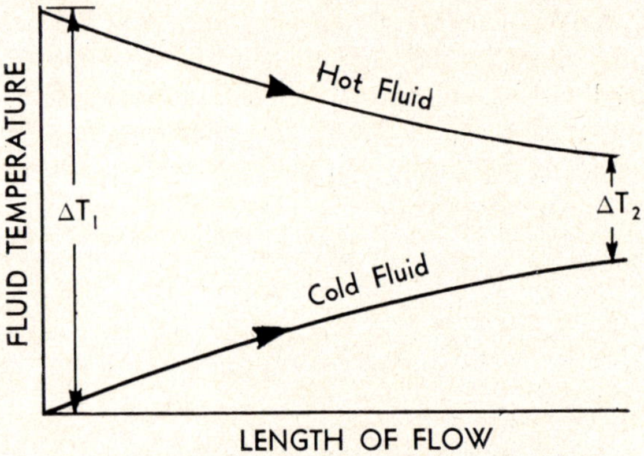

FIG. 5-3. Temperature Gradients through Parallel Flow Heat Exchanger.

logarithmic mean of ΔT_1 and ΔT_2, given by the following relation

$$\Delta T_m = \frac{\Delta T_1 - \Delta T_2}{\ln \frac{\Delta T_1}{\Delta T_2}} \cdot \quad (7)$$

(See *Elements of Applied Thermodynamics* by Johnston, Brockett and Bock.)

The situation shown in Figure 5-3 is called *parallel flow*, in which the two fluids flow in the same direction. *Counter flow* occurs when the two fluids flow in opposite directions. The temperature difference during the passage of the fluids through the apparatus is shown in Figure 5-4. Note two important points: (1) ΔT does not vary over so wide a range during the length of flow, and (2) the exit temperature of the cold fluid can be made higher than the exit temperature of the hot fluid, which is not possible with parallel flow. The counter flow arrangement is superior to the parallel flow. The following example is illustrative.

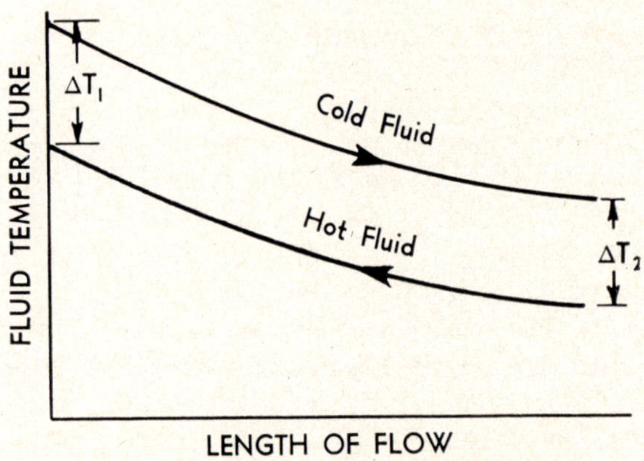

FIG. 5-4. Temperature Gradients through Counter Flow Heat Exchanger.

Example 52: It is desired to cool oil flowing at 300 gpm from 180°F to 120°F. Its density is 54 lb/ft³ and specific heat is 0.50 BTU/lb-°F. 225 gal/min of sea water (sp.ht. = 0.94) will be supplied at 75°F. (a) At what temperature must the water leave the exchanger? (b) Assuming U to be constant, compare the heat transfer areas necessary for parallel and counter flow heaters.

Solution:

(a) From heat balance

$$w_0 c_0 \Delta T_0 = w_w c_w \Delta T_w .$$

$$w_0 = 300 \frac{\text{gal}}{\text{min}} \times \frac{1 \text{ ft}^3}{7.48 \text{ gal}} \times 54 \frac{\text{lb}}{\text{ft}^3} = 2160 \text{ lbs/min.}$$

$$w_w = 225 \frac{\text{gal}}{\text{min}} \times \frac{1 \text{ ft}^3}{7.48 \text{ gal}} \times 64.0 \frac{\text{lb}}{\text{ft}^3} = 1925 \text{ lbs/min.}$$

$$2160 \times 0.5 \times (180 - 120) = 1925 \times 1 \times (T_w' - 75).$$

$$T_w' = 108.7°\text{F} \approx \underline{\underline{109°\text{F}}}.$$

(b) Under the conditions stated, the quantity of heat transferred by the two coolers must be the same. If the subscript p designates parallel flow, and c counter flow, then

$$q_p = q_c$$

or

$$U_c A_c \Delta T_{mc} = U_p A_p \Delta T_{mp}$$

but

$$U_c = U_p$$

so

$$\frac{A_c}{A_p} = \frac{\Delta T_{mp}}{\Delta T_{mc}} .$$

But

$$\Delta T_{mp} = \frac{\Delta T_1 - \Delta T_2}{\ln \frac{\Delta T_1}{\Delta T_2}}$$

$$= \frac{(180 - 75) - (120 - 109)}{\ln \frac{180 - 75}{120 - 109}}$$

$$= \frac{105 - 11}{\ln \frac{105}{11}} = \frac{94}{2.26} = 41.6°.$$

and

$$\Delta T_{mc} = \frac{(180-109)-(120-75)}{\ln \frac{180-109}{120-75}}$$

$$= \frac{71-45}{\ln \frac{71}{45}} = \frac{26}{.458} = 56.8°.$$

$$\therefore \quad \frac{A_c}{A_p} = \frac{41.6}{56.8} = .732.$$

or the counter flow exchanger will need only 73.2% the area of the parallel flow exchanger.

Another common heat exchanger involves changing the temperature of one fluid while the other fluid changes state at constant temperature (evaporates or condenses). A typical graph is shown in Figure 5-5. The direction of flow of either fluid has no meaning in this case, but ΔT_m can still be computed from terminal temperatures, ΔT_1 and ΔT_2. The name of the unit, *condenser* or *evaporator*, usually characterizes this type.

In a more complex arrangement, one fluid passes through a set of tubes, while another passes perpendicularly, or nearly so, to the direction of the tubes. This is called *cross flow*. Its temperature differences cannot be simply depicted as shown in Figures 5-3, 5-4, and 5-5.

504. THE MAIN CONDENSER.

The main condenser is the heart of the condensate system, and it accomplishes the two main functions of the system as given previously.

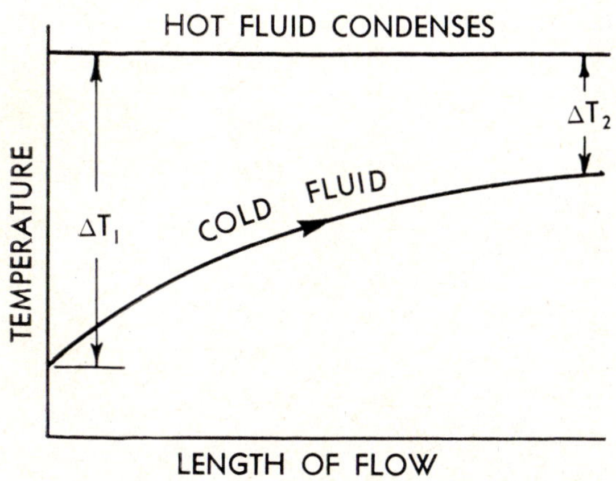

FIG. 5-5. Temperature Gradients through Cross Flow Heat Exchanger.

It is a shell and tube type heat exchanger. The exhaust steam passes around the outside of the tube bundle through which sea water (termed *circulating water* in this connection) flows. The relatively cool circulating water condenses the exhaust steam to form the condensate. The sea water enters the condenser through a scoop which picks up the water due to the motion of the ship; and it flows into the *inlet header* of the condenser, from which it passes through the tubes into the outlet header and then overboard. When the ship is traveling at very low ahead speeds, or is stopped, or backing, a large circulating pump (described in more detail later) pumps sea water through the condenser. Since the condenser is situated low in the ship, the headers, tubes, and sea connections must be constructed to withstand the submergence pressure. (This problem becomes more acute with condensers provided for steam-driven submarines.) The corrosive and erosive effects of sea water must also be considered. For this reason, liberal use is made in the construction of corrosion resisting materials such as bronze, composition brasses, and monel (a copper-nickel alloy). To minimize erosion-inducing conditions, the sea water velocity through the tubes is limited to 9 fps. Figure 5-6 is the schematic arrangement of a condenser installation. Note the non-return valve arrangement in the injection lines which permits supply from either the scoop or circulator without discharge through either.

Steam passes over the tubes from the top of the shell. The shell is made of steel, and must withstand the pressure difference between the atmosphere and the vacuum inside. On some installations, the condenser is mounted on a foundation, and itself supports the L.P. turbine. On others, the L.P. turbine is mounted on structural foundations and the condenser hangs from it. The tube sheets are the two end plates into which are fastened the ends of the tubes, and they are rigidly fastened to the flanged end of the condenser shell. Since the coefficient of expansion of the tube material is greater than that of the shell material, an expansion joint in the shell is provided to equalize the strain between the tube bundle and shell in order to prevent the tube connections to the tube sheet from working loose. Figure 5-7 shows a typical arrangement for this purpose.

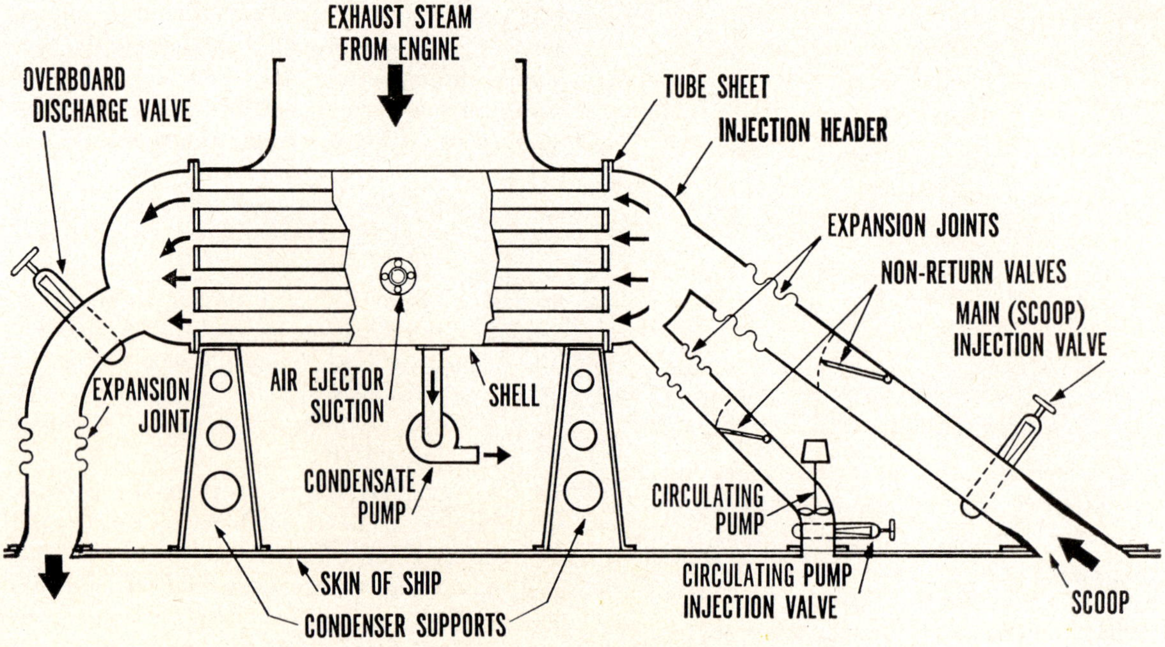

FIG. 5-6. Schematic Arrangement of a Condenser Installation.

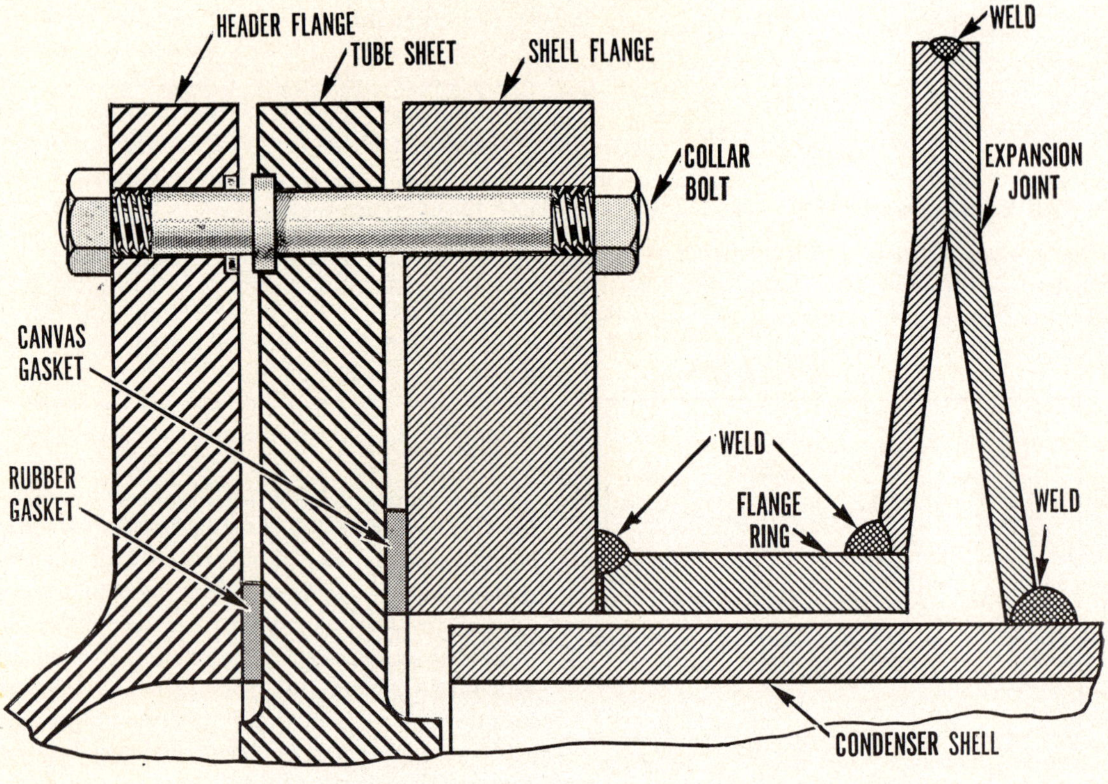

FIG. 5-7. Typical Condenser Shell Expansion Joint.

If the condenser were perfectly free of air, the pressure in the shell would exactly equal the vapor pressure of the condensate at the operating temperature established by the circulating water. Assume a reasonable operating temperature to be 95°F, then this pressure would be 0.815 psia, or 1.66″ Hg. The condensation of steam would provide the vacuum by itself. However, air is admitted into the condenser through leaks in the turbine gland seals, turbine casing drain lines, dissolved air in drains coming from various parts of the plant, and perhaps other sources. Furthermore, when starting up a cold plant, the condenser is filled with air. Therefore, condensation of the steam can neither create the initial vacuum nor alone maintain it.

In addition to contributing its proportional part to the pressure in the condenser, air tends to collect at the colder tube surfaces, causing an insulating effect on the flow of heat. This increases the temperature difference between the steam and water, and the condenser pressure correspondingly increases.

For the above reasons, it is necessary to provide air pumps called air ejectors, to purge air from the condenser. With effective air ejection, the condenser pressure is very close to the vapor pressure of water at the condensate temperature, and no significant error is caused by assuming the condenser pressure to equal the condensate vapor pressure. Removal of the air is discussed in Article 506.

The importance of vacuum on plant efficiency is demonstrated readily as follows. Suppose steam is available at 1200 psia and 900°F, to a perfect turbine which can extract work isentropically. If the turbine discharges to atmospheric pressure, it will produce 404 BTU of work per lb. of steam. If it discharges to a condenser with a pressure of 1.5″ Hg, it will produce 572 BTU/lb, an increase in work/lb. of over 41%. A real turbine does not show quite such improvement, but its increase is in the order of 30%.

505. CONDENSER PERFORMANCE.

The overall performance of the condenser may be computed by use of an energy balance. The energy entering the condenser is the kinetic and thermal energy carried in by the steam and the thermal energy carried in by the salt water. The energy leaving the condenser is the thermal energy of the condensate, and the thermal energy of the salt water leaving. The kinetic energies of the salt water entering and leaving are small, and about equal in any case. The kinetic energy of the condensate leaving is small, and that for the steam entering will be investigated, since it was significant as leaving loss from the turbine. Figure 5-8 is a diagram of the energy balance.

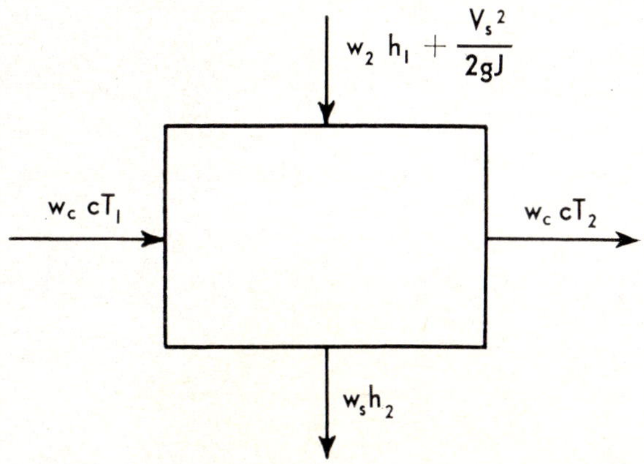

FIG. 5-8. Simplified Main Condenser Energy Balance.

The balance is

$$w_c c T_1 + w_s h_1 + \frac{w_s V_s^2}{2gJ} = w_s c T_2 + w_s h_2$$

or

$$w_s\left(h_1 - h_2 + \frac{V_s^2}{2gJ}\right) = w_c c(T_2 - T_1) \qquad (8)$$

where

w_s = weight rate of steam entering (and condensate leaving)
h_1 = enthalpy of steam entering
h_2 = enthalpy of condensate leaving
V_s = velocity of steam entering
w_c = weight rate of circulating water
c_{sw} = specific heat of circulating water (= 0.94 for sea water)
T_1 = injection temperature
T_2 = overboard discharge temperature.

Example 53: At full power on a destroyer, 4067 lbs/min of steam at 2.5 psia and $7\frac{1}{2}\%$ moisture are delivered to the condenser at a velocity of 198 fps. If sea water injection temperature is 75°F and discharge temperature is 92.7°F, what quantity of sea water is needed to condense the steam?

Solution:

$w_s = 4067$ lbs/min

$h_1 = 1043.4$ BTU/lb (from Steam Tables)

$h_2 = 102.3$ BTU/lb (for sat. water at 2.5 psia)

$c_{sw} = 0.94$ BTU/lb-°F, for sea water

$T_1 = 92.7$°F

$T_2 = 75.0$°F

Computing the kinetic energy for each pound of the entering steam,

$$\frac{V_s^2}{2gJ} = \frac{(198)^2}{64.4 \times 778} = 0.78 \text{ BTU/lb.}$$

This is insignificant compared to $h_1 - h_2$, so will be dropped. Therefore

$$w_c = \frac{4067(1043.4 - 102.3)}{(.94)(92.7 - 75.0)}$$

$$= \underline{\underline{230{,}000}} \text{ lb/min.}$$

Since sea water weighs about 8.5 lbs/gal this is equivalent to 27,000 gpm.

The above example demonstrates that the value of the kinetic energy of the steam entering the condenser is negligible, so the term may be dropped from Equation (8) in all practical cases, leaving

$$w_s(h_1 - h_2) = w_c c_{sw}(T_2 - T_1). \qquad (9)$$

Equation (9) equates heat given up by the exhaust steam to heat received by the sea water. Therefore, the total amount of energy transferred is equal to either side. In the preceding example, this amounts to

$$230{,}000 \times .94 \times 17.7 = 3{,}820{,}000 \text{ BTU/min.}$$

The rate at which heat can be transferred through the tube walls is proportional to the temperature difference between the fluids on each side of the tubes, the area, and the overall conductivity as discussed before and expressed by Equation (4). Now that the required rate of heat flow is known, and the terminal temperatures are known, the total heat transfer area can be closely estimated. Recall, from Example 51, that U for a combination of a water film, brass conductor, and steam film was computed to be 595 BTU/hr-ft²-°F. This is very close to 10 BTU/min-ft²-°F. This value is used together with the temperatures from the previous example to estimate the total tube area for the destroyer condenser of Example 53. Note that the refinement given by Equation (5) is not used in the example. It is left as an exercise for the student to use this more accurate equation for the below computation.

Example 54: Find the heat transfer area necessary for the conditions given in Example 53.

Solution: The temperature of the entering steam is 133.4°F (from Steam Tables for $2\frac{1}{2}$ psia). If the condensate is not subcooled, this is also the condensate temperature. Then

$$\Delta T_1 = 134.4 - 75.0 = 59.4$$

$$\Delta T_2 = 134.4 - 92.7 = 41.7$$

$$\Delta T_m = \frac{59.4 - 41.7}{\ln \dfrac{59.4}{41.7}} = \frac{17.7}{.358}$$

$$= 49.4°\text{F}$$

and

$$A = \frac{Q}{U \Delta T_m} = \frac{3{,}820{,}000}{10 \times 49.4} = \underline{\underline{7740}} \text{ ft}^2.$$

In this installation, the area is provided with 4504 tubes, 0.652″ O.D., 0.527 I.D., and 11′4″ long.

506. CONDENSER ARRANGEMENT.

Figure 5-9 is a cross section of one half of a typical condenser. The steam enters at the top, penetrating the tube nests, condensing as it does so. The water formed drips into the heating tray near the center of the drawing, or to the top of the air baffle whence it drains into the receiving tray, eventually overflowing into the heating tray. Note the wider spacing of the tubes on the outer periphery of the bundle. This allows better penetration into the bundle and helps to distribute evenly the heat transfer load among all the tubes. The steam lane in the center of the drawing permits direct contact between the steam and the condensate about to leave the condenser, in order to keep the latter as hot as possible. Cooling below the saturation temperature is undesirable since it represents a needless waste of heat to the sea water.

It is obvious that any air in the condenser will be completely saturated with water vapor (i.e. have relative humidity of 100%). The removal of

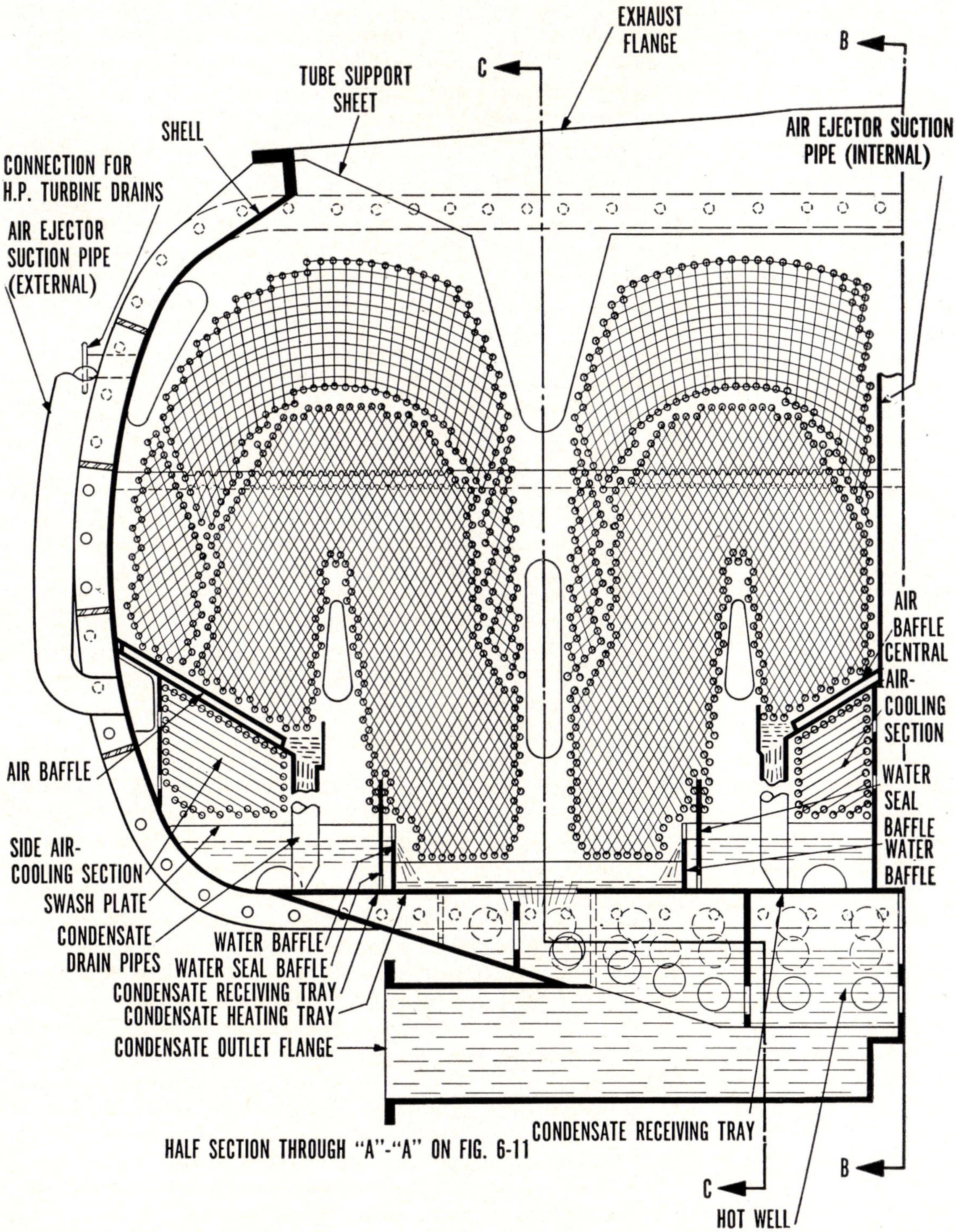

FIG. 5-9. Transverse Half-Section of a Typical Main Condenser.

the air must be accompanied by the removal of the water vapor. At temperatures near the saturation temperature at the condenser pressure, an extremely large quantity of vapor is present with each pound of dry air. In order to decrease the load necessary for the air ejector to handle (and thus to permit a reasonable size of air ejector), it is desirable to decrease the proportion of vapor to air. This can be done by cooling the mixture well below the saturation temperature. For this reason, certain sections of tubes in the condenser are well protected from the direct steam flow, and thus operate with surface temperatures more nearly approaching that of the circulating water. The air-vapor mixture surrounding these tubes is much "richer" in air content than in the rest of the condenser. The air ejector suction is taken from this section. The arrangement is well depicted in Figure 5-9. Normally, about 10% of the condenser surface is incorporated in the air cooling sections. The effectiveness of the cooling is shown in the following example.

Example 55: A condenser pressure is measured at 1.5″ Hg (.737 psia). The temperature of the condensate is 87°F. Find the proportion of the air-vapor mixture at 87°F and its proportion if cooled to 80°F.

Solution: The saturation pressure of water corresponding to 87°F is 0.635 psia. Since the condenser pressure is 0.737 psia, air is present to make up the difference. Therefore the pressure of the vapor, p_w is 0.635 psia and that of the air is 0.102 psia. The volume of the vapor for each pound, v_w, is 511.7 ft³/lb, as found from Steam Tables. The volume of the air is

$$v_a = \frac{R_a T_a}{p_a} = \frac{(53.3)(547)}{(.102)(144)}$$
$$= 1985 \text{ ft}^3/\text{lb}.$$

The proportion of air, by weight, is

$$\frac{511.7 \times 100\%}{511.7 + 1985} = 20.5\%$$

and of water, by weight

$$\frac{1985 \times 100\%}{511.7 + 1985} = 79.5\%.$$

Therefore, approximately 4 lbs of water vapor is pumped for every lb of air.

If the mixture is cooled to 80°, still at condenser pressure, the pressure of the vapor becomes 0.507 psia (from Steam Tables) and that of the air is $0.737 - 0.507 = 0.230$ psia. Now the specific volume of air is

$$v_a = \frac{(53.3)(540)}{(.230)(144)} = 870 \text{ ft}^3/\text{lb}$$

and that of steam is $v_w = 633.1$ (from Steam Tables). The proportion of air present, by weight, is

$$\frac{633.1 \times 100\%}{870 + 633.1} = 42\%$$

and of water, by weight, is

$$\frac{870 \times 100\%}{870 + 633.1} = 58\%.$$

This is a ratio of about 1.4 lbs. of water pumped for every lb. of air. Cooling 5 more degrees (if possible) would reduce the ratio to 0.85 lbs. water per lb. of air.

507. MISCELLANEOUS.

Below the heating tray at the bottom of the condenser, the condenser shell forms a collecting tank called the *hot well.* The condensate piping leads from here to the condensate pump. Two pumps are usually provided, one steam driven and one electric driven. These will be discussed later. It is noted at this point however, that the condensate is practically at its boiling temperature, and any slight decrease in pressure will cause it to "flash" into steam. For this reason the condensate pump is located below the condenser in order to keep a positive head of water between the surface of the condensate in the hot well and the pump suction. The condensate pump discharge pressure is usually about 45 psig, varying somewhat in different plant arrangements. This pressure drops somewhat as the condensate passes through the heat exchangers between the condensate pump and the D.F.T.

It is common practice to speak of condenser pressure conditions in terms of *vacuum.* Thus, a 25″ vacuum corresponds to a 5″ Hg pressure (based on a 30″ Hg barometric pressure). A 5″ Hg pressure is the pressure created by a 5″ vertical column of mercury, and is equal to $5/30 \times 15 = 2.5$ psia, approximately.

The low pressure maintained in the condenser makes it a convenient point for the introduction of make-up feed to the cycle, to compensate for fresh water lost as condensate to the bilges, vapor to the atmosphere, or other leaks from the system. A feed line is connected from the reserve feed tanks to the condenser shell. Opening a valve allows atmospheric pressure to force fresh feed water into the condenser.

Fresh water drains from several auxiliary units in the plant are collected in a Fresh Water Drain Collecting Tank at atmospheric pressure. The drain system will be discussed later in more detail. These drains collect continuously while the plant is in operation. In order to conserve them as feed water and to conserve the thermal energy which they contain, it is desirable to re-introduce them into the feed circuit. This can be done either by admitting them from the Drain Collecting Tank to the condenser, or by pumping them into a higher pressure point in the cycle. Both arrangements are provided. The first alternative results in the loss of a large portion of the thermal energy of the drains because of cooling in the condenser. Because of this, the second alternative is preferable.

The Drain Collecting Tank is fitted with an automatic float valve, or similar device, which opens a valve in the tank. When the drains are routed to the condenser, atmospheric pressure on the surface of the water in the tank forces the drains into the condenser. As the tank level drops, the float valve closes to prevent entry of air into the condenser. When the drains are pumped into the condensate line at a higher pressure point, the float valve operates a switch to control the F.W. Drain Pump when the tank is filled, and shuts off the pump when the tank is emptied. In the latter event, the drains join the main condensate just prior to entry into the D.F.T.

508. AIR EJECTOR ASSEMBLIES.

The function of the air ejector in removing the air from the condenser has already been discussed. The air ejector is a jet pump. A jet of steam, expanding through a nozzle, acquires a high velocity (supersonic). It entrains the air-vapor mixture surrounding the nozzle exit and accelerates it to the stream velocity. The jet then enters a diffuser tube where the kinetic energy is converted into pressure, so that the pressure at the diffuser outlet is higher than at the air-vapor entrance. Figure 5-10 is a sketch of the arrangement.

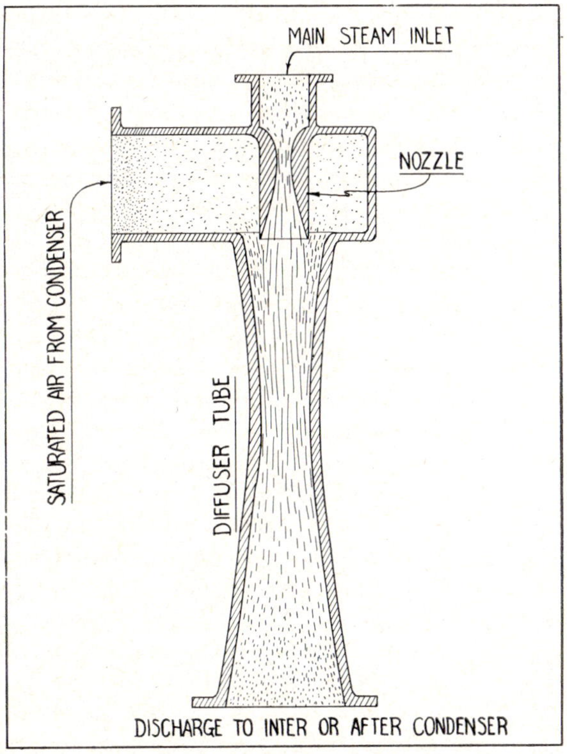

FIG. 5-10. Air Ejector Cross-Section.

Experience has shown that the air ejector has a very poor efficiency when the pressure at its outlet, p_0, is greater than about 8 or 9 times its suction pressure, p_i. Yet the purpose of the air ejector is to raise the pressure of the air from condenser pressure to atmospheric pressure, a ratio of 15 or 20 to 1. The problem is solved by placing two ejectors in series, permitting them to operate at pressure ratios, p_0/p_i, of 4 to 6. For instance, the first stage ejector raises the pressure from 1.5″ Hg to about 7″ Hg and the second from 7″ Hg to about 32″ Hg, or 1 psig, slightly higher than atmospheric.

The actuating steam for the first stage mixes with the air and vapor pumped, and is a large proportion of the outlet volume. In order to circumvent the necessity for providing a second stage ejector large enough to handle the entire first stage outlet, the latter is cooled in a heat exchanger called the *air ejector inter-condenser.* The cooling fluid is the condensate which has just been delivered from the main condenser. Thus the thermal energy used in the motive steam is largely recovered. It is also desirable to conserve the condensate which forms in the inter-condenser. Since its pressure is below that of the main condensate or the atmosphere, a pump would have to be provided to raise its pressure. It is very simple, however, to lead it back to the main condenser which has a lower pressure. This is done through the loop seal. See Figure 5-11. The leg of water in the loop seal maintains the desired pressure difference between the inter-condenser and the main condenser, preventing the pump from being "short-circuited."

From the shell of the inter-condenser, the air and uncondensed vapor is pumped by the second stage ejector into another heat exchanger called the *air ejector after-condenser*. The pressure now has been raised to slightly higher than atmospheric. The steam-air mixture is again cooled by the main condensate. The fresh water as produced is led to a low pressure drain tank, which collects fresh water drains from other points in the plant. Modern arrangements provide for the L.P. drain tank to discharge directly into the condensate line.

The air and uncondensed vapor from the shell of the after condenser is then led through a gland exhaust condenser and from here discharged to the atmosphere. The function of the gland exhaust condenser is to conserve the thermal energy and the working fluid of the turbine gland seal steam system. The gland exhaust steam gives up its heat to the main condensate and flows to the low pressure drain tank after it is condensed.

The actuating steam for the air ejectors is taken from the auxiliary steam line through a reducing valve. In current installations, operating pressures are 125 to 135 psig, but in older installations, they may be found as high as 250 psig. 15 to 20 lbs. of steam are required for every pound of air removed. This would be extremely inefficient if the energy of the steam were not largely recovered in heating the condensate. Roughly 1000 BTU/lb are so recovered. In modern naval plants, an air ejection capacity is provided to eject approximately one pound of air per hour per thousand rated shaft horsepower. This is not intended to imply that maximum air leakage occurs at full power; the maximum actually occurs at low power.

In modern arrangement, the air-ejectors, the inter and after condensers, and the gland exhaust condenser are all assembled in one unit. In the unit, two first stage and two second stage ejectors are provided, either set being capable of handling the full load under normal operation. The ejectors are so arranged that either set can be disassembled for maintenance while the other is operating.

The gland seal steam, discussed briefly in Chapter 4, is supplied to the turbine glands at very low pressures. After being utilized in the glands, it is collected in the gland exhaust system, and is now virtually at atmospheric pressure. It is then led to the gland exhaust condenser ,together with considerable air mixed with it. Most of the vapor is condensed, giving heat to the main condensate, and the air and remaining vapor led off to atmosphere, along with the air pumped from the main condenser. The condenser is designed to handle roughly 20 lb. gland steam and 10 lb. air for each 1000 rated SHP.

509. HEAT EXCHANGE RELATIONS.

An approximate analysis of the heat exchange taking place in the condensate system from the condensate pump outlet to the gland exhaust condenser outlet is relatively simple. The energy entering the system is (1) the energy in the condensate leaving the pump, (2) the energy in the actuating steam entering the air ejectors, (3) the energy in the air-vapor mixture entering the air ejectors, (4) the energy in the gland exhaust steam-air mixture entering the gland exhaust condenser. The energy leaving the system is in (1) the actuating steam condensate from the inter-condenser, (2) the actuating steam condensate from the after condenser, (3) the gland exhaust condensate from the gland exhaust condenser, (4) the main condensate from the gland exhaust condenser. Since the quantity of air-vapor mixture entering from the main condenser is small compared to the actuating steam, and the energy in it and the air entering the gland exhaust condenser are partially compensated for by the energy in the air-vapor mixture leaving the system, only steam and water flow into and out of the system need be considered for a reasonably accurate energy balance. The example below will illustrate this and provide an idea of the amount of heating of the main condensate.

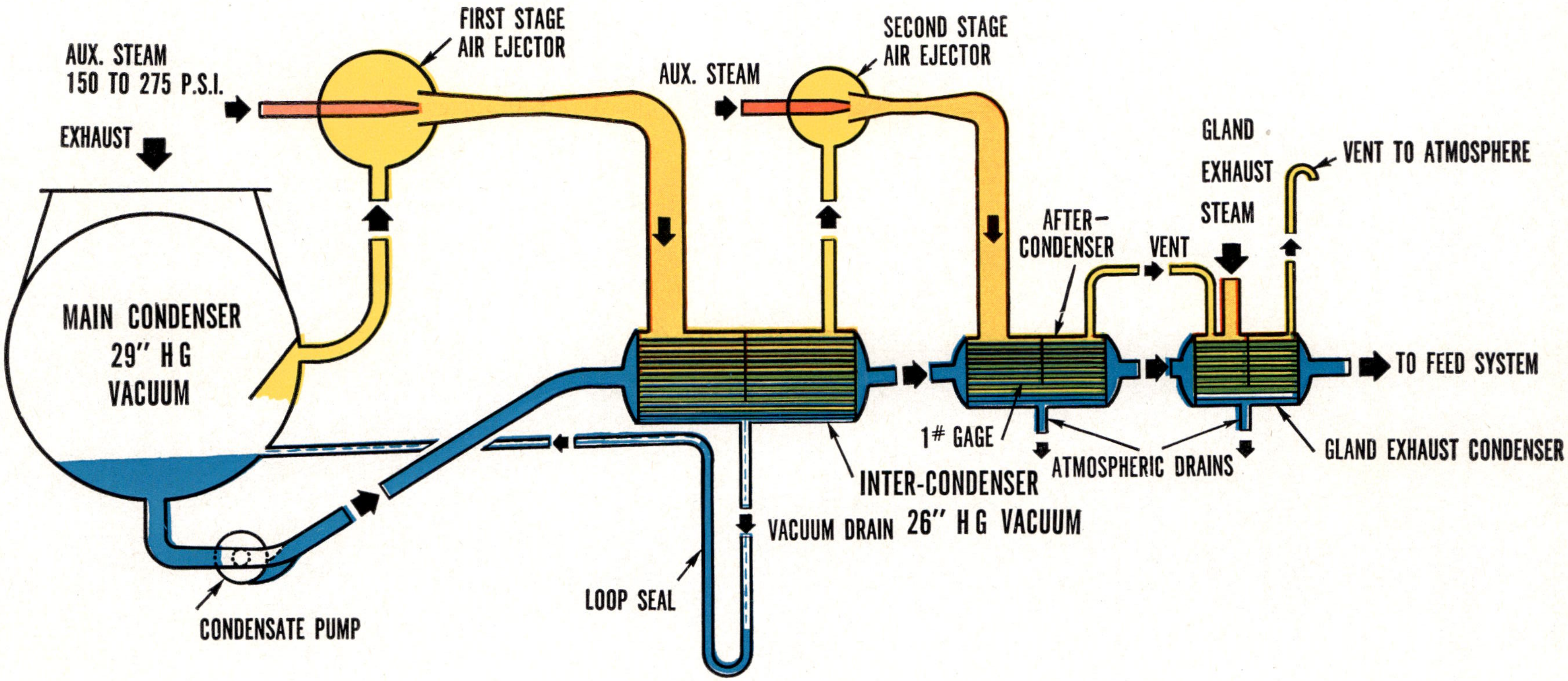

FIG. 5-11. Main Condensate Flow in Air Ejector.

Example 56: A destroyer turbine throttle flow at a given speed is 485 lbs/min of steam. The condensate leaving the condenser also includes 28 lb/min of water drained into the condenser from the L.P. drain tank and the inter-condenser. The condenser pressure is 1″ Hg and the condensate leaves the condensate pump with a temperature of 74°F. Steam at 130 psig and 400°F is supplied at the rate of 5.5 lbs/min to the first stage ejector and 4.4 lbs/min to the second stage ejector. Gland sealing exhaust steam is estimated at 6 lb/min at approximately saturated conditions. The inter-condenser shell pressure is 6″ Hg. The after condenser and gland exhaust condenser pressures are at very nearly atmospheric pressure. Find the increase in condensate temperature through the heat exchangers.

Solution: The energy entering is:

(a) Main condensate entering

$$Q = w_c c_w (T_1 - 32)$$
$$= 513 \times 1.0 \times (74 - 32) = 21550 \text{ BTU/min.}$$

(b) From 1st stage act. steam

$$Q = w_{s1} h_1$$
$$h_1 = 5.5 \times 1222.7 = 6730 \text{ BTU/min.}$$

(c) From 2nd stage act. steam

$$Q = w_{s2} h_1$$
$$= 4.4 \times 1222.7 = 5380 \text{ BTU/min.}$$

(d) From gland exhaust steam (sat. at 212°F)

$$Q = w_g h_g$$
$$= 6.0 \times 1150.4 = 6910 \text{ BTU/min.}$$

$$\text{Total entering} \quad = \underline{\underline{40570}} \text{ BTU/min.}$$

The energy leaving is:

(a) From inter-condenser drain

$$Q = w_{s1} h_1 \text{ (sat. liquid at 6″ Hg)}$$
$$= 5.5 \times 109 = 599 \text{ BTU/min.}$$

(b) From after-condenser drain

$$Q = w_{s2} h_1 \text{ (sat. liquid at 212°)}$$
$$= 4.4 \times 180 = 792 \text{ BTU/min.}$$

(c) From gland exhaust condenser drain

$$Q = w_g h_g' \text{ (sat. liquid at 212°)}$$
$$= 6.0 \times 180 = 1080 \text{ BTU/min.}$$

(d) Main condensate leaving

$$Q = w_c c_w (T_2 - 32)$$
$$= 513 \times 1.0 \times (T_2 - 32).$$

Total leaving

$$= 513(T_2 - 32) + 2471 \text{ BTU/min.}$$

Equating energy leaving to energy entering and solving for T_2:

$$513(T_2 - 32) + 2471 = 40{,}570$$
$$T_2 = \frac{40{,}570 - 2471}{513} + 32$$
$$= 106.4°\text{F.}$$

If the rough assumption is made that each lb. of steam entering the system gives up 1000 BTU to the condensate (approximately its heat of condensation), the total heat transferred is $(6.0 + 4.4 + 5.5) \times (1000)$, or 15,900 BTU. Each pound of condensate receives 15,900/513 = 31 BTU, which will raise its temperature 31°, from 74°F to 105°F. This is in good agreement with the above figure of 106.4°F.

510. THE DEAERATING FEED TANK.

The deaerating feed tank is at the junction of the condensate and boiler feed systems. The D.F.T., in carrying out its functions of deaerating and heating the condensate, converts the condensate to boiler feed. As previously mentioned, it also serves as a storage for this feed. It will be discussed here as the final component of the condensate system, but it will be necessary to consider it later to some extent in connection with the boiler feed system.

Most of the air in the system has already been separated from the condensate in the main condenser. However, the quantity remaining, even though small, can cause a significant amount of corrosion in the high pressure, high temperature boilers in present use; and means must be found to remove as much as possible. There are two mechanisms by which the air is retained by the condensate. First, the condensate is capable of holding in solution a certain amount of the various gases in the air. Secondly, the process of adsorption causes molecules of the gases in the air to cling to the condensate particles in an intimate physical mixture. The deaerating process then must serve the dual purpose of driving the

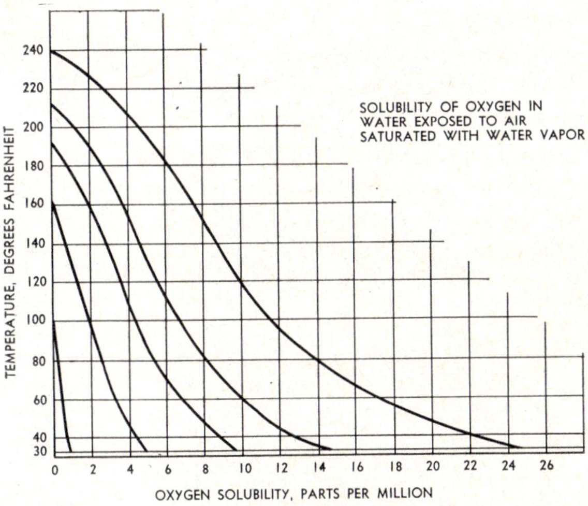

FIG. 5-12. Solubility of Oxygen in Water.

gases out of solution and of separating the adsorbed gas molecules from the liquid. This is accomplished by two processes, heating, and mechanical agitation.

The two most harmful gases from a corrosion standpoint, which are present in significant quantity, are oxygen and carbon dioxide. Figure 5-12 shows the solubility of oxygen in water as a function of temperature and pressure. It is noted that the *solubility becomes zero at the boiling temperature for any given pressure.* The curves for CO_2 and other gases are similar. Heating the condensate to the boiling point corresponding to its pressure provides the means for releasing dissolved gases. Mechanical agitation to separate the adsorbed particles is accomplished by two spray processes which will be apparent from the description of the D.F.T.

Condensate enters the D.F.T. at the water inlet shown in Figure 5-13. It flows through a nest of tubes called the *vent condenser* and serves to cool the air and vapor surrounding these tubes, discussed later. The water then flows into a spray head from which it is discharged through spray valves by the pressure of the condensate pump. This spray action breaks the water up into fine droplets which releases some of the adsorbed gases and allows the water to rapidly receive heat from the surrounding steam atmosphere, and thus to release the dissolved gases. The water droplets fall into the conically shaped tray and are led to the central port. Here the water is picked up by the incoming steam and is hurled in a helical path outward and upward through the helical baffles. This action again atomizes the water. The scrubbing and heating actions of the steam blast remove the remainder of the dissolved and entrapped air. The water droplets collect on the lower surfaces of the conical baffle and drop to the bottom of the tank, remaining in an atmosphere of steam. The steam carrying the gases rises up around the outside of the baffle to the top of the tank where it heats the incoming water spray. Since the steam is giving up heat to the condensate, it is continually being condensed. The gases tend to concentrate in the coolest portion of the tank (just as they did in the main condenser). This portion is around the tubes of the vent condenser containing the incoming condensate. Some of the water vapor carried by the air is condensed at this point; the remaining air and vapor pass out of the tank through the air outlet.

To insure effective air removal, sufficient vapor is permitted to pass out of the tank through the air vent to sweep out all of the freed air. The amount of steam so discharged is limited by an orifice installed in the vent line. In most installations, the vent line is led to the gland exhaust condenser to recover the water and some of the heat thus carried out of the D.F.T. air vent.

The steam supply is auxiliary exhaust steam. It enters the tank through a check valve which will open when the D.F.T. pressure is less than that of the auxiliary exhaust line, but prevents the backflow of water to the steam line should the tank be flooded or should the tank pressure rise higher than the steam pressure. Pressure control is automatic, since the tank pressure will tend to drop as the steam is condensed, causing the valve to open to admit more steam. Some tanks are provided with an external operating mechanism which permits adjusting the pressure difference between the steam supply and the tank. These are normally set so that the tank pressure is maintained at $\frac{3}{4}$ to $1\frac{1}{2}$ psi below the steam supply pressure.

Steam also flows into the tank from the high pressure drain connection. As the liquid drains from high pressure connections are throttled through the traps to a lower pressure, they flash into steam, the energy of which is conserved by utilization in the D.F.T. In the event of lack of sufficient heating steam from the above two

IRELAND, K.

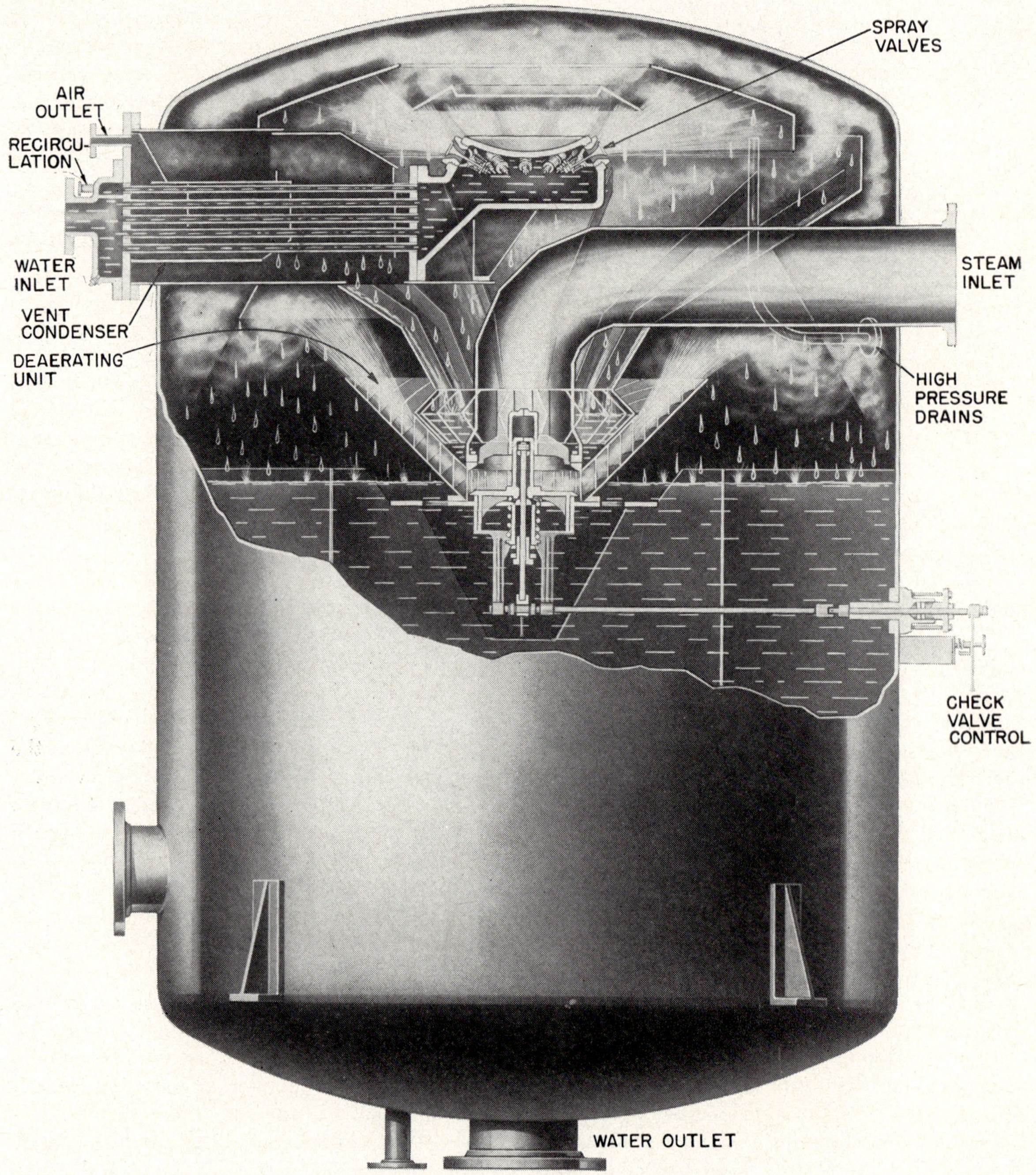

FIG. 5-13. Typical Deaerating Feed Tank.

sources, the necessary steam may be supplied from the auxiliary steam line through a reducing valve. Since the throttling process is wasteful, this is uneconomical practice. Normally, the number of steam and electric driven auxiliaries may be so balanced that the auxiliary exhaust steam supply is just sufficient to operate the D.F.T. and the evaporators.

A relief valve is provided on the tank to protect the tank against excess pressure resulting from abnormal operating conditions. A *vacuum breaker* is installed to admit air to the tank in the event too rapid cooling tends to create a vacuum, which the tank is not built to withstand. Condensate and boiler feed recirculating connections will be discussed in following pages. General specifications require that the feed water delivered to the boiler contain less than 0.01 parts per million of air.

511. HEAT EXCHANGE IN THE FEED TANK.

In Example 56 the temperature of the condensate leaving the gland exhaust condenser was computed to be 106°F. In actual practice this temperature will vary from 100°F to 125°F, depending on the power being developed and the individual characteristics of the plant. The pressure of this condensate is reduced to that of the D.F.T. as it passes through the spray nozzles. The D.F.T. pressure is normally between 14 and 16 psig in most modern plants, the corresponding saturation temperature being about 250°F. It is necessary, then, for each pound of condensate to be raised about 140°F. This is a requirement of about 140 BTU/lb of condensate which must be furnished by the auxiliary exhaust steam. The quantity of exhaust steam per pound of condensate is then readily calculated, provided heat from high pressure drains is neglected.

512. SUMMARY AND OPERATION OF THE CONDENSATE SYSTEM.

It has been shown that the exhaust steam from the turbine gives up heat to the sea water, being condensed in the process. The condensate pump raises the condensate pressure to about 45 psig and sends it through the air ejector condensers, the gland exhaust condenser, and to the D.F.T. where the pressure is throttled to about 15 psig. During its passage through the air ejector and gland exhaust condensers, the condensate is heated (from 74°F to 106.4°F in our example). In the D.F.T., the condensate is deaerated and further heated to about 250°F, the boiling temperature corresponding to its pressure at this point. Sources of energy were (1) the energy possessed by the condensate in the condenser, (2) the energy added by the condensate pump (small and to be discussed in Chapter VI), (3) the energy added by the air ejector actuating steam, (4) the energy added by gland exhaust steam, (5) the energy added by the L.P. drains, (6) the energy added by H.P. drains, (7) the energy added by the auxiliary exhaust steam. Energy escaping from the system was contained in (1) actuating steam condensate from 1st and 2nd stage air ejector condensers, (2) air and vapor vented from 1st and 2nd stage air ejector condensers, from the gland exhaust condenser and from the vent condenser of the D.F.T. Figure 5-14 indicates the above points.

In starting up a cold plant, one of the first requirements is the deaeration of the water going to the boiler. A condensate flow is necessary for this. To provide this flow, water from the D.F.T. may be recirculated from the tank discharge to the tank water inlet. Normally, the emergency feed booster pump, to be discussed later, may be so connected as to provide this recirculation. Also a flow of condensate is required through the air ejector condensers when the air ejectors are started up to create the condenser vacuum. Water from the D.F.T. may be circulated back to the main condenser to provide this function When operating at low plant loads, it is possible that normal condensate flow is insufficient to provide adequate cooling to the air ejector condensers, since the air ejector actuating steam requirement is independent of plant load. Under these conditions, the condensate temperature entering the D.F.T. will rise above its normal value. A thermostatically operated valve opens to allow condensate to recirculate to the main condenser which, in effect, increases the flow rate of condensate through the air ejector condensers. The above three recirculating schemes are shown diagrammatically in Figure 5-15. It is emphasized that these measures are used in starting up, or at low loads. Under normal operating conditions, no recirculating is necessary.

513. FLOW OF BOILER FEED.

The boiler feed system comprises the system of pumps and piping leading from the D.F.T. to the feed stop-check valve at the entrance to the economizer, which is here considered as a part of the boiler. (See *Naval Boilers* by Latham.)

Two major considerations exist in connection with the pumps. First, the boiler feed as it leaves the D.F.T. is practically at the saturation temperature corresponding to the D.F.T. pressure. Any slight reduction in this pressure will cause the water to flash into steam and to "vapor-lock" the pump. Therefore, the first pump in the line from the D.F.T. must be located below the D.F.T. in order to provide a positive head of

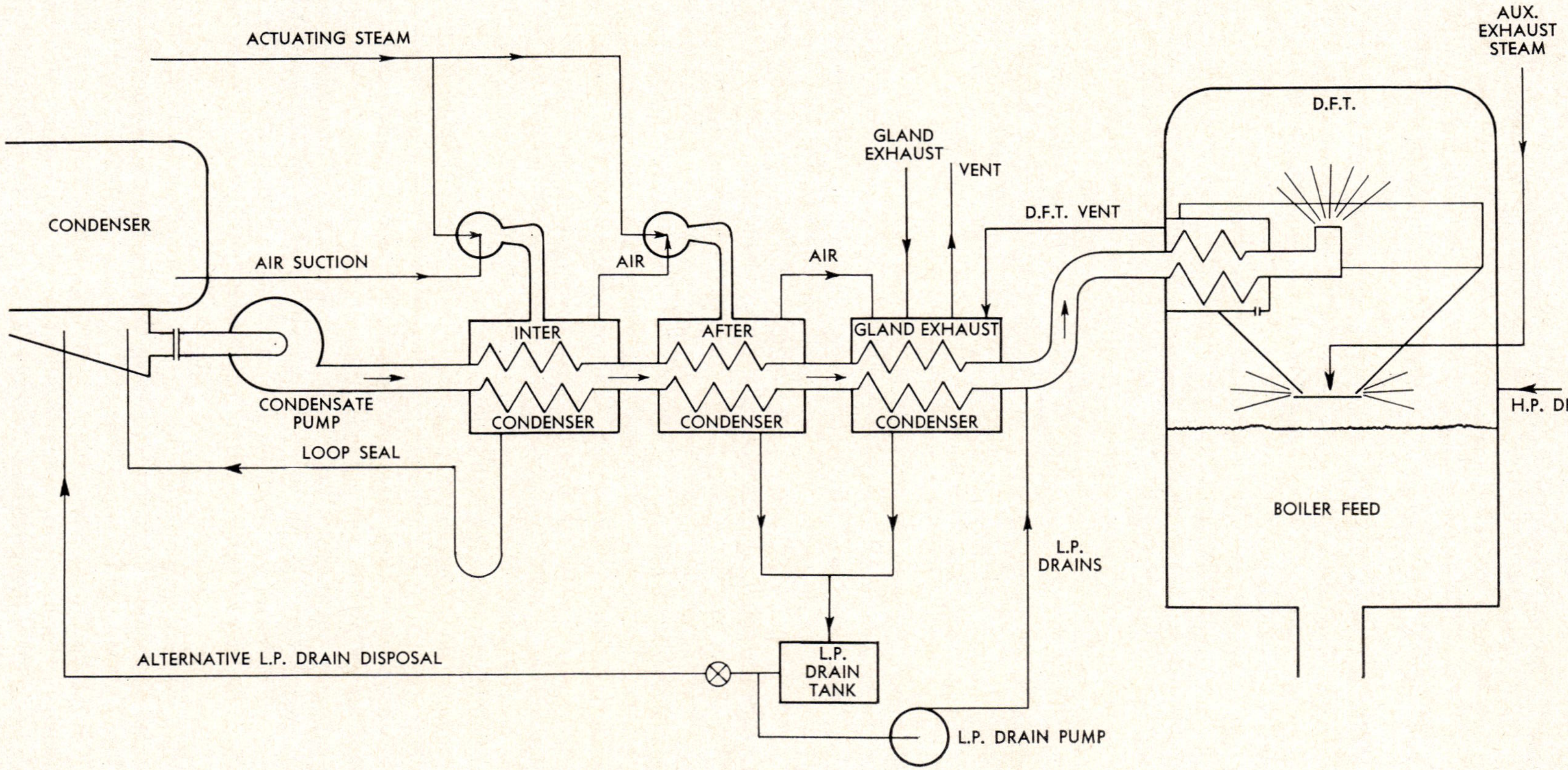

FIG. 5-14. Main Condensate Flow Diagram.

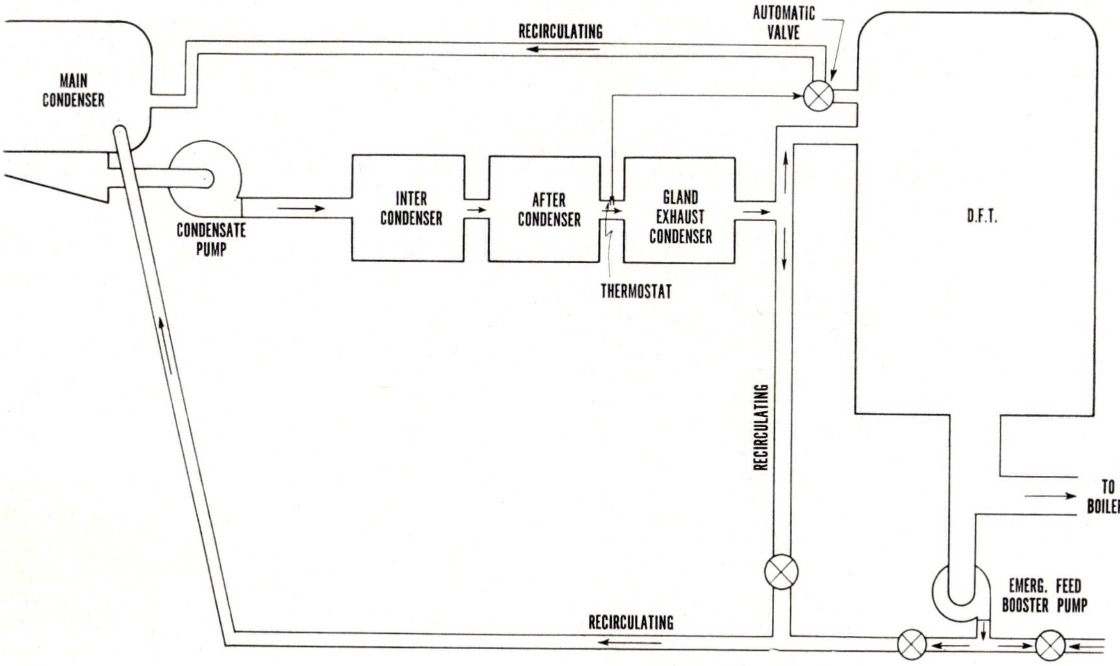

FIG. 5-15. Main Condensate Recirculation.

water on the pump suction to prevent this occurrence. In the crowded machinery spaces of combatant ships, where it is desirable to keep the center of gravity of all large equipment as low as practicable, it is the usual practice to locate the D.F.T. between the upper and lower levels. This makes it convenient to use a relatively small pump as the first pump in the boiler feed circuit. This pump is called the *main feed booster pump.* Another reason for use of a separate smaller pump is that it is desirable that it operate at a lower speed than the subsequent high pressure pumps, to minimize the possibility of developing a pressure less than saturation pressure at its suction, with the accompanying vapor lock discussed above. These characteristics will be discussed more fully in Chapter VI.

The second major consideration is that some feed water flow through the pumps must be assured at all times. This is to prevent overspeeding of the pumps under no-flow conditions, which will cause overheating of the pumps and consequent damage. This characteristic of the pumps used will also be discussed in the next chapter. To insure a continuous flow through the pumps, even when there is no demand for feed water at the boiler inlet, each pump is fitted with a recirculating line leading back to the D.F.T. This line contains an orifice, sized to permit a minimum of 5% of the rated pump capacity to recirculate continuously while the pump is in operation. In some installations, an automatic recirculating valve is provided in lieu of the orifice.

The main feed booster pump, then, takes suction from the D.F.T. and discharges at about 65–75 psig (depending on the installation) to the main feed pumps. The main feed pump raises the pressure of the boiler feed to about 20% to 25% above the boiler pressure.

In a typical modern arrangement, three main feed booster pumps are provided (for each propulsion plant) two of which are turbine driven and one is motor driven. Three main feed pumps are provided. In addition a motor driven emergency feed booster pump is provided, which can also be used to transfer feed from the system to the storage tanks, and to recirculate from the D.F.T. to the condensate system while starting up the plant, as previously discussed.

The arrangement is indicated schematically in Figure 5-16. The system serves the two boilers in this plant which supply steam to one main propulsion turbine and its auxiliaries.

The total capacity of the main feed pumps is

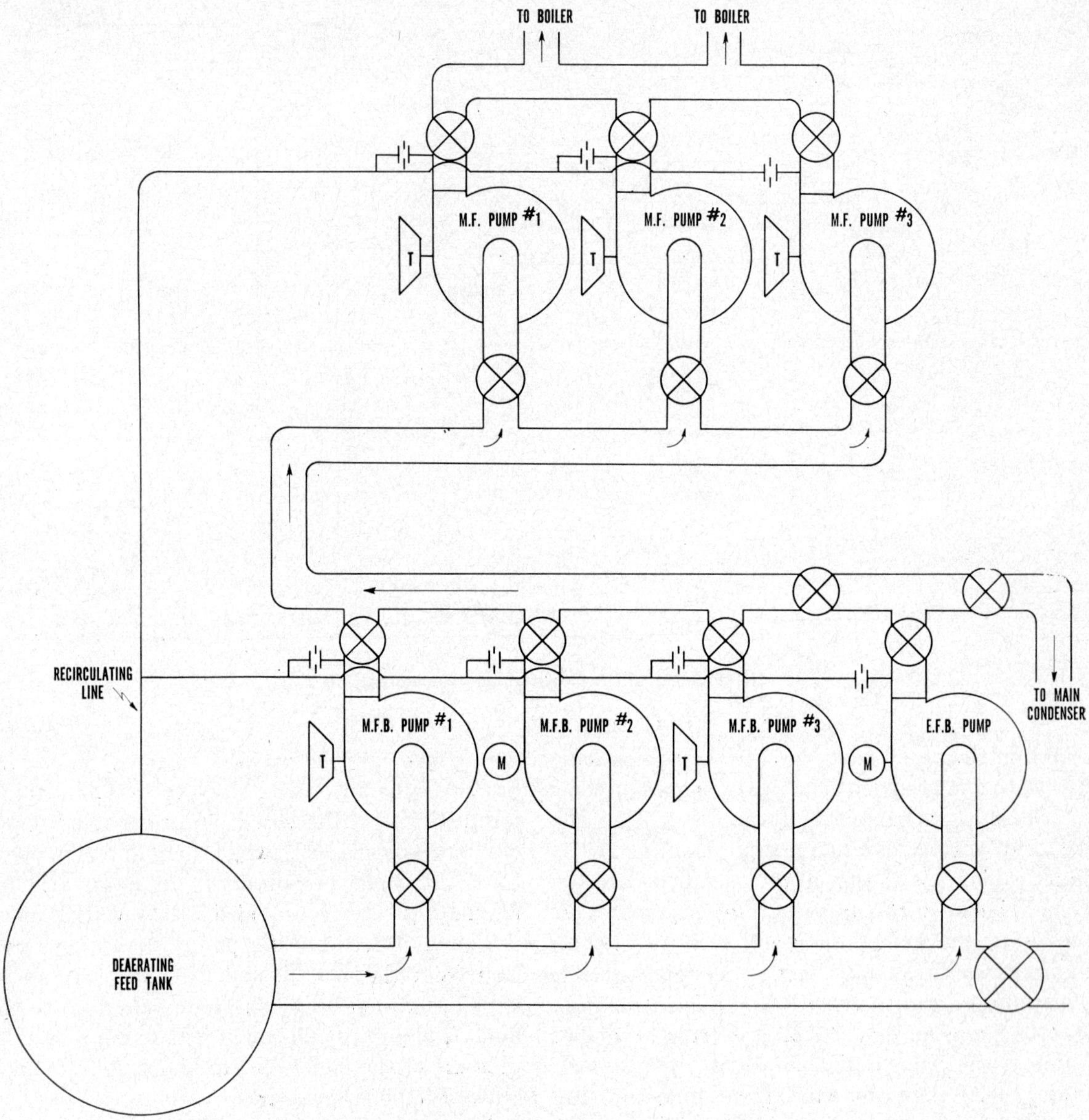

FIG. 5-16. Typical Boiler Feed System Arrangement.

required to be 150% of the boiler requirement at full power plus the capacity required for recirculation. Each main feed booster pump must have a capacity of 30 gallons per minute in excess of that required by each main feed pump in order to insure adequate supply to those pumps.

In some modern arrangements, the main feed booster pump has been combined in the same casing with the main feed pump. In such cases, the main feed booster pump rotor is geared to the main feed pump shaft, in order to permit its operation at a lower speed, as previously mentioned. Such units are called *main feed and booster pumps.*

CHAPTER VI
PUMPS

600. GENERAL.

In the previous discussion of the main steam system, pumps have been mentioned many times. Specifically, the functions of the lube oil, main circulating, main condensate, main feed booster and main feed pumps have been discussed. Many other pumps are necessary for operation of the ship. Fresh water pumps, fire and flushing pumps, drain pumps, fuel oil service pumps, fuel oil transfer pumps are some examples. Because of their importance it is deemed appropriate to discuss the more common types in some detail.

601. CLASSIFICATION.

Pumps are classified in three main groups: rotary, reciprocating and centrifugal. The first two classifications are positive displacement pumps. This term conveys the fact that in these pumps each operation cycle (one revolution of the shaft or one complete reciprocating stroke) causes the pump to deliver essentially a fixed quantity of liquid, regardless of the time required for the cycle, or the pressure change the pump causes. Therefore, the rate of discharge of such a pump is directly proportional to its speed, except for small corrections which will be discussed later. The centrifugal pump, on the other hand does not have this characteristic. Its rate of discharge is a function of speed, but it is not directly proportional, and it is also a function of the pressure difference across the pump.

602. DEFINITIONS AND ENERGY RELATIONS OF PUMPS, GENERAL.

In well designed pumps having normal efficiencies, the friction and turbulence losses within the pump have a negligible heating effect on the fluid pumped. Therefore, the thermodynamic energy terms in the General Energy Equation may be ignored and the equation conveniently rearranged to apply to pumps.

$$Z_1+\frac{p_1}{w}+\frac{V_1^2}{2g}+H_p=Z_2+\frac{p_2}{w}+\frac{V_2^2}{2g}$$

or

$$H_p=\frac{p_2-p_1}{w}+\frac{V_2^2-V_1^2}{2g}+Z_2-Z_1 \qquad (1)$$

where

H_p = head developed by pump (ft-lbs/lb or ft)
p_2 = outlet pressure, lbs/ft²
p_1 = inlet pressure, lbs/ft²
w = specific weight of fluid, lbs/ft³
V_2 = pump exit velocity, ft/sec
V_1 = pump inlet velocity, ft/sec
Z_1 = static inlet head, ft
Z_2 = static outlet head, ft.

In analyzing a pump, the vertical height of the inlet and outlet are nearly the same, if sections 1 and 2 are taken close to the pump. Therefore $Z_2 \approx Z_1$. The average velocity V_2 is different from the inlet velocity V_1 if inlet and outlet pipe diameters are different. For the moment we shall assume that $V_1=V_2$. Then the pump head is expressed by

$$H_p=\frac{p_2-p_1}{w}. \qquad (2)$$

Note that H_p, as a measure of pressure rise across the pump, is conveniently expressed in feet of head. *It is also a measure of energy acquired by the fluid being pumped*, and in this sense its units can be assigned as ft-lbs/lb. Therefore, multiplying it by the weight rate of flow, W, in lbs/sec will give the power added to the fluid. This power is the useful output power of the pump, P_0. Thus

$$P_0=\frac{WH_p}{550} \qquad (3)$$

where P_0 is expressed in horsepower. The mechanical efficiency of the pump is then the output power divided by the input power (commonly called the brake horsepower).

$$E_p=\frac{P_0}{P_i}. \qquad (4)$$

The difference in input and output power is lost in friction and turbulence.

Example 61: A pump operating at 80% efficiency is supplied with water at 75 psig and delivers it to a boiler at 1500 psig at the rate of 400 lbs/min. Find the output power, input power, and the temperature rise in the water.

Solution: Using equation (3):

$$P_0 = \frac{400}{60} \times \frac{(1500-75)144}{62.4} \times \frac{1}{550}$$
$$= 40 \text{ HP.}$$

From Equation (4):

$$P_i = \frac{40}{.8} = 50 \text{ HP.}$$

Power loss to friction $= 50 - 40 = 10$ HP

or

5500 ft-lbs/sec

or

$$\frac{5500}{778} = 7.07 \text{ BTU/sec.}$$

But water flows through at

$$\frac{400}{60} = 6.67 \text{ lbs/sec}$$

and carries away the friction heat. Therefore each pound of water gains

$$\frac{7.07}{6.67} = \underline{\underline{1.06 \text{ BTU}}}$$

which raises its temperature about 1°F.

Consider Figure 6-1. When the fluid is not flowing, the pressure at the pump suction, p_1, is the sum of the reservoir pressure head, p_r/w and the height of the leg of liquid, h_1, between the liquid surface and the pump suction. When the fluid is flowing, the pressure p_1 is less than the sum of these two components by the amount of friction head, h_f, opposing the flow in the suction pipe. The net suction head, h_s, then is

$$h_s = \frac{p_1}{w} = p_r + h_1 - h_f \qquad (5)$$

It is important that this head be greater than the equivalent vapor pressure head of the liquid being pumped, otherwise the liquid will flash into vapor at the pump suction, and the pump will not function. When h_s is greater than the vapor pressure head, the pump is said to have a *net positive suction head* (NPSH). Note that h_1 may be positive or negative in equation (5), depending on whether the pump is above or below the reservoir surface.

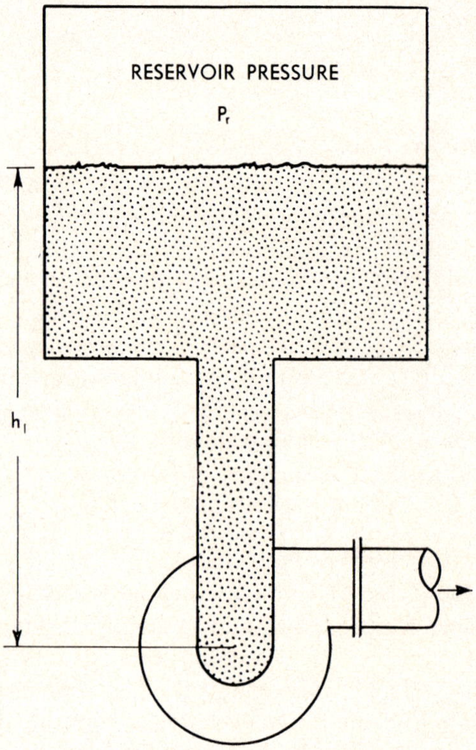

FIG. 6-1. Static Pressure on Pump Inlet.

A NPSH is a requirement for all pumps. Its designed value is usually large for cold liquid service where vapor pressures are low. With liquids near saturation temperature, such as encountered in the main feed system, it is difficult to provide large values, especially in the condensate and main feed booster pump locations.

603. ROTARY PUMP TYPES.

Perhaps the simplest type of rotary pumps is the spur gear pump, consisting of two meshing spur gears enclosed in a casing with close tolerances between the casing and gears. Figure 6-2 illustrates this type. Fluid is trapped in the space between the teeth and carried around to the discharge opening, into which it is forced as the teeth mesh. As they mesh, each tooth forces an amount of liquid, equal to its own volume, out of the discharge opening. One gear shaft is driven from an external power source, and the other idles. If clearances are small, this pump, as well as other rotary types, can pump air, and thus can "take a suction" on the fluid being pumped. The viscosity of the fluid is relatively unimportant in the functioning of the pump, and it is therefore

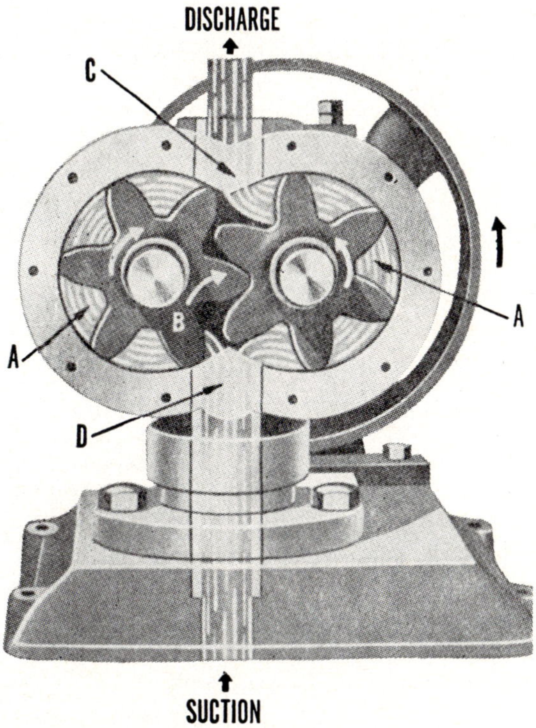

FIG. 6-2. Simple Gear Pump—End View.

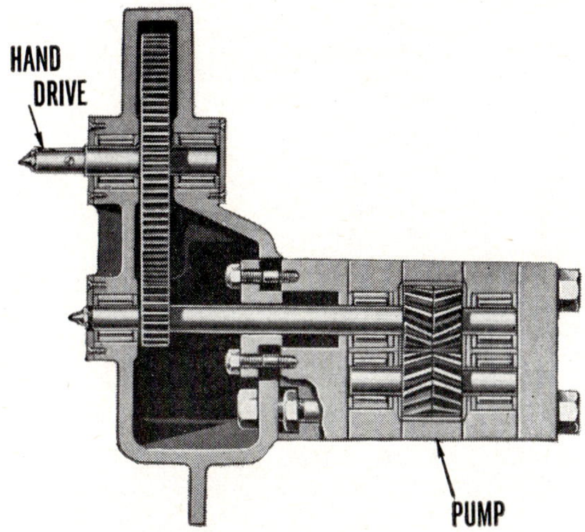

FIG. 6-3. Herringbone Gear Pump.

suited for pumping fuel and lubricating oil. The simple gear pump will cause some discharge pressure pulsation due to its intermittent discharge.

Modifications of the gear pump are the herringbone gear pump and helical gear pump shown in Figures 6-3 and 6-4 respectively. Note that as the helix angle of the gear increases, it becomes nec-

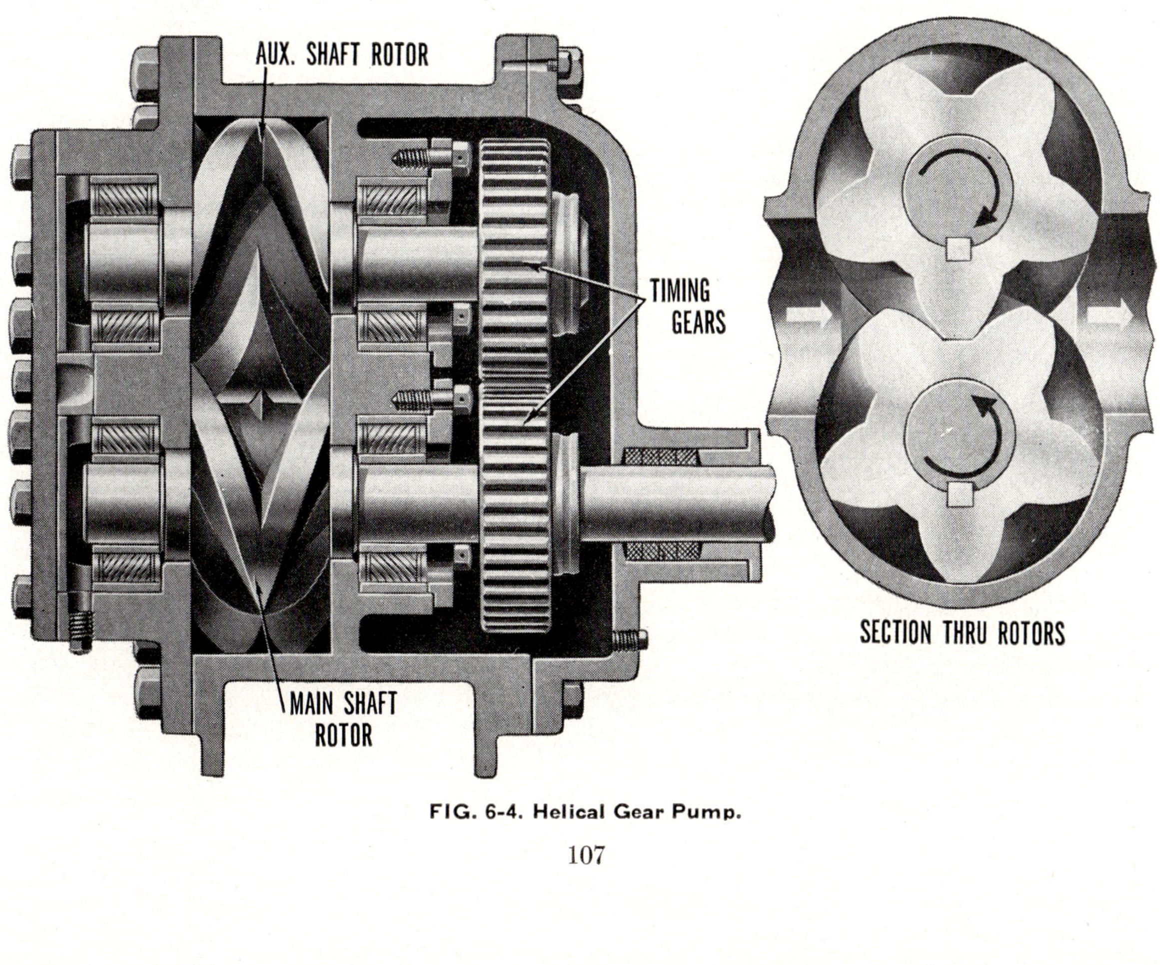

FIG. 6-4. Helical Gear Pump.

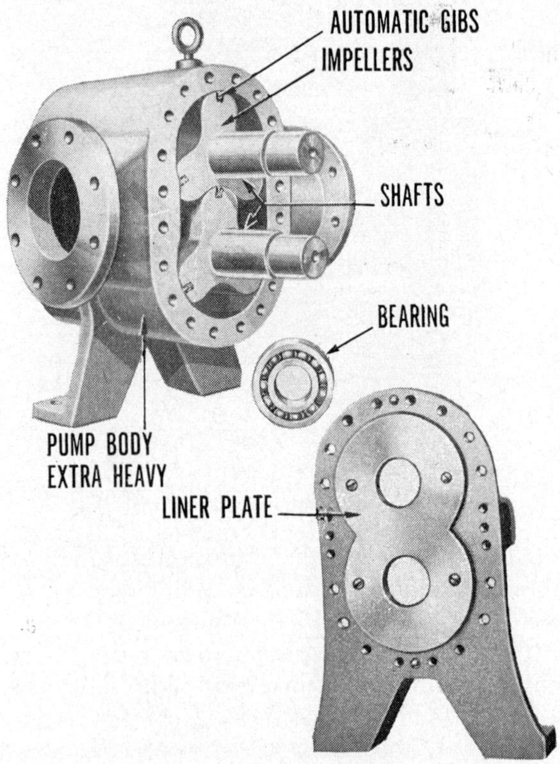

FIG. 6-5. Lobe Pump—End View.

essary to provide a driving gear, called a *timing gear* for the idler, as shown in Fig. 6-4.

Figure 6-5 shows an evolution of the gear pump, called the lobe pump, which is particularly adapted for pumping air. It is most commonly used in naval service as a low pressure compressor for providing combustion air for diesel engine use. The gibs on the lobe tips provide a close fit to minimize leakage and are replaceable as they wear. The liner plates at the ends of the lobes are also replaceable, as they are subject to wear.

Another important sub-type of the rotary class is the screw pump. Several variations of this type are common. Figure 6-6 is an example. The two closely-matched, helically threaded shafts fit in a casing with close tolerances. The threads trap the fluid at the suction ends and carry it to the discharge in the center. Each revolution of the shaft delivers a finite quantity of the fluid. Taking the suction at both ends and discharging at the center has two advantages: it obviates the necessity for providing high pressure packing at the shaft ends, and it balances the end thrust on the rotors caused by the pressure of the discharge fluid acting in the direction of the low pressure

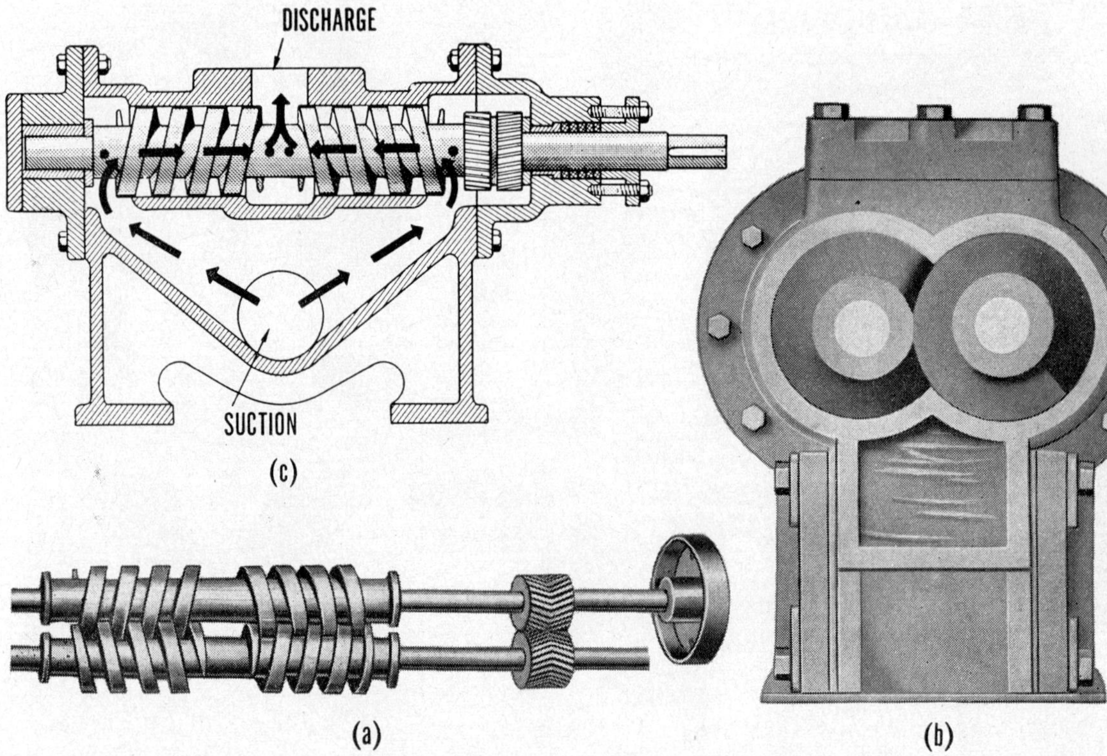

FIG. 6-6. Double Low Pitch Screw Pump.

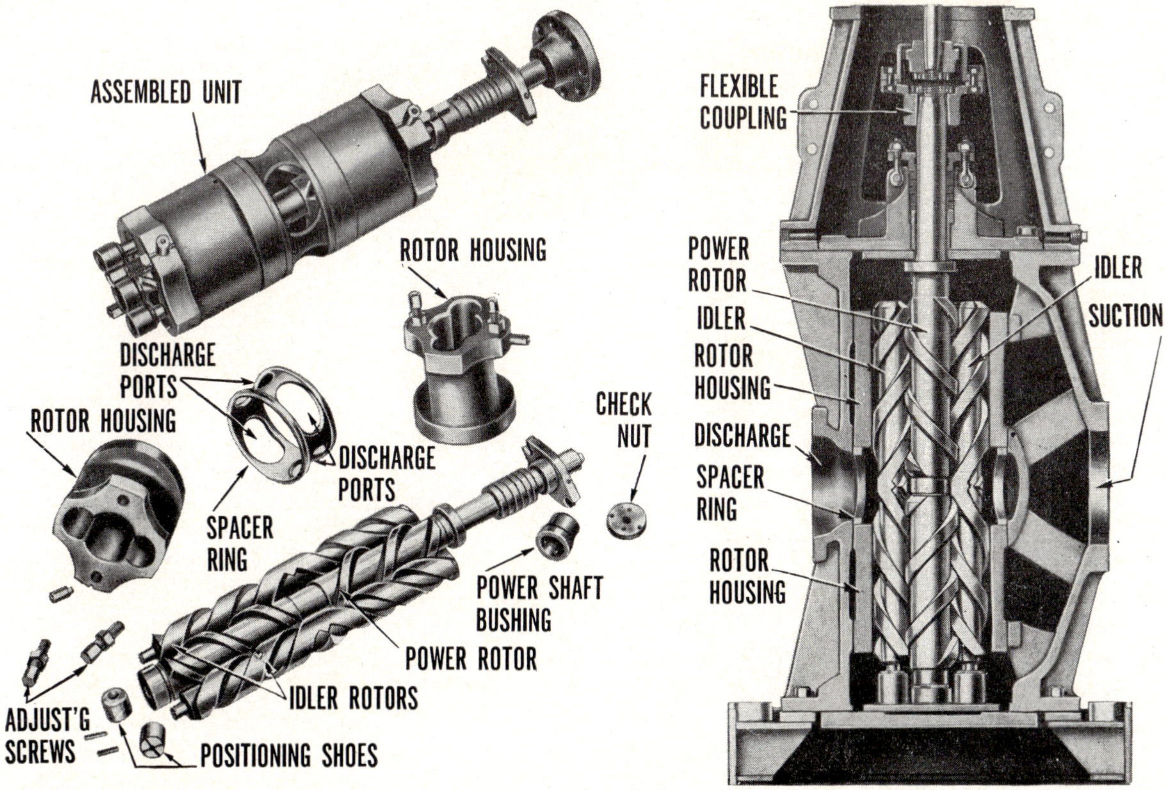

FIG. 6-7. Triple High Pitch Screw Pump.

fluid entrance to the rotor. Helical timing gears are provided in the pump to drive the idler, a necessity with rotors which are the equivalent of low-pitch worm gears.

A high pitch screw pump with two idling rotors is shown in Figure 6-7. This pump has the same general features as the preceding one, except that no timing gears are required due to the high pitch of the rotors.

Another type of rotary pump will be discussed in connection with hydraulic gears at the end of this chapter.

It is noted that all rotary pumps require close mechanical tolerances. For this reason they are not particularly suitable for handling fluids of a corrosive or abrasive nature, since corrosion in the pump would necessitate frequent maintenance. This, and the fact that they are adaptable for handling high viscosity fluids, makes them ideally suitable for oil service.

604. ROTARY PUMP CHARACTERISTICS.

The *displacement* of the rotary pump is the theoretical volume displaced per revolution. It depends only on the geometry of the pump.

The *capacity* of the rotary pump is the quantity of fluid actually delivered per unit of time at stated operating conditions of speed, inlet and outlet pressure, temperature and characteristics of pumped fluid.

The existence of a pressure differential across the pump causes flow from the high pressure side back to the low pressure side through the working clearance areas of the pump. This flow causes the *capacity* to be less than the *displacement* per unit of time, the difference being termed *slip*. The slip for a given pump is increased as the pressure differential increases, and decreases as the viscosity of the pumped fluid increases. It is seldom greater than 5% to 10% within the range of speeds and pressures for which the pump is designed. The two curves in Figure 6-8 are representative of rotary pump characteristics.

605. CENTRIFUGAL PUMPS.

The classification of *Centrifugal Pump* is now commonly used to include all pumps operating with a non-positive displacement rotating *im-*

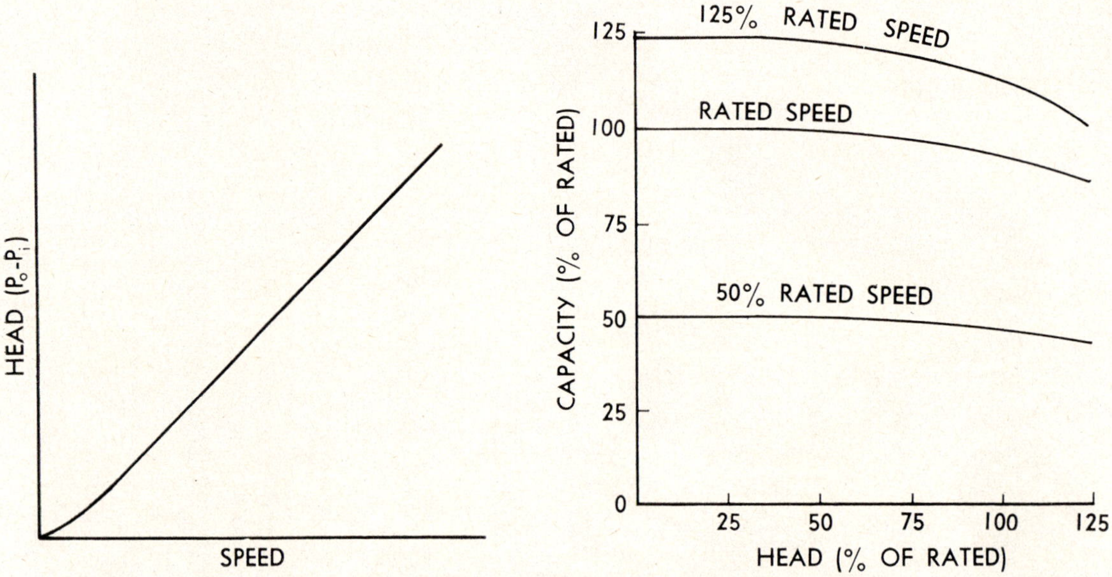

FIG. 6-8. Rotary Pump Characteristics.

peller. They may be divided into three general sub-classes: (1) Radial flow, (2) Axial flow, (3) Mixed flow.

In the radial flow pump, fluid is admitted into the center of a rotating impeller having radial channels. Rotation of the impeller imparts kinetic energy to the fluid, hurling it outward with centrifugal force. For the pump to develop a pressure head, it is necessary that this kinetic energy be converted into pressure energy by means of a diffuser of some type. The simplest arrangement for this, used almost exclusively in modern naval pumps, is shown in Figure 6-9. The volute shape of the casing provides the necessary diffuser. In present volutes, the conversion from K.E. to pressure is almost complete, very little energy being lost to thermal energy changes in the fluid.

In the axial flow pump, the impeller is in fact a *propeller*. This pump develops most of its head by the lifting action of the blades, or vanes, on the fluid. It receives and discharges fluid along the propeller axis. The conversion of mechanical energy to pressure energy takes place with little intermediate production of kinetic energy. The axial flow arrangement is sketched in Figure 6-10.

The mixed flow pump is one in which the head is developed partly by centrifugal force and

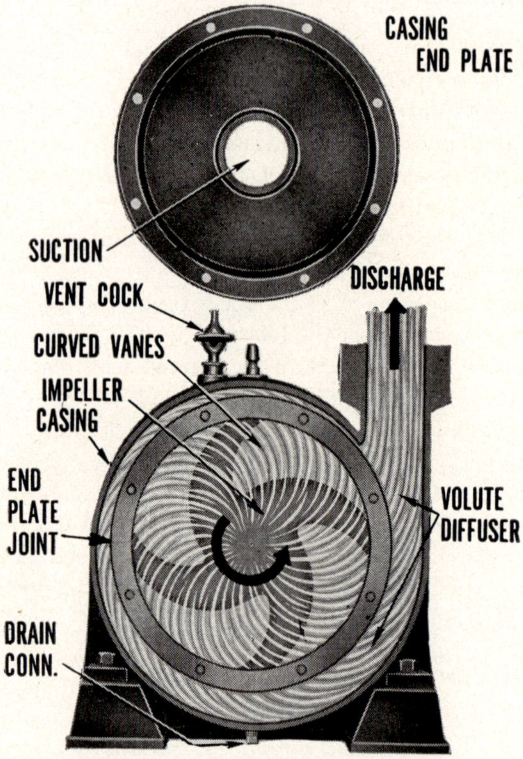

FIG. 6-9. Volute Casing Centrifugal Pump.

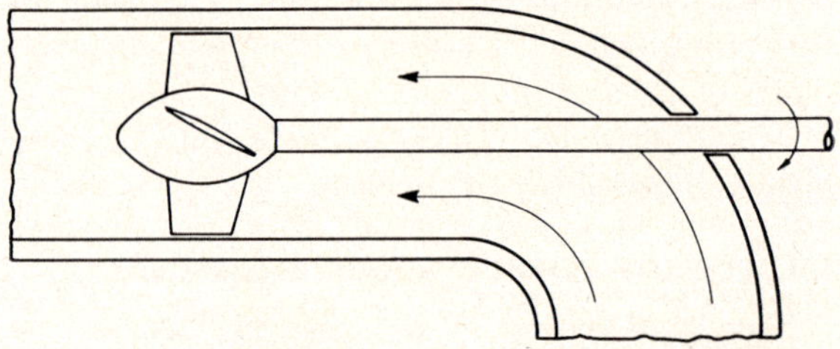

FIG. 6-10. Schematic of Axial Flow Pump.

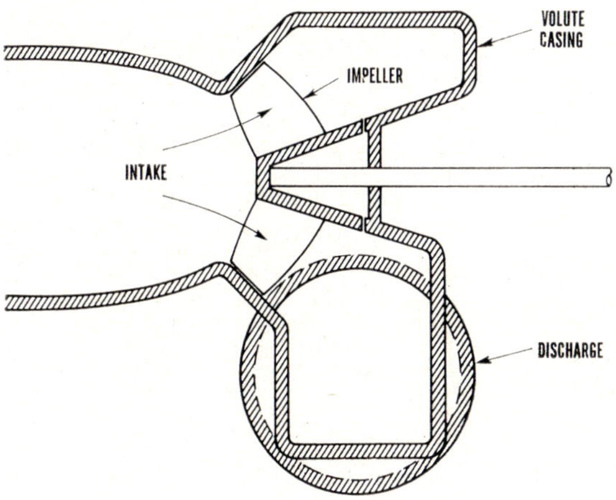

FIG. 6-11. Schematic of Mixed Flow Pump.

partly by the lift of the vanes on the liquid. It happens that the characteristics of the centrifugal pumps and propeller pumps more closely fit naval power plant requirements. The characteristics of the mixed flow pumps, useful in many industrial applications are seldom required for naval practice, and therefore such pumps are found infrequently aboard ship. Figure 6-11 is a sketch of a mixed flow pump.

606. THE RADIAL FLOW PUMP.

Many arrangements of the radial flow centrifugal pump are in use. This type is by far the most common type found in the machinery plant.

Figure 6-12 illustrates a horizontal, single-stage, volute type centrifugal pump with a single-suction impeller. To avoid the necessity of costly renewal of pump casings due to erosion of the casing by water leakage from the high pressure to the low pressure side through the close clearance space between the impeller hubs and the casing, removable soft metal wearing rings are threaded into the casing at the close-clearance points. These rings, in providing the small clearance between the discharge and suction sides of the pump, suffer the erosion, but are easily and economically replaced.

It should be noted, in Figure 6-12, that the impeller contains balancing holes, which connect the space around the hub to the suction side of the impeller. The balancing holes have a total cross-sectional area which is considerably greater than the cross-sectional area of the annular space between the left wearing ring and the hub. This results in the suction pressure being on both sides of the impeller hub, thus maintaining a hydraulic balance of axial thrust. Any unbalanced thrust existing is absorbed by a ball type thrust bearing (not shown).

A lantern ring spacer, or seal cage, is shown inserted between the rings of packing in the stuffing box. Seal piping leads water from the discharge side of the pump to the annular space formed by the lantern ring. The web of the ring is perforated so that water can flow in either direction along the shaft between the shaft and the packing rings. The cooling water entering the lantern ring is thus enabled to serve three purposes, viz.:

(1) To cool the shaft and packing,

(2) To lubricate the packing, and

(3) To seal the joint between the shaft and packing against leakage of air into the pump in the event the pump suction pressure is less than that of the atmosphere.

In Figure 6-13 is shown a horizontal, single-stage pump, having a double-suction type impeller, whereby hydraulic end-thrust is eliminated.

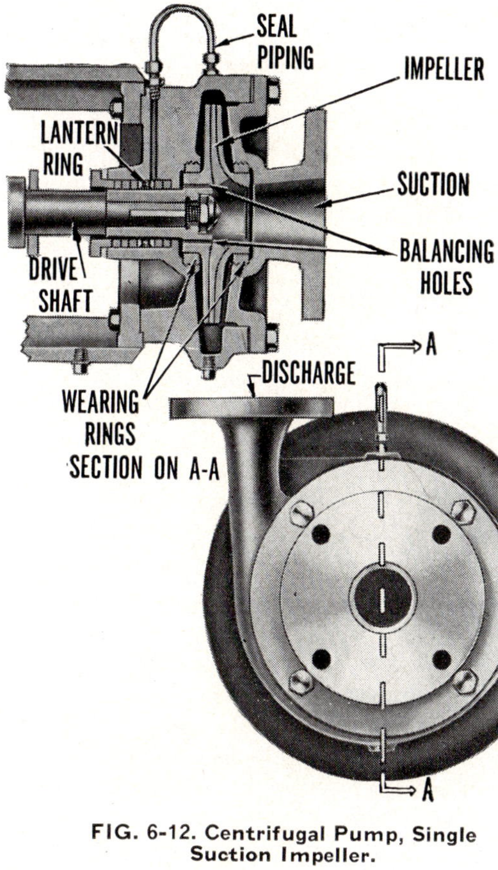

FIG. 6-12. Centrifugal Pump, Single Suction Impeller.

The peculiar cross-sectional shape of the casing wearing rings should be noted. In this construction the impeller wearing ring rotates within a groove in the casing wearing ring; thus the clearance passage forms a labyrinth which further restricts flow from the high to the low pressure sides. Water seal piping (not shown) supplies the lantern ring shaft gland seals. This pump has approximately twice the capacity of pump having a single suction impeller of the same diameter, with the same size eye, and operating under the same speed and head conditions.

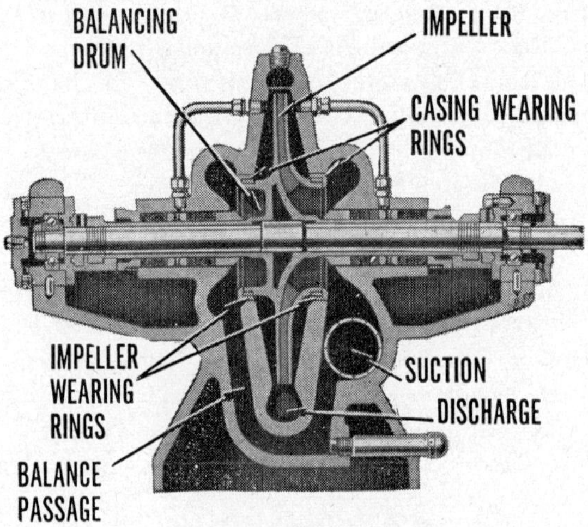

FIG. 6-13. Centrifugal Pump, Double Suction Impeller.

There is a practical limit in discharge head beyond which it is not economical to use a centrifugal pump having a single impeller. For the highest discharge heads, either a very high rotative speed or a large diameter of impeller must be used. Both of these lead to high mechanical stresses and to lowered efficiency, due primarily to disk friction and leakage losses. Disk friction increases very rapidly with increased speed or increased impeller diameter, the horsepower lost varying approximately with the cube of the speed and the fifth power of the diameter. Thus a higher speed and smaller impeller would give less disk friction for the same head. However, leakage loss, due to liquid passing back from discharge to suction through clearance spaces, increases with the smaller diameter. At high heads, with the greater pressure difference between discharge and suction, the leakage loss is an important item. As a result of these effects, the efficiency of a high-head pump of the single impeller type is likely to be comparatively low. While single stage pumps have been built which have delivered a head of more than 1500 feet, practical designs limit the head in most cases to about 1000 feet or less per impeller. For the higher heads, two or more impellers are connected in series, the discharge from one impeller being connected to the suction of the next. The total head is the head of one impeller multiplied by the number of impellers. For the sake of economy such an arrangement is built in one casing and is known as a multi-stage pump, each impeller constituting a stage. Ordinarily not more than six stages can be built in one pump casing on account of the length of the shaft. Room for a shaft of large diameter is difficult to obtain because of the space required to lead the liquid into the eye of the impeller. The shaft of a multi-stage pump must be carefully designed in order to prevent undue vibration, such as would occur if the deflection of the shaft, due to its length, were such as to cause the pump to operate at a critical speed.

The essential features of the two pumps shown in Figures 6-14 and 6-15 are characteristic of modern high pressure pumps. The casings of these two pumps are made in two halves, split at a horizontal flange (though in the views shown, the casing is cut away in a vertical plane). Some pumps have casings built up in a series of cylindrical segments which must be disassembled from one end. This provides extra strength to withstand high casing pressures by avoidance of axial joints.

In another arrangement, the entire pump casing is enclosed in another cylindrical casing which has no circumferential or longitudinal flanged joints. The discharged liquid from the pump surrounds the inner pump casing, and flows out through pipe connections from the outer casing. The single-piece outer casing can be designed to withstand high pressures without danger of leakage. The inner casing flanges are kept tight by the discharge pump pressure acting on the outside of the inner casing. Pumps using this type of construction, suitable for very high pressure pumps, are called *barrel* pumps.

Note the sequence of arrangement of stages in Figure 6-14. This helps to balance thrust produced and places the highest pressure stage inboard of other stages so that the shaft packing does not have to seal the maximum pump pressure. Also note, on both pumps, the auxiliary

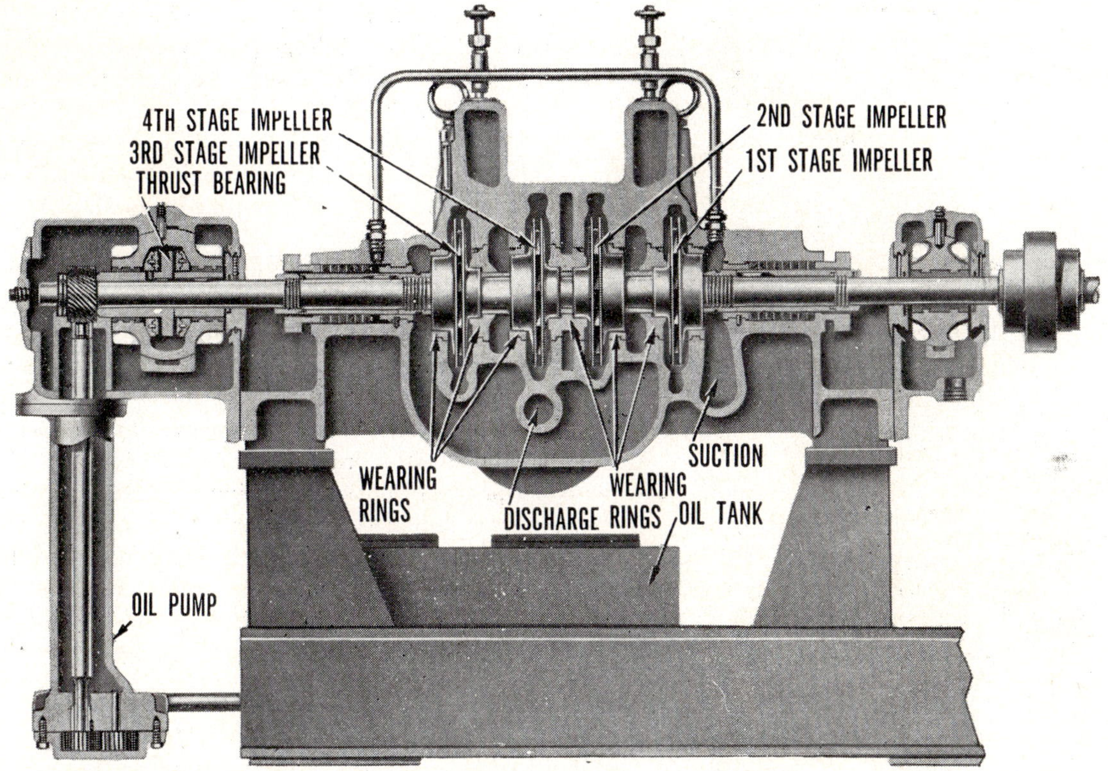

FIG. 6-14. Multistage Centrifugal Pump.

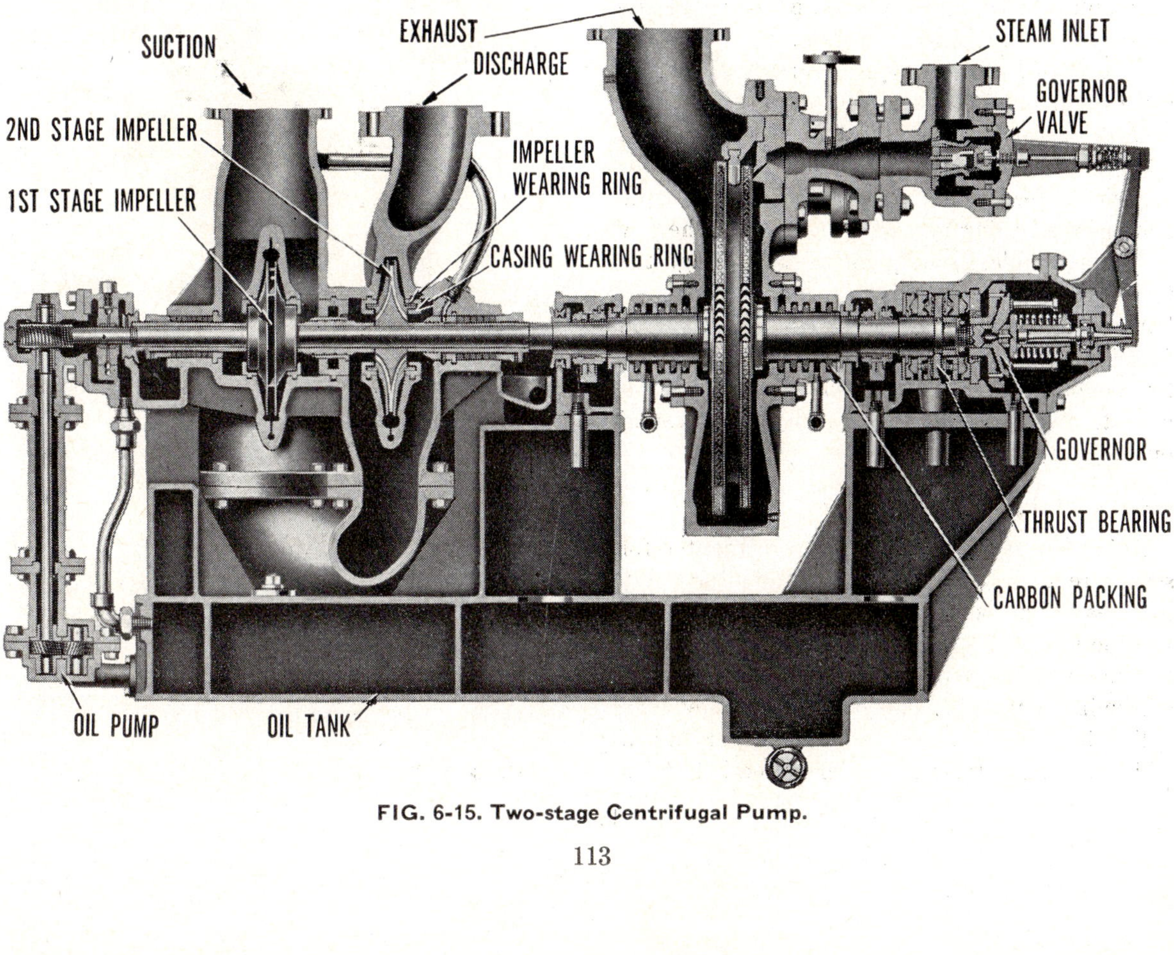

FIG. 6-15. Two-stage Centrifugal Pump.

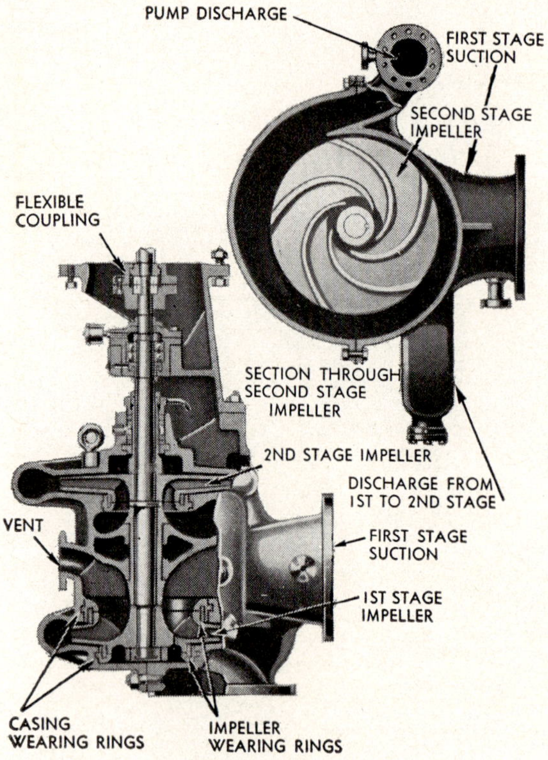

FIG. 6-16. Typical Main Condensate Pump.

gear pumps used to provide oil pressure for the self-contained lubrication system.

Figure 6-16 shows a typical vertical two stage centrifugal pump used as the main condensate pump. This pump operates at a speed of about 1000 RPM (as opposed to about 10,000 RPM for the main feed pumps described above)—the slower speed being necessary because of the low NPSH available. Consequently, even though the total pressure rise required is small, two impellers are necessary under these conditions. Main feed booster pumps are similar, but may have a large enough NPSH so that only one stage is required. Two stage main feed boosters are found on the latest destroyers.

607. AXIAL FLOW PUMPS.

Axial flow, or propeller pumps, are adapted for service requiring a large capacity and a relatively low head. The only service aboard ship having this requirement is the main circulating system of the main condenser. A typical arrangement is shown in Figure 6-17. This pump is mounted on a cylindrical stool, or sea-chest, the lower end of which is open to the sea, and the up end of which is fitted with a valve. (See Figure 5-1.) Just above this valve, a large pipe joins the sea chest. The other end of the pipe is open to the engine room bilge. This drainage system can be used to pump out the engine room quickly in the event of damage. The circulating pump is thus sometimes referred to as the *main drain pump*.

608. RADIAL FLOW PUMP CHARACTERISTICS.

In order to get a qualitative understanding of the characteristics of the centrifugal pump, consider Figure 6-18. Two simplifying assumptions are made. First, the cross section area of flow at the inlet to the channels in the eye of the impeller is equal to the cross section area of flow at the periphery, or outlet of the impeller. This requires that the thickness of the channel, or its dimension perpendicular to the page, be greater at the inlet than at the outlet, since the circumference of the eye is smaller at the inlet than at the outlet. Stated analytically, ignoring thickness of the vane ends:

$$A_1 = A_2 = 2\pi R_1 t_1 = 2\pi R_2 t_2$$

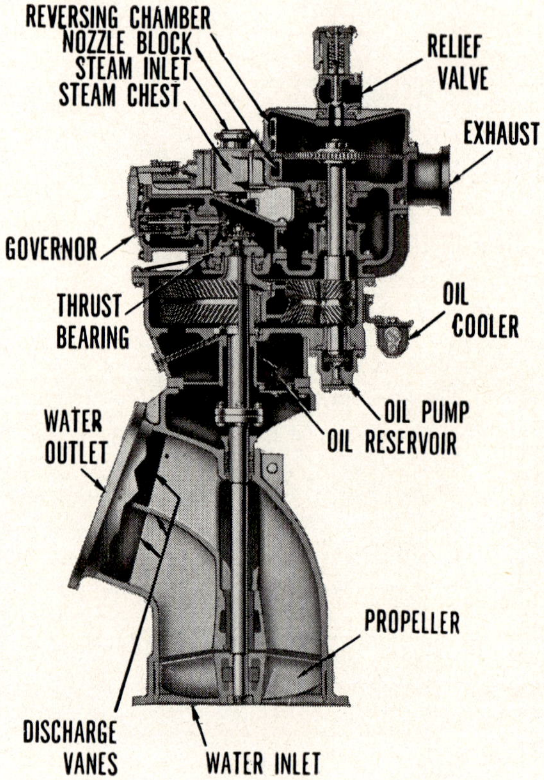

FIG. 6-17. Typical Main Circulating Pump.

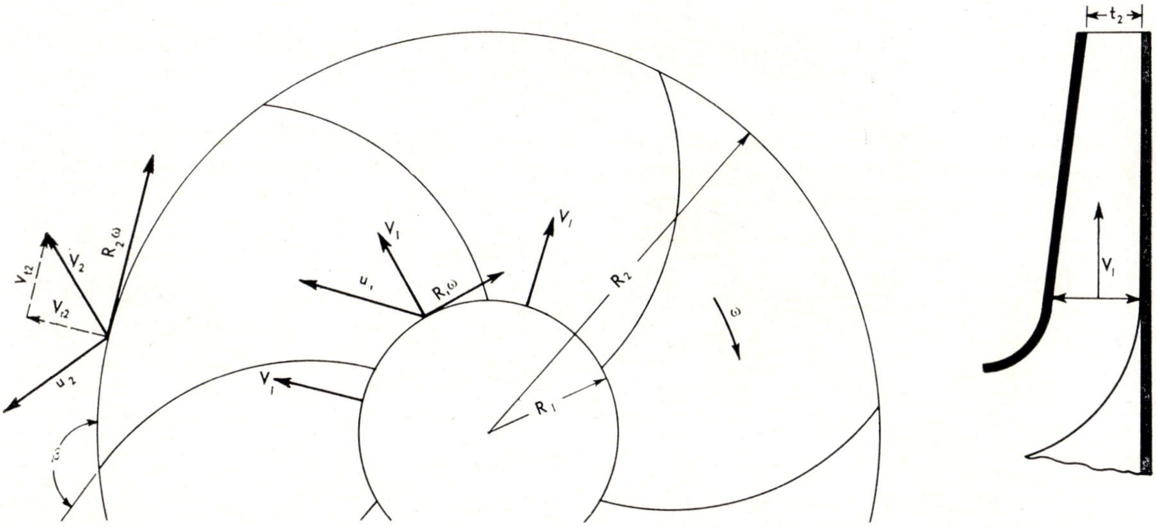

FIG. 6-18. Vector Relations, Centrifugal Pump.

The second assumption is that fluid enters the channel inlets in the eye with an absolute velocity exactly radial as shown by the vector V_1.

The velocity of the fluid relative to the impeller, u_1, at entrance section 1, is the absolute velocity of the fluid minus the linear velocity of the impeller at this point. If the impeller is rotating with an angular speed, ω, the impeller velocity is $R_1\omega$. Then, vectorially, as shown

$$u_1 = V_1 - R_1\omega \tag{6}$$

If the first assumption holds, the area of flow is constant and the magnitude of u does not change as the fluid flows to the outlet section. However, its direction does change, as determined by the curvature of the blade, and the relative velocity at the exit is represented by u_2. The absolute velocity, V_2, at the exit is the vector sum of the relative velocity and the linear speed of the impeller exit, $R_2\omega$. The vector addition is shown. This velocity represents kinetic energy and is converted to pressure in the volute (or diffuser) leading to the pump outlet.

In the study of Fluid Mechanics, it is shown that a change in the moment of momentum is caused by, or produces, a torque. Note that V_2, the absolute exit velocity of the fluid, may be resolved into two components: V_{t2} tangent to the impeller, and V_{n2} normal to the impeller. The product of the tangential component, V_{t2}, the mass rate of flow, and its radius, R_2, is the moment of momentum of the fluid leaving. The moment of momentum of the fluid entering is zero, by the assumption that V_1 is radial, thus having no tangential component. The change in the moment of momentum is equal to the torque, and is

$$\begin{aligned} T &= mR_2V_{t2} - mR_1V_{t1} \\ &= mR_2V_{t2} \quad \text{(Since } V_{t1}=0) \\ &= \frac{w}{g} R_2V_{t2} \end{aligned} \tag{7}$$

where

T = torque, ft lbs

w = weight rate of flow, lbs/sec

R_2 = outer radius of impeller, ft

V_{t2} = tangential component of absolute exit velocity, ft/sec

g = gravitational constant, ft/sec^2.

The power delivered to the fluid is equal to the torque times the angular speed, or

$$P_0 = \frac{\omega T}{550} = \frac{\omega w R_2 V_{t2}}{g\ 550}. \tag{8}$$

Equating this expression to the expression for power of equation (3), and eliminating common terms, an expression for pump head is obtained:

$$H_p = \frac{\omega R_2 V_{t2}}{g}. \tag{9}$$

From an examination of Figure 6-18, it is seen that

$$V_{t2} = R_2\omega + u_2 \cos \beta.$$

Therefore, equation (9) is rewritten

$$H_p = \frac{\omega R_2}{g}(wR_2 + u_2 \cos \beta). \qquad (10)$$

Equation (10) is useful for evaluating the effect of the angle of the impeller blades on pump performance. Three cases will be considered.

(*A*) *When* $\beta > 90°$. In this case the term, $u_2 \cos \beta$, is negative. At constant speed, pump head is then a maximum when the discharge is shut off, and u_2 is zero. As the discharge is opened, permitting flow, the value of the pump head falls off in proportion to the magnitude of u_2. This is shown in Curve *A* of Figure 6-19.

(*B*) *When* $\beta < 90°$. In this case the term, $u_2 \cos \beta$, is positive. As the discharge is opened, pump head rises fro n the shut off head, at constant pump speed, in proportion to the magnitude of u_2. Curve *B* of Figure 6-19 shows this.

(*C*) *When* $\beta = 90°$. The term, $u_2 \cos \beta$, is zero and independent of discharge. Therefore the pump head remains constant at all speeds.

The performance indicated above is theoretical; it neglects the fact that the fluid flow actually conforms to the blade shape at only one set of discharge conditions. There are generally fluid friction losses, eddies, and shock losses not considered in the above discussion. Actual head capacity curves always drop off at high discharge rates, as indicated by the dotted curves of Figure 6-19.

The operating point is determined by the characteristics of the piping system into which the pump discharges. The piping system head generally increases as more liquid is pumped, representing increased frictional resistance to flow. The dashed curve, shown in Figure 6-19, is typical of the parabolic shape of the system head curve usually encountered. The point where this curve crosses the pump Head-Capacity curve determines the discharge *pressure* at which the pump will operate. Note that Pump "*B*" can operate at this pressure with two capacities, and may therefore tend to "hunt" from one toward the other. Pumps "*A*" and "*C*" would give more stable operation. All shipboard centrifugal pumps will have backward curved impeller blades. However, wrong rotational direction is easily possible

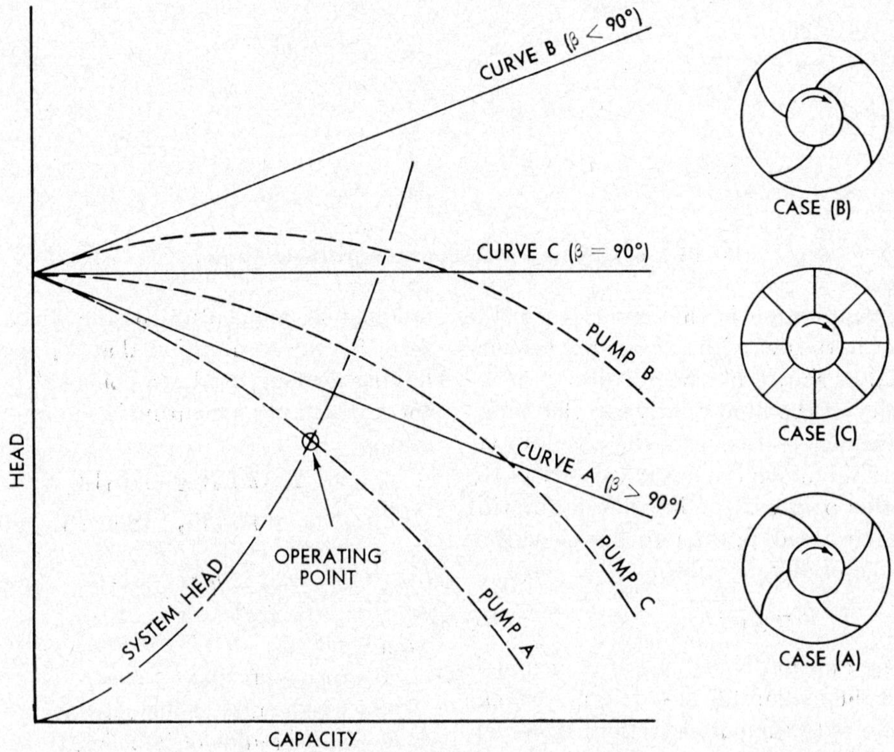

FIG. 6-19. Centrifugal Pump Characteristics.

when reinstalling electric drive pumps after repairs, and this will give the characteristics of a forward curved impeller.

It is noted from equation (8) that the power requirement is directly proportional to flow, *w*. Therefore, the power requirement, at constant speed, falls off as the pump discharge is throttled. If the pump is driven by a steam turbine, the turbine tends to deliver constant power, as long as the weight rate of steam flow is constant. Therefore, the turbine (and pump) will tend to speed up as its torque decreases. This must be prevented to avoid damage to the pump and turbine. Use is made of the fact that pump head increases with the square of the speed (equation 10). A governor which responds to pump pressure change and controls the turbine steam input is used. The governor is described in the later discussion on auxiliary turbines. Electric motors used to drive pumps are essentially constant speed machines, the power drawn varying with torque requirements. Therefore, no governor control is necessary when the pump is motor driven.

609. AXIAL FLOW PUMP CHARACTERISTICS.

Axial flow, or propeller, pumps are best analyzed by airfoil theory. No attempt will be made to give such an analysis. However, Figure 6-20 shows ideal and actual Head-Capacity curves. The very steep curve is characteristic. As stated before, this type of pump is suitable for large capacities at low heads, which the curve indicates.

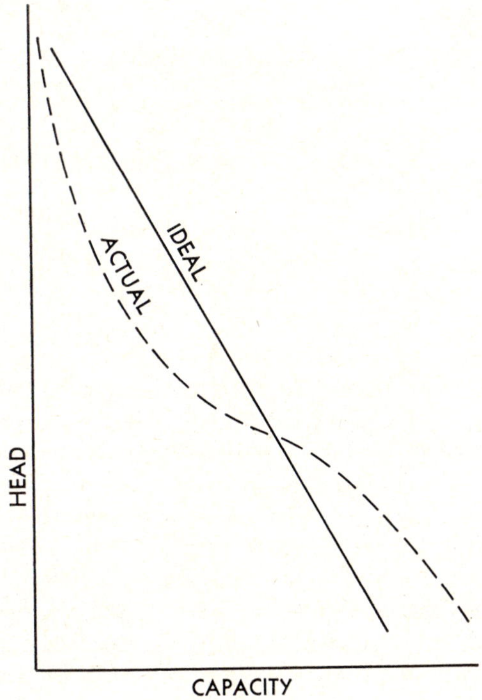

FIG. 6-20. Typical Characteristics of Axial Flow Pump.

610. MAIN FEED SYSTEM PUMPS, SPECIAL CONSIDERATIONS.

When the local pressure within a pump falls to the vapor pressure of the liquid being pumped, vapor bubbles form in the impeller. Since a centrifugal pump designed for liquid use has very poor vapor pumping characteristics, the pump will not discharge the vapor and allow liquid to continue to enter. The pump is said to be *vapor-locked* or *vapor-bound* in this circumstance. By Bernoulli's relation, low local pressures are caused by high local velocities. Therefore, when pumps are operating at pressures and temperatures near the saturation point for the pumped fluid, low impeller speeds are desirable. This is the case for condensate pumps and main feed booster pumps. As seen from Equation (10), low impeller speeds are associated with low pump heads. To minimize the chance for undesired pressure drops, the piping runs from the condenser and D.F.T. to their pumps are kept short, straight and continuously leading downward to the pump. In addition, the pump casings are vented back to the vapor sections of these units to permit any vapor that may form in the casing to flow back to the condenser or D.F.T.

The main feed pump is required to deliver a high head. This requires high pump speed and multi-staging. Therefore, it cannot accept fluid near its saturation temperature. The function of the booster pump is to raise the pressure of a nearly saturated liquid, so that the temperature of the fluid discharged is well below its saturation temperature, a condition that will minimize its flashing to vapor in the main feed pump. Data are given below for pumps on the DL Class destroyers.

Table 6-1

Pump	Speed, RPM	Capacity, GPM	Pump Head	Stages
Condensate	1,170	325	120 ft	2
Booster	2,000	430	136 ft	1
Main Feed	10,000	430	3290 ft	4

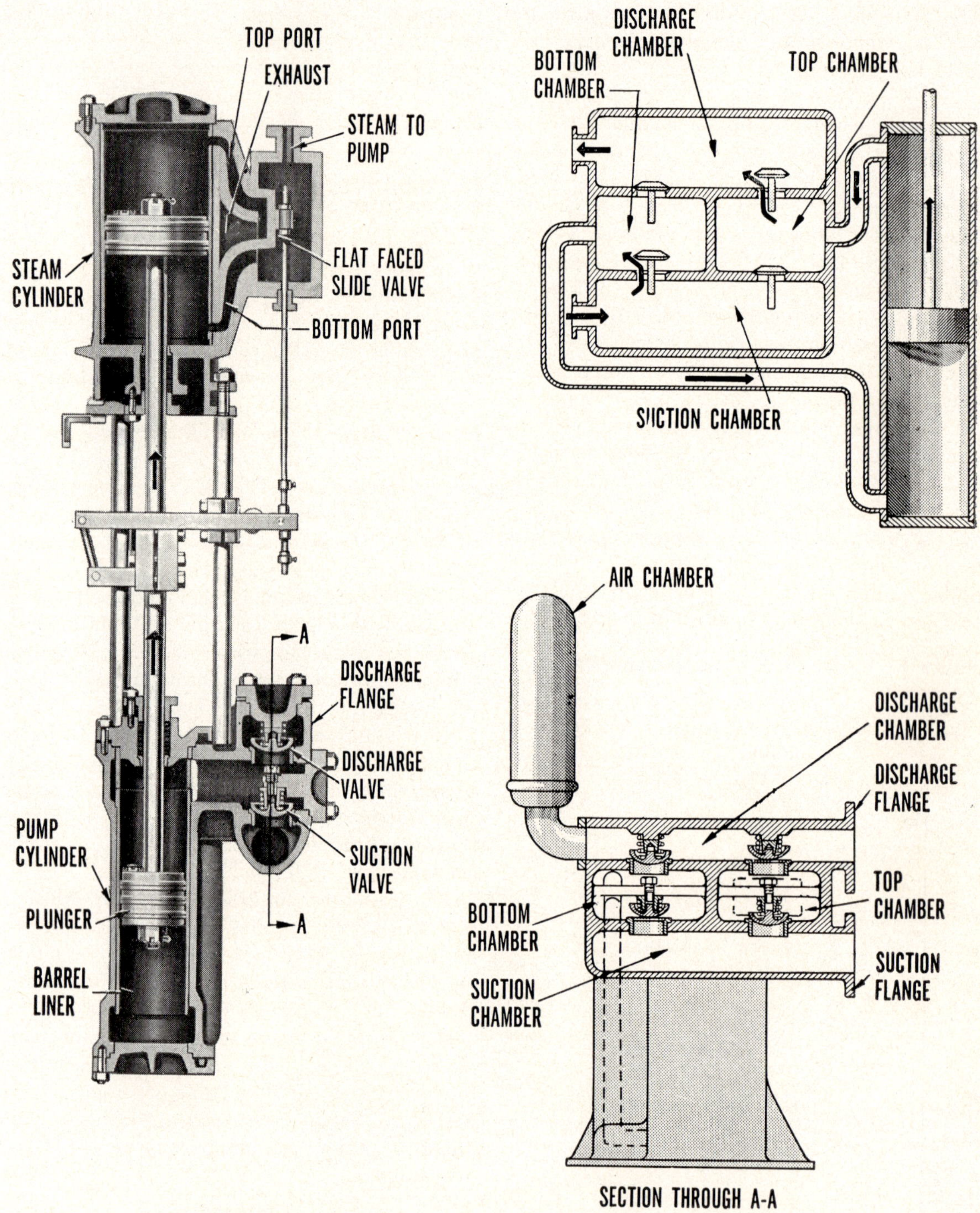

FIG. 6-21. Reciprocating Pump Details.

Local cavitation is also harmful to the pump physically in that it may cause damage to the pump due to overheating, and it causes severe erosion on the impeller itself. Therefore, no attempt should be made to increase the capacity of a pump by increasing its speed beyond the rated value, when the normal capacity falls off due to wear of the scaling rings, packing leaks, etc.

611. RECIPROCATING PUMPS.

In nearly all shipboard pump applications, reciprocating pumps have been replaced by the rotary or centrifugal types. The latter have the advantages of smaller size in a given application (due to higher speed); smoother discharge; and easier adaptation to a high speed prime mover, such as an electric motor or auxiliary turbine. Reciprocating pumps are rugged, reliable, and relatively easy to operate. Reciprocating pumps are still used as emergency feed pumps on many combatant type ships, though in later designs the centrifugal pump is used even for this purpose.

Figure 6-21 is a typical arrangement of a reciprocating pump and its steam drive end. It is called a *direct-acting* pump because the steam and water pistons are directly connected. It is called a *double-acting* because the water piston (or plunger) discharges water on each stroke (see schematic in upper right corner of page). It is a *simplex* pump since only one steam and one water cylinder are used.

Reciprocating pumps are designated by three dimensions, i.e., *the diameter of the steam piston, the diameter of the water plunger, and the length of the stroke.* For example, an 8″×5″×10″ pump signifies a steam piston diameter of 8″, a pump plunger diameter of 5″ and a stroke of 10″. When the steam piston is larger than the water piston, the pump is called a high pressure pump, otherwise it is called a low pressure pump.

612. SPECIAL APPLICATIONS.

A rotary, positive displacement pump of considerable importance outside the propulsion plant is the variable-stroke pump. This pump is characterized by a very fine control of its discharge rate from zero to full power, and it is therefore useful in hydraulic control applications, such as the steering gear, anchor windlass, various gun mounts and turrets, etc.

The axial piston type of variable stroke pump, combined with an axial piston motor is shown in Figure 6-22. When the pump and motor combination are used like this, the assembly is called a *hydraulic speed-gear*. The electric motor drives the shaft which rotates the socket ring and the cylinder barrel. The socket ring rotates inside a

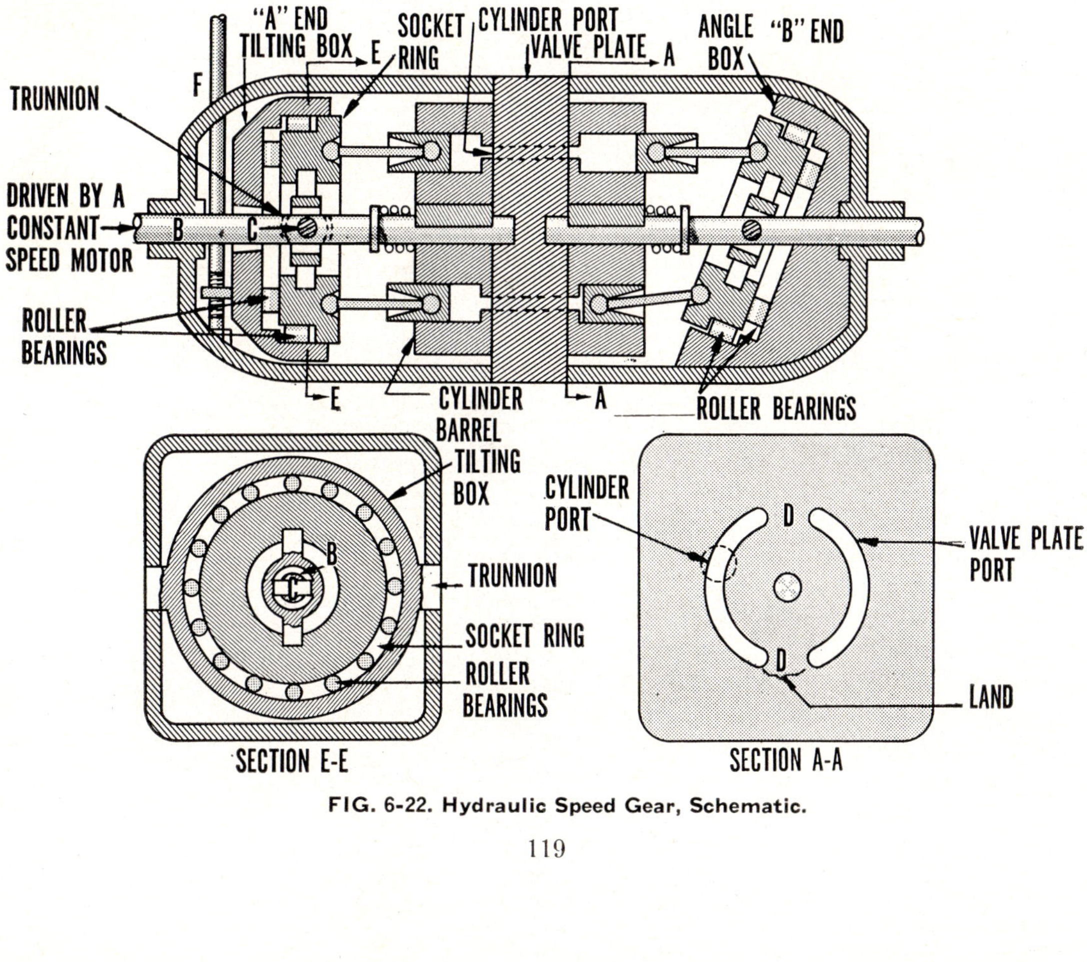

FIG. 6-22. Hydraulic Speed Gear, Schematic.

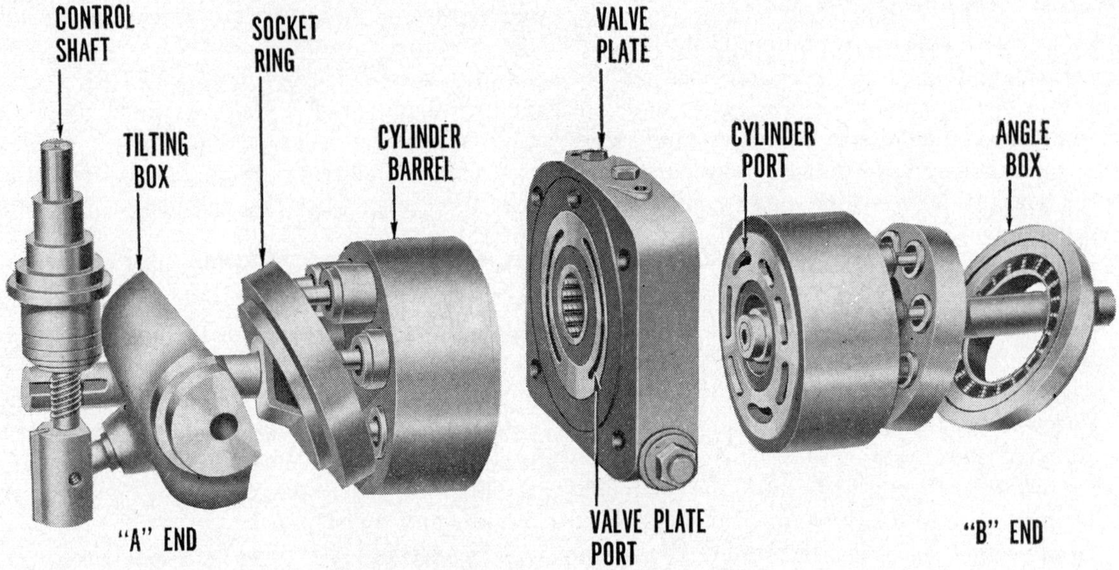

FIG. 6-23. Hydraulic Speed Gear, Exploded View.

tilting box which can be tilted so that its top moves to the right or left in the figure, by the action of the control shaft. Assuming that the top moves to the right, it tilts the socket ring, attached to the shaft by a universal joint. This forces the pistons into the cylinder block at the top and out at the bottom. Thus, with each revolution of the cylinder block, the piston makes one reciprocating stroke in its cylinder. The reversal points of the stroke are at the extreme top and extreme bottom. The cylinder block rotates with very small clearance against the fixed valve plate, which has two circular ports milled in it. Oil is discharged through one port and enters through the other, depending on the rotational direction of the pump and direction of tilt of the socket ring. In the position shown in Figure 6-22, no oil is being pumped. In the exploded view in Figure 6-23, viewing the pump from the left end, and assuming clockwise rotation oil is entering from left side valve port and being discharged in the right side port. The quantity of flow is controlled by the angle of tilt of the tilting box, since the pump is turned by a constant speed motor. In its application is a pump, the valve ports are connected to inlet and discharge piping.

In the speed-gear application, a hydraulic motor unit is mounted, as shown in the two figures, to complete the oil circuit. The motor end is identical to the pump end, except that the socket ring has a fixed tilt determined by the angle box. In Figure 6-23, still viewing from the left end under the conditions described above, oil will be forced into the right hand set of motor cylinders, tending to force the pistons out. This will cause the motor socket ring to rotate counter

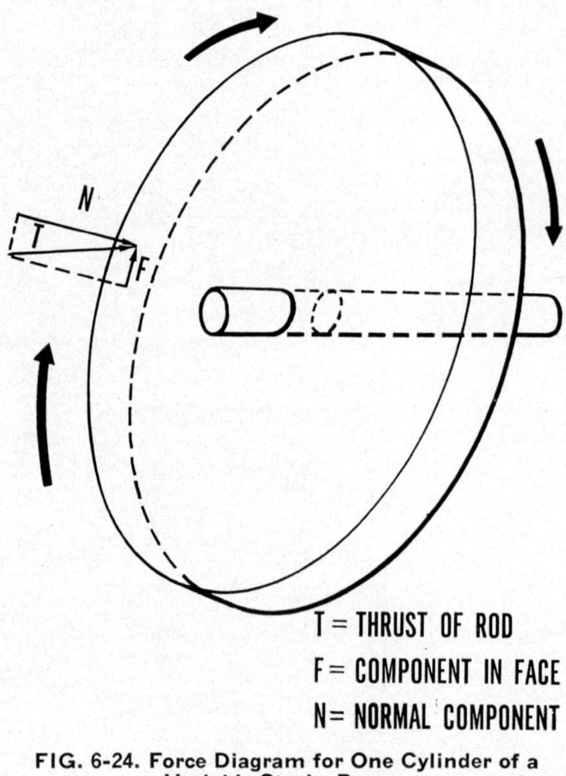

FIG. 6-24. Force Diagram for One Cylinder of a Variable Stroke Pump.

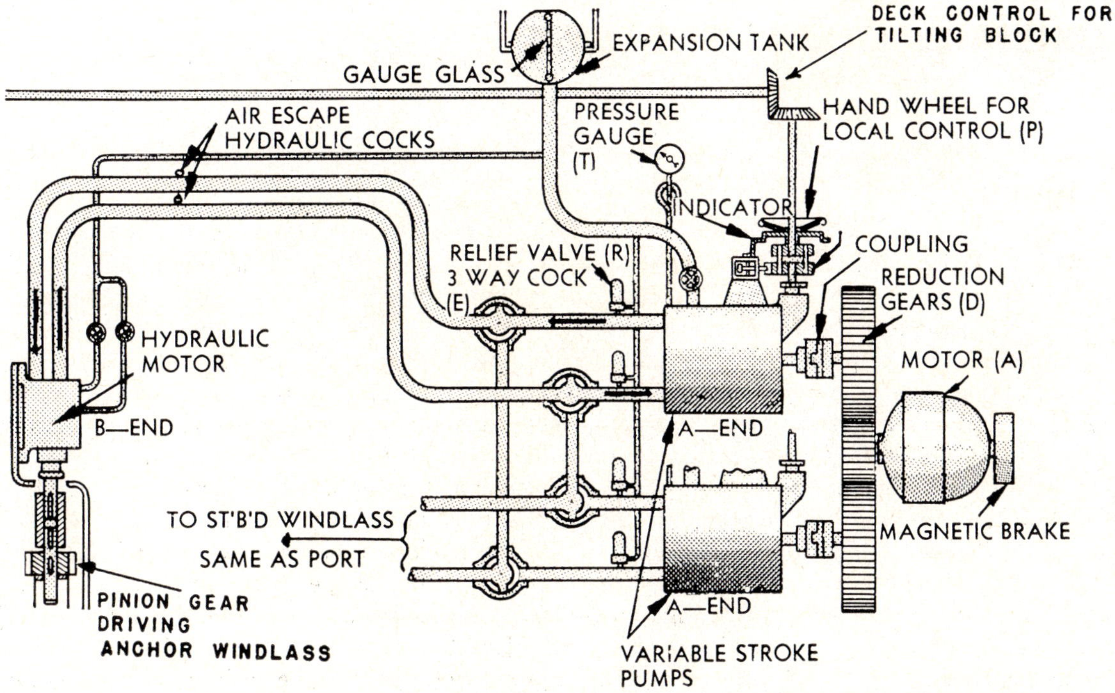

FIG. 6-25. Anchor Windlass Control Arrangement.

clockwise, rotating its attached shaft as it does so. The force diagram in Figure 6-24 shows this. The thrust of the piston rod, T, consists of two components, one normal to the socket ring, N, and one tangential to the ring, F. It is the latter which causes rotation.

The speed of rotation is determined by the amount of oil being pumped, which is a function of the amount of tilt of the pump, or A-end, socket ring. The direction of rotation is determined by the direction of oil flow, which is a function of the direction of tilt of the A-end socket ring. Thus it is seen that the *speed* and *direction* of a power output shaft may be controlled by the *position* of the control shaft, while the power is supplied by an electric motor turning in a constant direction with a constant speed.

One control application is shown in Figure 6-25, a schematic of an anchor windlass control system.

Occasionally, a variable stroke pump will be found in use in the engineering plant as an in-port fuel oil service pump, but this is not common.

A variation of the axial piston arrangement is a radial piston arrangement. It possesses the same characteristics although its mode of operation is somewhat different. Its frequency of use does not warrant a detailed description here.

CHAPTER VII
AUXILIARY STEAM SYSTEMS

700. GENERAL.

The auxiliary steam system has been briefly discussed in Article 202. It will be recalled that this system furnishes steam to drive the propulsion auxiliaries, and for other functions necessary to the operation of the plant. The pressure and steam conditions of this system vary depending on the installation. Until recent years, the auxiliary system carried 600 psig saturated steam. With the advent of higher boiler pressures, many of the major auxiliaries are driven by superheated steam at the higher boiler pressure. In some other instances, a portion of the main steam is desuperheated and reduced in pressure to 600 psig to provide steam for driving certain auxiliary units.

As an example, the CVA-60 machinery plant provides *boiler pressure auxiliary steam* at 1200 psig for power supply to the main feed pumps and the turbo-generators (ship's service and aircraft-starting). In addition, a quantity of 1200 psig steam emerging from the superheater is led back through desuperheater coils in the boiler steam drum, thence to reducing valves where it is reduced to 600 psig. The 600 psig system then supplies steam to the condensate pumps, main feed booster pumps, fuel oil service pumps, forced draft blowers, main circulator, lubricating oil service pumps, fire pumps, and to reducing stations for lower pressure systems.

In contrast, all of the major auxiliaries of the DL-5 are powered by auxiliary steam coming directly from the superheater outlet, with no reduction in pressure except that caused by the line pressure drop.

As a consequence of these variations, standard nomenclature is somewhat difficult, and the two terms, *boiler pressure auxiliary steam system* and *auxiliary steam system*, used herein, will refer to the steam supply used for propulsion auxiliaries. Auxiliary steam used for other purposes will be designated by the system pressure.

701. AUXILIARY TURBINES, GENERAL CONSIDERATIONS.

Two types of auxiliary prime movers are commonly used: electric motors and steam turbines. The selection of the type of prime mover is subject to many considerations. The generation of electric power in an efficient multistage turbogenerator, which operates condensing, and its use to drive a pump with a well designed electric motor, is generally more efficient than driving the same pump with a small, few-stage turbine which operates noncondensing. However, there are several considerations which make the choice of steam turbine drive preferable. First, above about 50 horsepower, steam turbines begin to have a decided advantage over electric motors in weight and space per horsepower. Second, even for smaller sizes, the necessity for plant reliability makes the use of steam power advantageous to prevent stoppage of the entire plant in the event of an electrical casualty of even short duration. For instance, if the fuel oil service pumps were electric driven, the ship may have to stop, or at least slow down, for such a minor casualty as a circuit breaker trip caused by a temporary electrical overload. Because of this, all auxiliaries vital to propulsion either are normally steam driven, or they are provided in duplicate or triplicate, at least one of which is independent of electric power. Third, some auxiliary units are required to operate at speeds above those feasible for electric motors or at varying speeds which may be difficult to obtain with electric motors. For example, main feed pump speeds may be as high as 10,000 RPM, and forced draft blower speeds of over 5000 RPM may be required. Finally steam at low pressure for D.F.T. heating and deaerating and other propulsion plant uses is required. It is thermodynamically more efficient to use steam in these instances in which the pressure has been decreased by the accomplishment of work than to use steam in which the pressure has been decreased by a straight throttling process.

Auxiliary turbines are constructed to be rugged and reliable. These two considerations lead to a simplicity in design at the expense of efficiency. Practically all auxiliary turbines installed prior to 1950 except the ship's service generator turbine were single stage units. More recent plants incorporate one, two, or three-stage turbines.

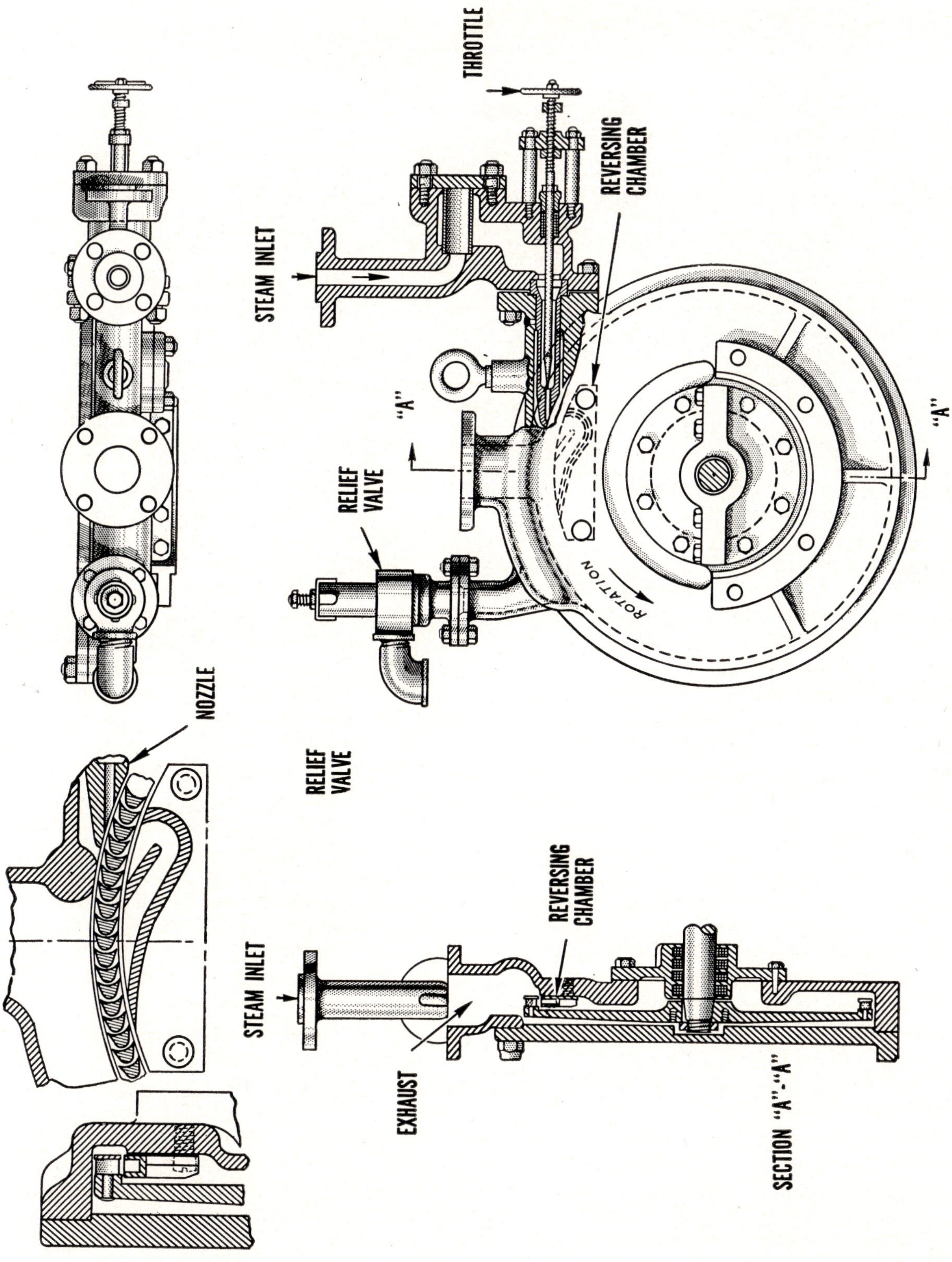

FIG. 7-1. Radial Flow, Single Stage, Re-entry Turbine.

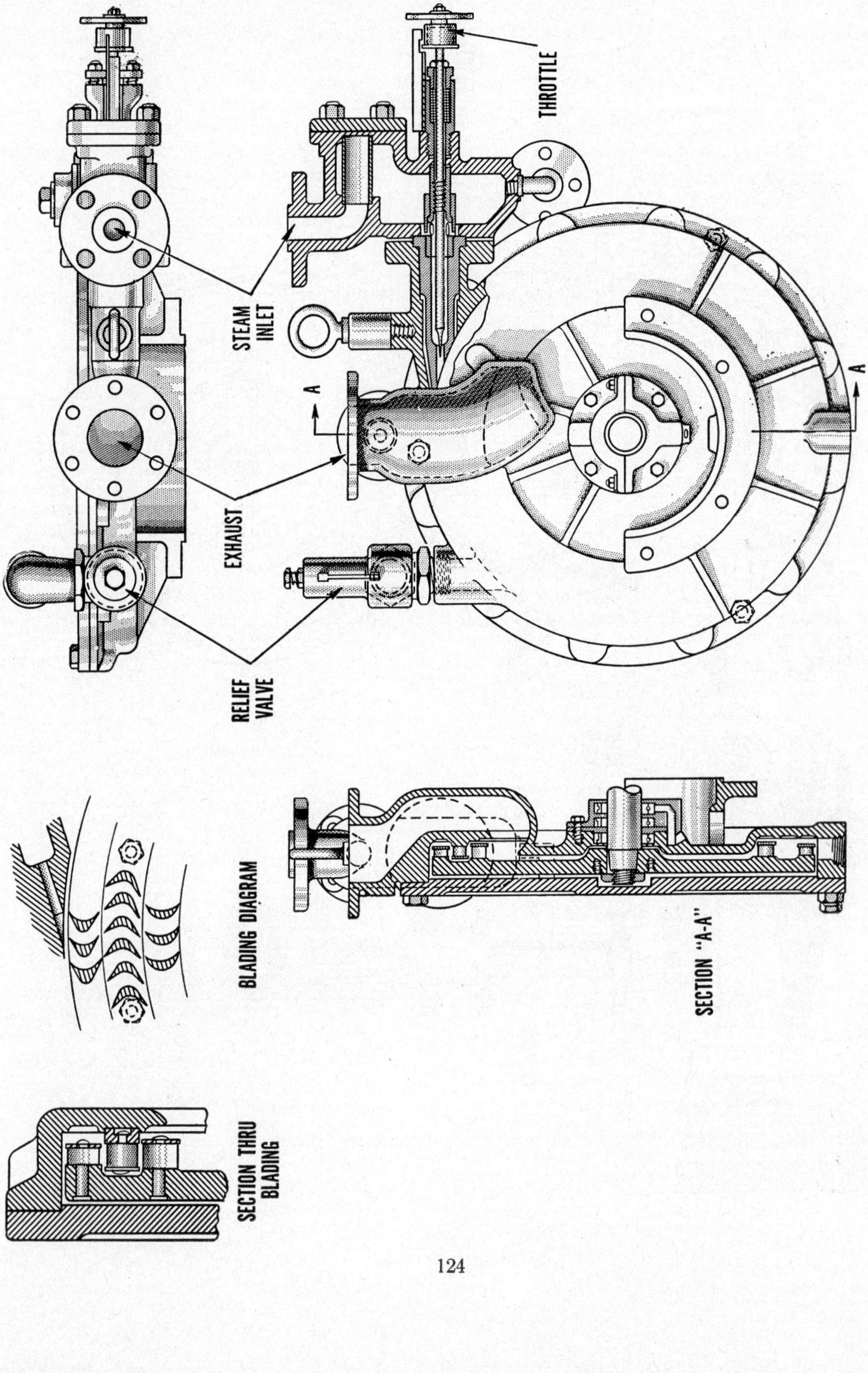

FIG. 7-2. Radial Flow, Velocity Compounded Turbine.

They are all of the impulse type. They are characterized by a large pressure drop across the stages, high rotor velocities to keep the blade-speed steam-speed ratio in a reasonable range, and high leaving losses.

Older auxiliary turbines are fitted with a casing relief valve to prevent the casing from being subjected to full line pressure in the event the throttle valve is inadvertently opened before the exhaust valve is opened. This valve usually relieves directly to the machinery space atmosphere and is usually set to open at 25 to 30 psig. It is the latest practice to provide a spring loaded exhaust valve which cannot be positively closed by hand. The valve is designed to open automatically when the steam pressure in the turbine casing exceeds the exhaust line pressure by more than about 2 psi. Whatever the arrangement, all auxiliary turbines are equipped to prevent the turbine casings from being subjected to line pressure. This permits much lighter construction of the turbine casing than would otherwise be possible and prevents damage to the shaft seals.

702. AUXILIARY TURBINE CLASSIFICATION AND TYPES.

Auxiliary turbines are usually of the radial, helical, or axial flow type. They may be one, two, or three-stage turbines, except for the ship's service generator turbine, which will be discussed separately. Often the first stage will be velocity compounded to permit a large pressure drop across the first stage nozzles without an inordinately high wheel speed. The particular type turbine used in any given application will vary according to the choice of the supplier of the unit. Normally, only operating characteristics are specified; within these limits the contractor has a wide choice of types. It may be found that the same auxiliary components on sister ships will be driven with different type turbines.

Figure 7-1 shows a radial flow, single stage, re-entry turbine. The re-entry feature effects a velocity compounding. Note the needle valve control, a feature found on many auxiliary turbines. The needle valve arrangement approaches the efficiency of a nozzle in the conversion of thermal to kinetic energy and avoids much of the thermodynamic loss associated with pure throttling. The needle serves, in effect, to vary the flow through the nozzle by changing the throat area, thus varying the flow without decrease of pressure at the nozzle inlet, as would be caused by an upstream throttle valve. The single penetration of the casing by the shaft is also typical; this simplifies the shaft sealing problem, solved in this case by four rings of carbon packing. Figure 7-2 is another arrangement of the radial flow turbine. This one has two rows of moving buckets to effect the velocity compounding.

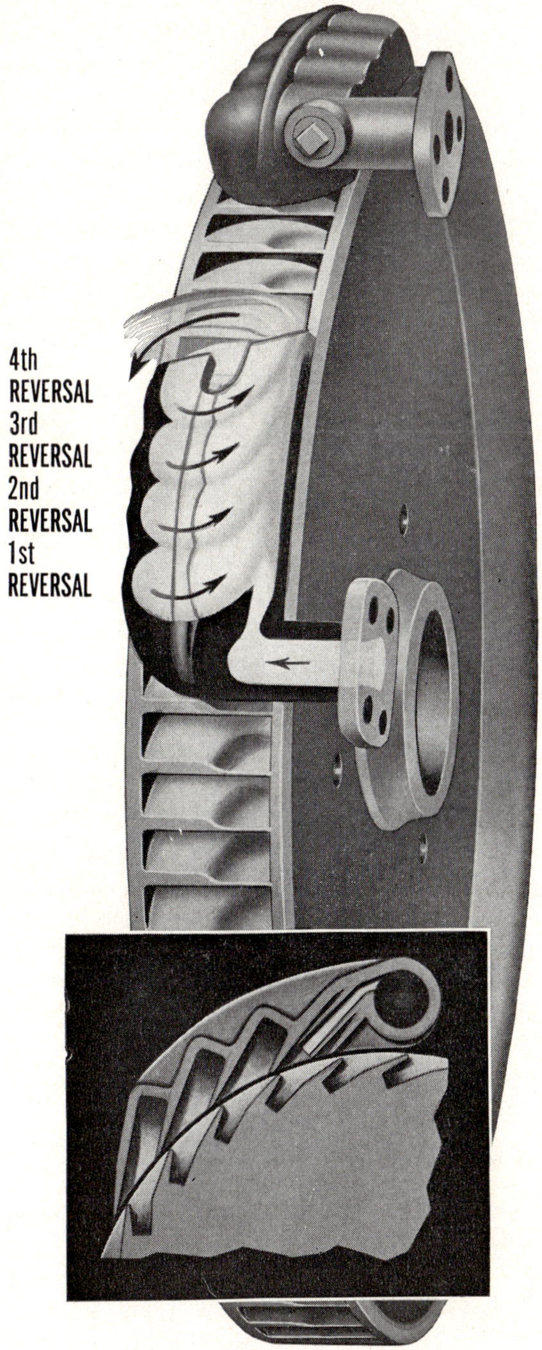

FIG. 7-3. Helical Flow Turbine.

Helical flow, or tangential flow, turbines are often found in various applications. Figure 7-3 is a phantom view of this type. Note the four reversals, which is equivalent to a velocity compounded stage with four sets of fixed blades. Several sets of nozzles may be mounted around the periphery of this type, giving a wide range of power for the same size wheel. Another view, showing the whole unit is presented in Figure 7-4.

A typical velocity compounded, axial flow turbine is shown in Figure 7-5. This is a vertically mounted turbine (with reference to the shaft) whereas the others discussed have been horizontally mounted. In its application here as a main circulating pump drive, the thrust on the pump impeller is downward, as well as the weight of the turbine and pump rotors. Consequently a large thrust bearing is provided as shown to support the thrust and weight. Another type axial flow arrangement of the re-entry type is depicted schematically in Figure 7-6. An arrangement with two reversing chambers is given in Figure 7-7.

The above turbines are all single stage turbines, designed to operate with 600 psig saturated steam. A recent arrangement utilizing two stages, with the first stage velocity compounded is sketched in Figure 7-8.

703. AUXILIARY TURBINE EFFICIENCIES.

While the engine efficiency of the main propulsion turbine (energy at the turbine shaft divided by energy available to the turbine) may go as high as 75%, the engine efficiency of auxiliary turbines is seldom greater than 60% under the the best conditions, and may range as low as 17–20%. However, the failure of the turbines to utilize more of the energy available is not as serious as these figures indicate, since the thermal energy remaining in the exhaust steam is not normally lost to the plant, as in the case of the main propulsion and ship's service generator turbines, but is used for various purposes to be discussed in connection with the auxiliary exhaust system.

704. AUXILIARY TURBINE CONTROL.

Depending on their application, auxiliary turbines may have any one of several means of control. Detailed arrangements vary widely; only basic considerations will be discussed here. Descriptions of individual arrangements are available in the instruction books provided with each equipment installation.

A. Constant Pressure Control. Many pumps operate on the principle of providing constant output pressure, regardless of flow demand. The fuel oil service pump, the main feed pumps in many ships, and the fire and flushing pumps are examples of such applications. A scheme for sensing the output pressure and using its variation from the desired pressure to control the steam supply to the turbine is the basis for this type of control. Such a device, called a *constant pressure pump governor* is shown in Figure 7-9.

The desired pressure is set by the top handwheel and adjusting screw which controls the force on the control diaphragm by varying the spring tension. When the pump discharge pressure is at its correct value, the diaphragm is balanced. Suppose the pressure drops a little below normal. The reduced pressure will allow the diaphragm to drop a slight amount. This motion is transmitted by the crosshead to the spool in the center body. The spool forces the lower diaphragm down, pushing on the spindle of the small pilot valve, and causing it to open. This allows steam to flow from the supply line to the top of the main piston, forcing it down and opening the main valve wider. More steam now flows to the pump turbine, causing it to speed up and raise the pump discharge pressure. As the pressure increases the sequence is reversed, closing the main valve as necessary to limit the turbine speed.

It can be visualized that such a system could easily be unstable; that is, the characteristics of the governor could cause each correction to "overshoot," causing the pump to alternately slow and speed up.

The pressure chamber, with its passages to the diaphragms at each end of the spool piece, throttled by a needle valve, provides a damping device to retard the action of the pressure signal on the valve and prevent overshooting. This system allows the crosshead to control the pilot valve easily with slow changes in the pressure signal, but introduces a lag caused by the time required for steam (or condensate) to flow from one side of the diaphragm to the other when large changes in the pressure signal are received.

B. Constant Speed Control. Pumps, such as the main condensate pump and main feed booster

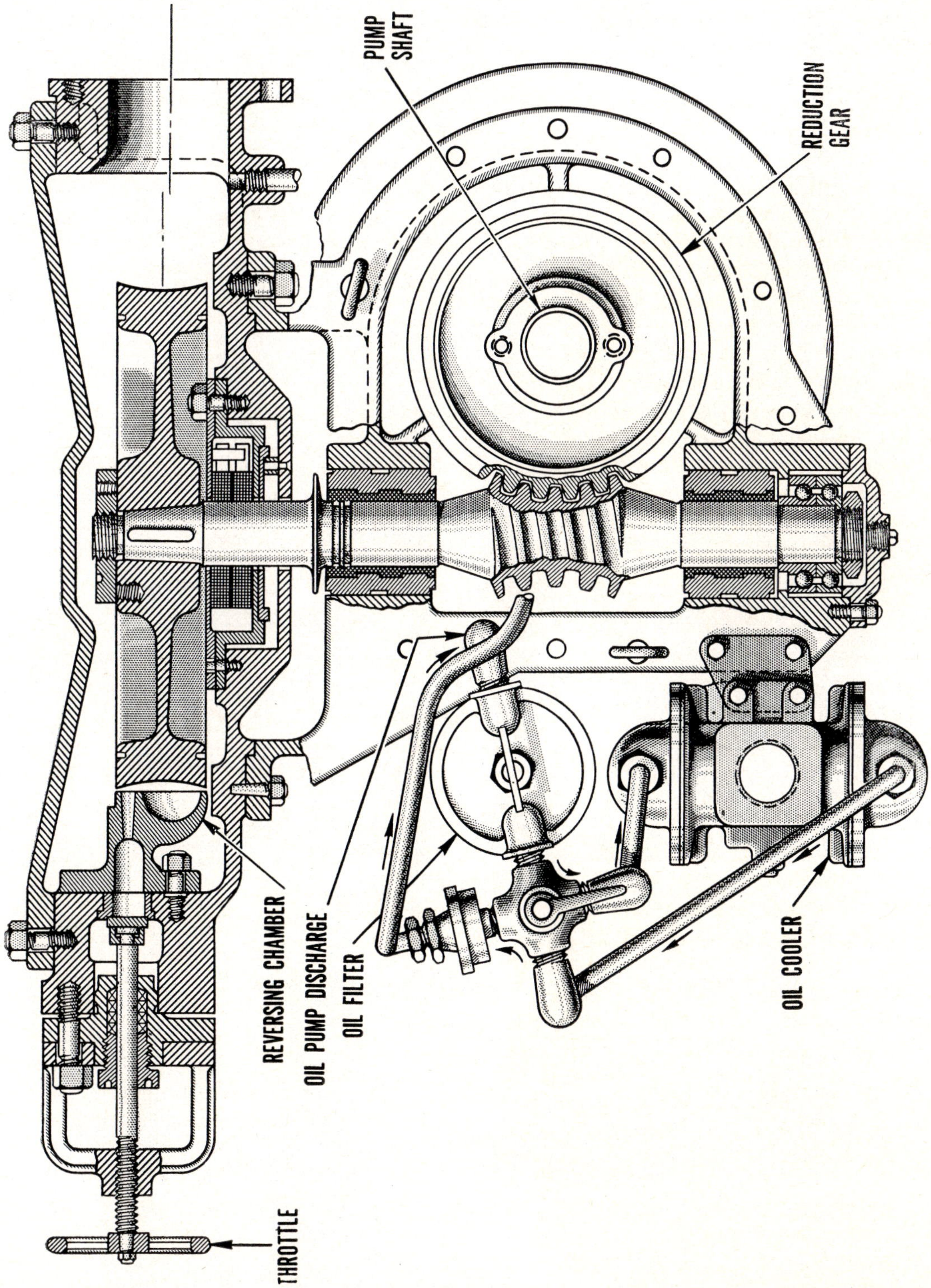

FIG. 7-4. Helical Flow Turbine, Sectional View.

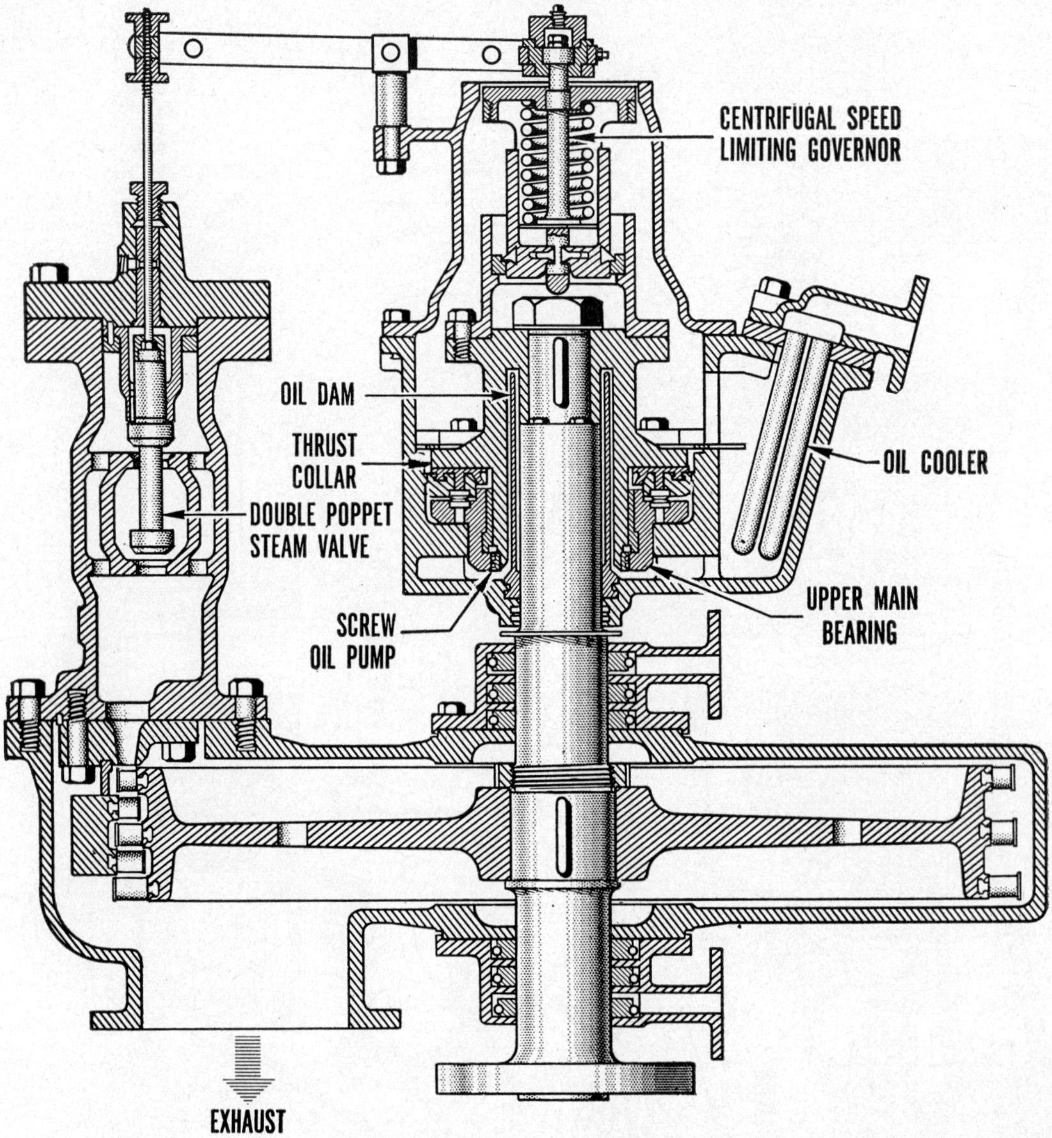

FIG. 7-5. Axial Flow, Velocity Compounded Turbine.

pump (when separate) are often controlled by constant speed governors. In these applications, discharge pressure is not as important as the prevention of too high a pump speed, which would result in cavitation due to the near saturation of the liquid entering the suction. If kept at a constant speed, the flow and discharge pressure will be regulated automatically by the Head-Capacity characteristic of the pump. For instance, at low ship speeds, when condensate flow is low, the condensate pump will lower the level of condensate in the condenser. This decreases the positive head available in the condenser, increasing the total pump head, and lowering its capacity. (See Figure 6-19.) The same effect occurs in the main feed booster pump. These two pumps, then, are essentially self-regulating as to flow, providing their speed remains constant.

Figure 7-10 shows a simplified constant speed control of the centrifugal weight type. The weights tend to fly outward due to centrifugal force. As they do so, they raise the weight seat, which pushes the housing upward, eventually causing the governor spring to compress until its force equals that produced by the weights. The movement of the housing tends to close the

FIG. 7-6. Axial Flow, Single Re-entry Turbine.

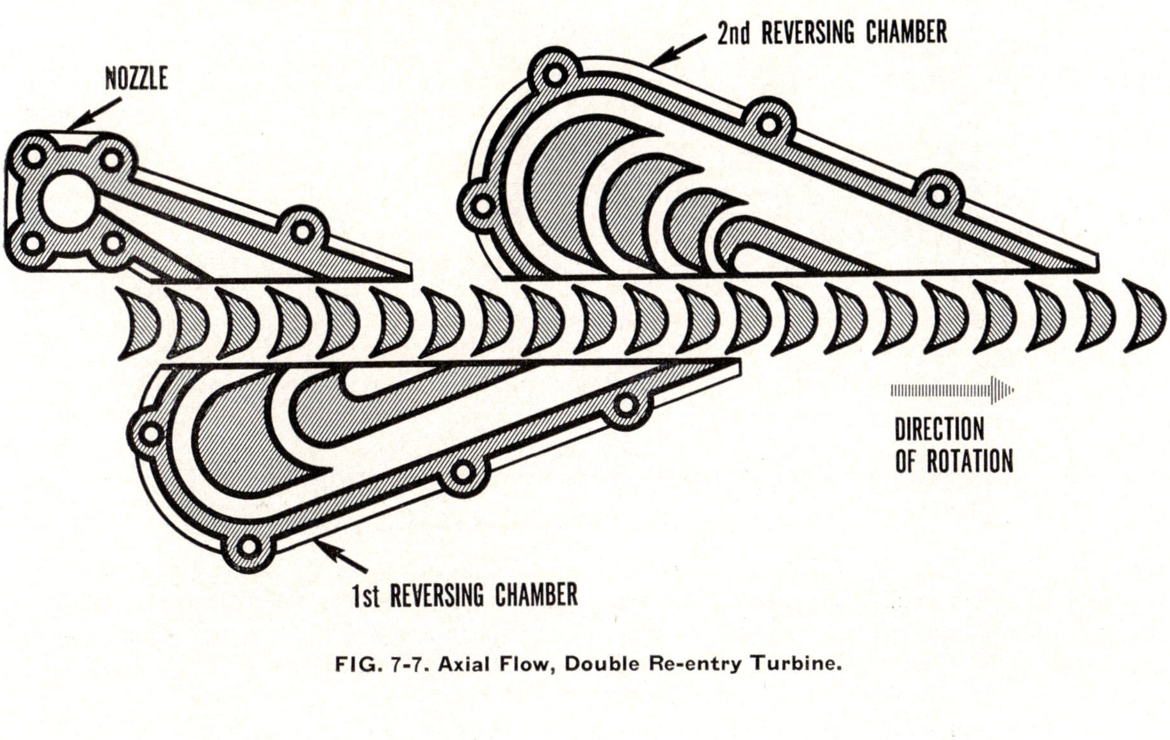

FIG. 7-7. Axial Flow, Double Re-entry Turbine.

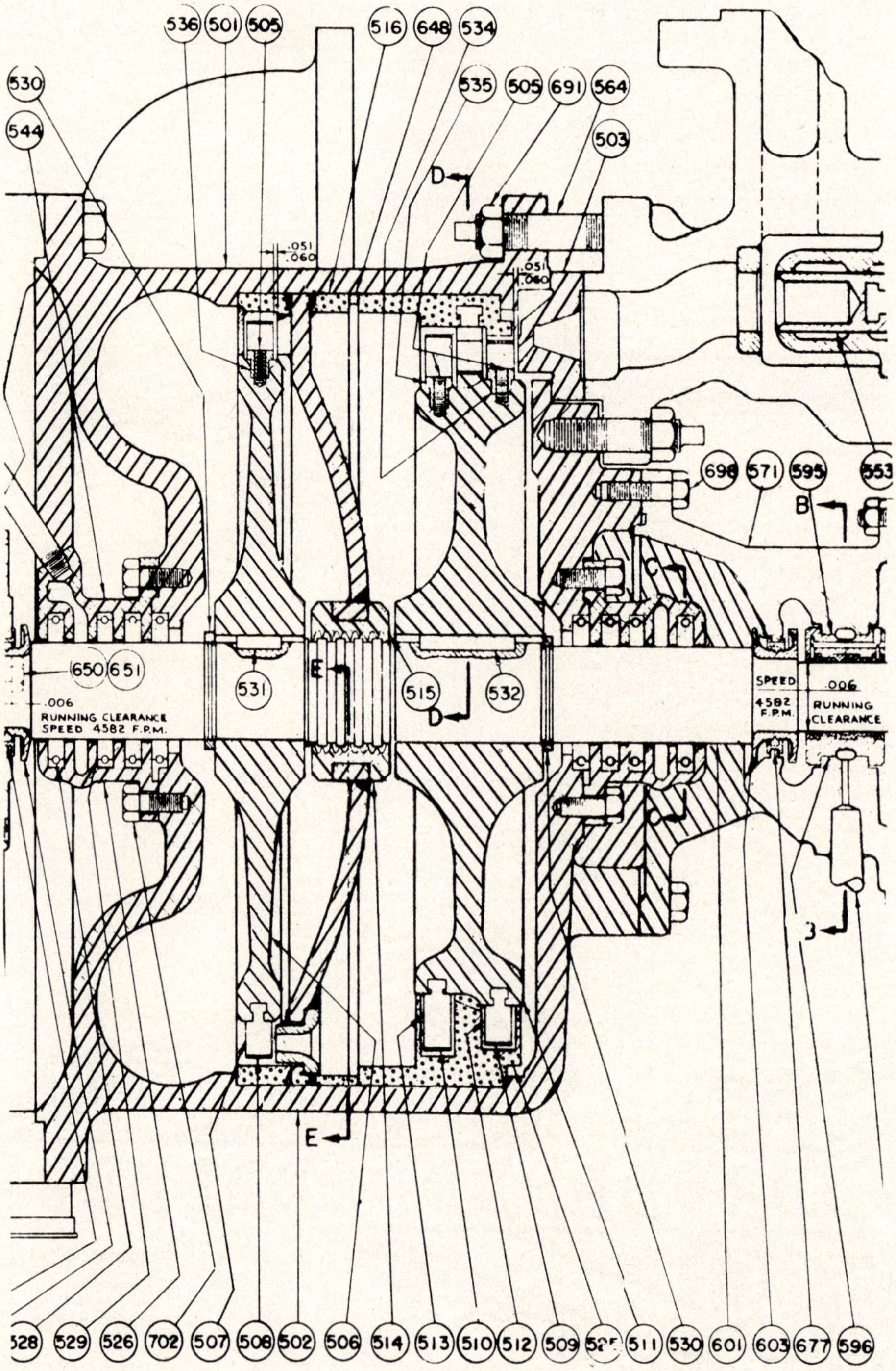

FIG. 7-8. Axial Flow, Two-Stage Turbine.

throttle valve, cutting down on steam supply to the turbine, until an equilibrium speed is reached. The governor can be set to the desired speed by adjustment of the relative position of the housing and sleeve, made possible by their threaded connection. Many variations of this arrangement are used, often involving the control of a hydraulic valve by the rotating weights, as shown schematically in Figure 7-11. One such system will be covered later in connection with the ship's service generator.

C. Pressure Differential Control. Often it is

necessary to control the turbine speed so that some pressure differential is kept constant. This is desirable where boiler feed water regulation is automatic. In this case, the main feed pump speed is controlled to provide a constant pressure differential across the boiler feed water regulating valve. (The actual flow through the valve is then controlled by the valve position, which is in turn controlled by a boiler drum level regulating device). In one system, the pressure differential is sensed by a device which operates a pneumatic system, which in turn operates a hydraulic system to position the turbine throttle valve. The details of these systems vary widely. Where installed, instruction books are provided giving complete descriptions. An older type excess-pressure pump governor for boiler feed control is shown in Figure 7-12 to illustrate the basic principles of pressure-differential controls. This particular arrangement is found infrequently now. In the figure, the upper spool, with its two diaphragms, is sensitive to the difference between pressure at the top, which equals the boiler pressure plus the spring pressure, and the pump pressure on the lower diaphragm. Thus the desired excess pressure of the pump discharge over the boiler drum is equal to the spring pressure. Variations in pump pressure cause the spool to rise or fall, transmitting such motion through the exterior connecting rods to the lower spool. The lower spool controls the pilot valve and causes the remainder of the assembly to function in the same manner as the constant pressure governor.

D. Manual Control. Manual control is necessary on very few auxiliaries, the most notable exception being the forced draft blowers on manually controlled boilers. Regulation of air by manual control is the practice on most existing installations, though recently automatic combustion control, which includes automatic control of air supply, has been adopted for new installations. Manual control consists of varying supply steam by a hand-operated nozzle control valve, or by a hand-operated throttle valve.

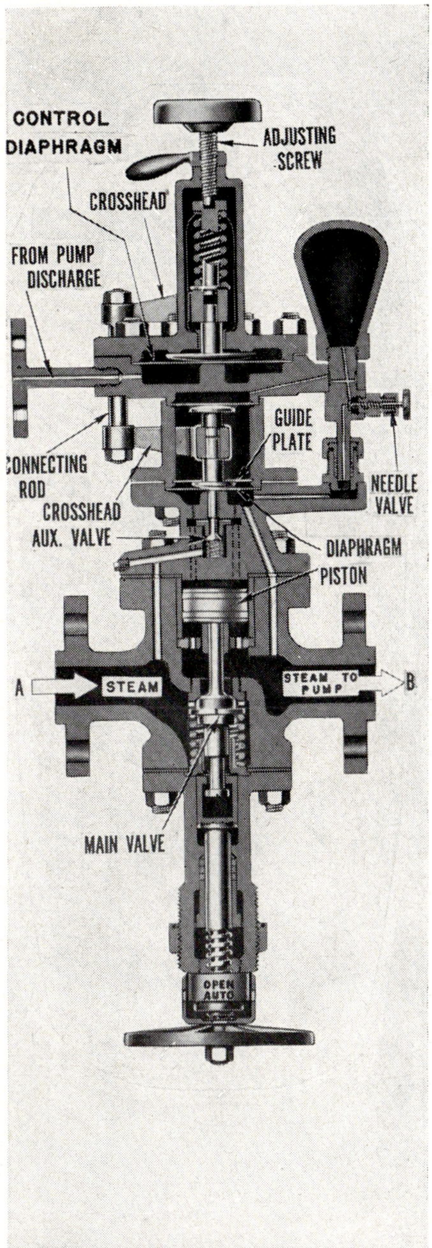

FIG. 7-9. Constant Pressure Pump Governor.

705. AUXILIARY TURBINE SAFETY DEVICES.

All pumps must be protected against too high an operating speed, to prevent overheating of the pump itself with consequent expansion and seizing of the impeller, or shafting, or to prevent damage to the drive unit. Examination of equation (7), Chapter 6, will indicate that the torque on a radial flow pump decreases with a decrease in flow, and thus, if constant power is available, an increase in speed will occur. On the other hand, the torque on a propeller type pump increases with a flow decrease, and such a pump will tend to slow if discharge is limited while constant power is supplied.

Electric motors approach constant-speed ma-

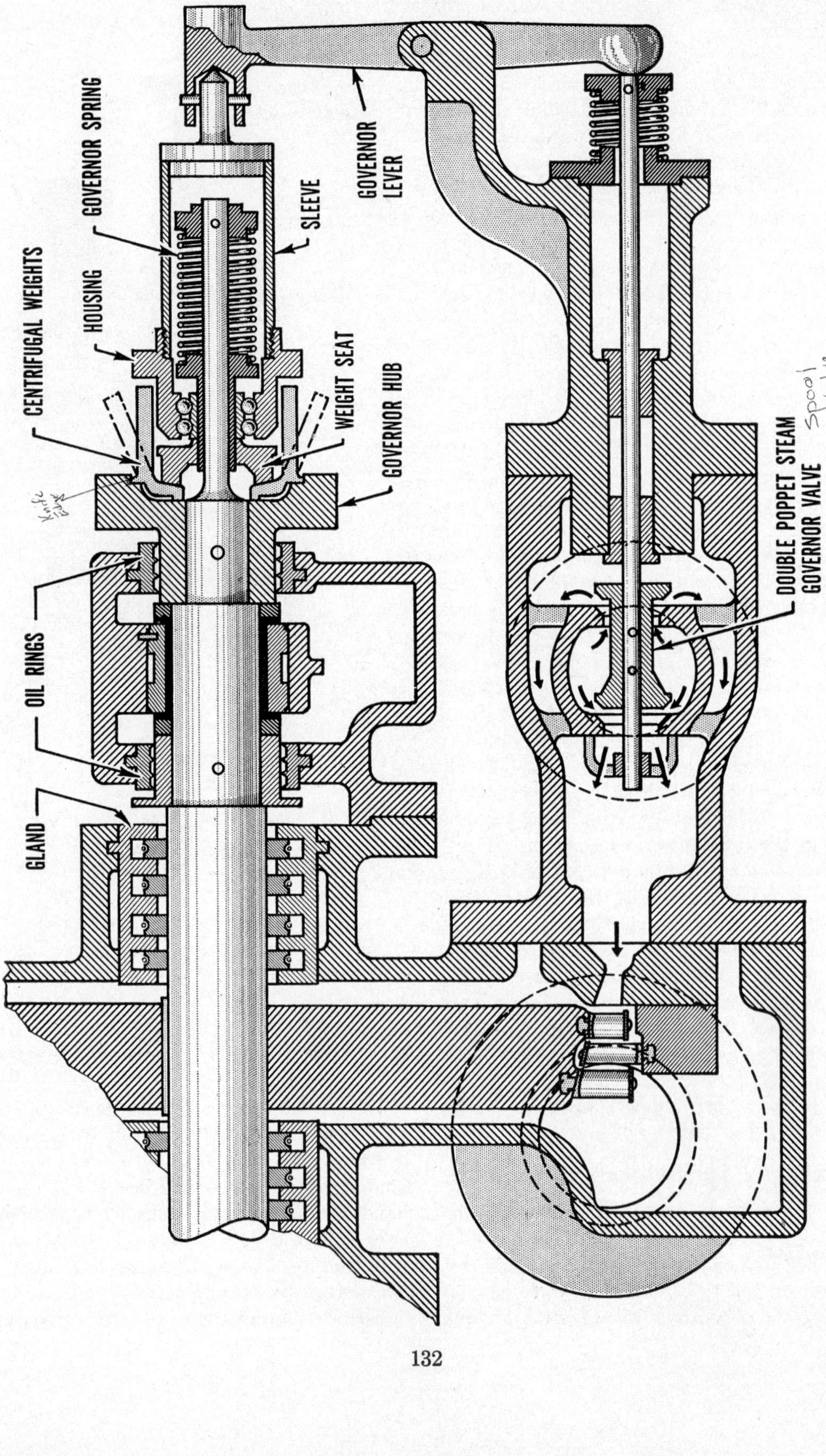

FIG. 7-10. Scheme for Constant Speed Control.

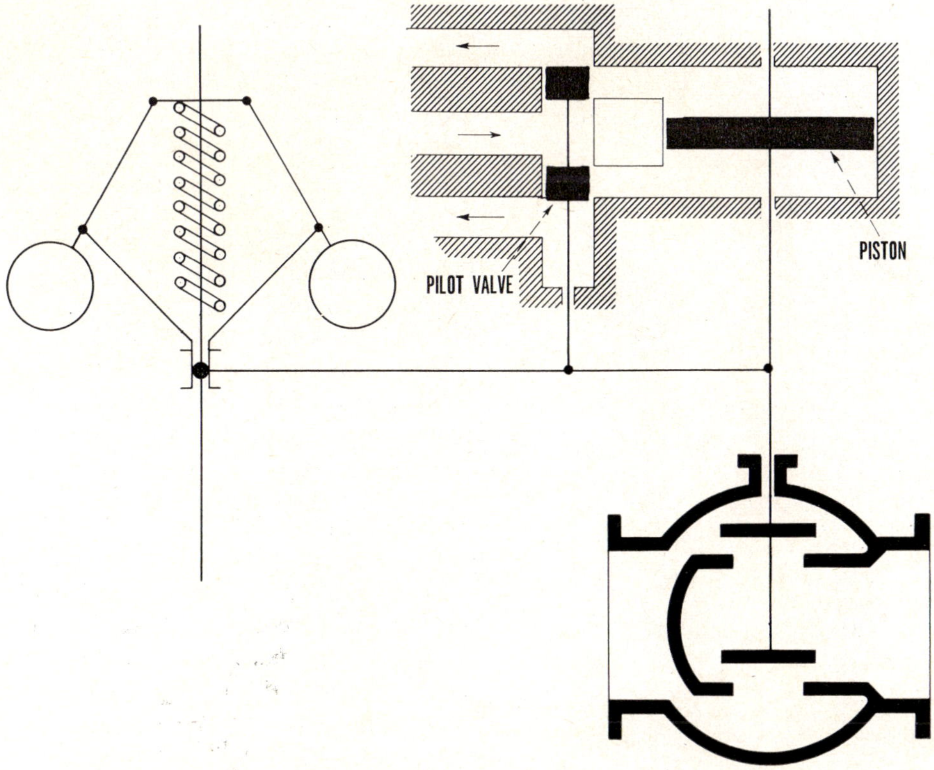

FIG. 7-11. Schematic of Hydraulic Governor System.

chines, i.e., their speed changes only slightly with changes in power demand. Therefore, centrifugal pumps driven by electric motors will not tend to overspeed should flow be stopped; the motor speed will simply rise slightly to its no-load speed, and remain constant. Thus electric driven pumps are essentially self-protecting.

If pump flow is shut off on a turbine-driven pump which has constant pressure control, both the control device and the power characteristic of the pump will tend to cause an increased speed. This can occur very rapidly and result in almost immediate damage. An inadvertent closing of the suction or discharge valve to the pump, a break in the discharge line causing a sudden discharge pressure drop, or a loss of suction by the pump due to cavitation could cause such a circumstance. To protect the pump and drive turbine, a speed-limiting governor is provided, which acts to limit the speed of the turbine to a safe maximum, regardless of the action of other control devices. This governor operates on the centrifugal weight principle discussed under constant-speed control, and in the same manner. Pump turbines which are provided with constant-speed control do not need an additional protective device.

706. OTHER AUXILIARY STEAM REQUIREMENTS.

In addition to providing steam for driving the auxiliary turbines and the ship's service generator turbine, the auxiliary steam system must provide steam for operation of the air ejectors, the fuel oil heaters, the boiler soot blowers, the fuel oil tank heating coils, the laundry, galley, ship's compartment heating and other miscellaneous services. This steam is provided by leading a portion of the superheated steam back through the steam drum desuperheater, where it gives up its superheat to the feed water, and becomes saturated at boiler pressure. Then it is led to a reducing valve where its pressure is reduced to that desired. Figure 7-13 is a schematic of this arrangement. As noted, a spray desuperheater is often fitted downstream from the reducing valve. This will be discussed below.

Operation of a typical reducing valve can be visualized by referring to Figure 7-14. This is called a spring loaded reducing valve. The dis-

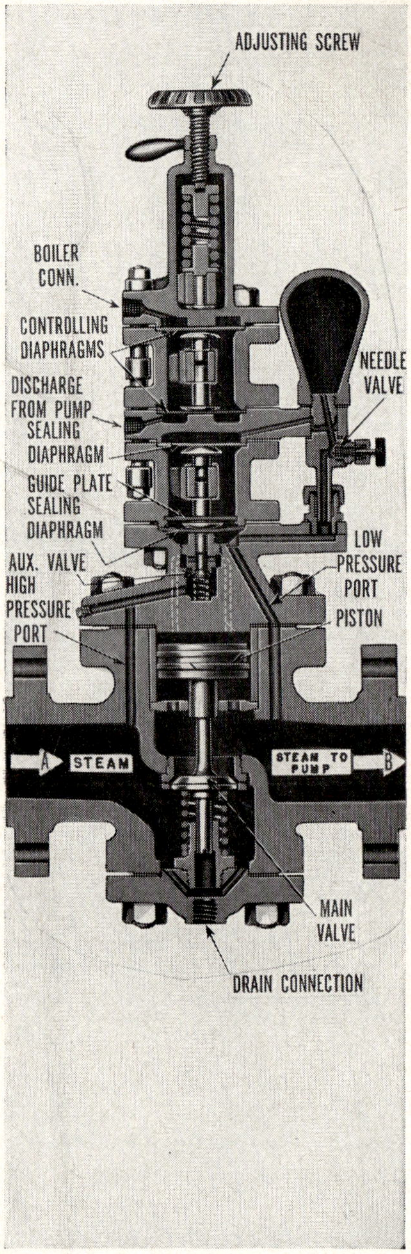

FIG. 7-12. Excess Pressure Pump Governor.

charge pressure is ported to the lower side of the diaphragm. The adjusting spring is set to balance the desired discharge pressure. Should the discharge pressure drop, the diaphragm moves downward opening the pilot valve, allowing inlet steam to force the piston down, opening the main valve wider, which raises pressure on the discharge side. This action, or its reverse, goes on continuously to maintain the discharge pressure at that required by the adjusting spring setting.

When the pilot valve is closed by an excessive discharge pressure, the main spring pushing up on the piston, and the pressure difference across the piston, causes steam to leak from the top chamber of the cylinder around the piston, allowing the valve to close.

Another type of reducing valve is shown in Figure 7-15. This is called a *gas-loaded valve*, since the desired discharge pressure is set by charging the dome with compressed air. In the valve shown, the loading chamber is charged to the desired pressure from the charging connection. Glycerine is forced through the orifice tube by this pressure to the lower side of a synthetic rubber diaphragm. On the other side of the diaphragm, steam pressure is led from the remote control connection, which is connected to the discharge line a few feet downstream from the valve. This distance, and the separation of the dome from the valve body, allows the steam to condense. The condensate protects the diaphragm from direct contact with the steam. The diaphragm bears against an upper plate which is connected to the main valve by a stem. The force tending to open the valve is that resulting on the diaphragm from the gas pressure in the loading chamber. The forces tending to close the valve are the spring pressure on the main valve and the force on the diaphragm from the discharge line. Should the discharge pressure drop, the diaphragm will move upward, opening the valve and thus restoring the discharge pressure. Should the discharge pressure increase, the diaphragm will move down, tending to close the valve, and reducing the pressure.

A simpler, integral unit is shown in Figure 7-16. This valve is used for air or other gases in low temperature service, since it does not have the high temperature protection features described above. The action is the same as that previously described. Note the provision for charging the loading chamber with the inlet gas.

707. RECOVERY OF USED AUXILIARY STEAM.

From the list of uses of auxiliary steam given in the first paragraph of Article 706 it is seen that nearly all involve the utilization of the steam as a source of heat, rather than one of kinetic energy. The major portion of this desired heat is obtained by condensation of the steam in these various uses. For instance, at 165 psia, one pound

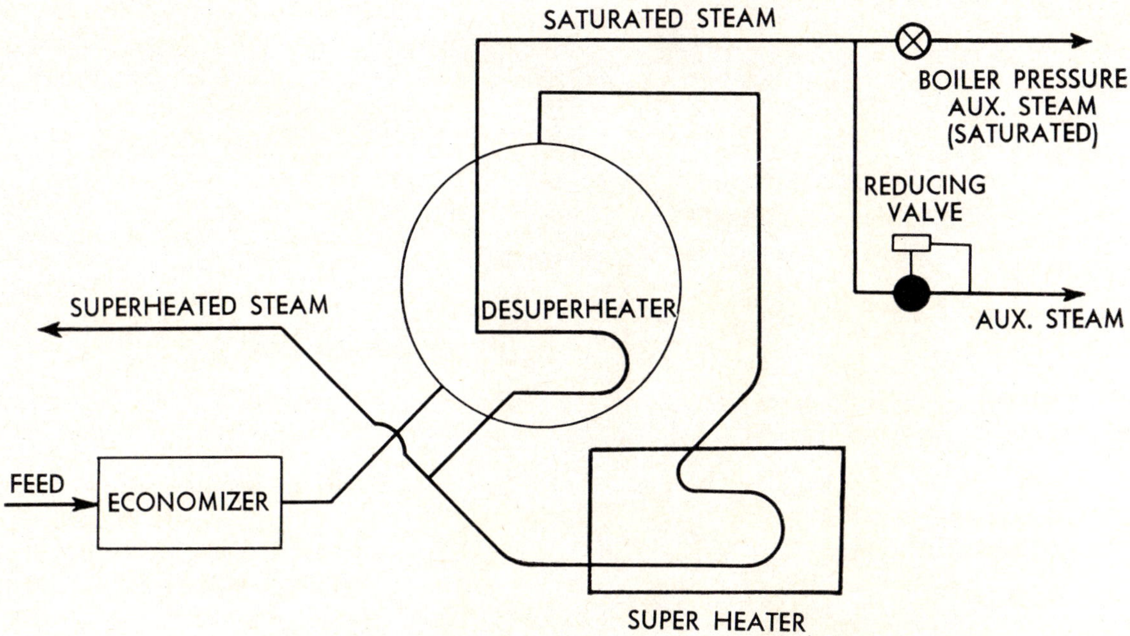

FIG. 7-13. Schematic Arrangement of Auxiliary Steam Supply.

of steam gives up 857 BTU in condensing from a saturated vapor. Superheating 300° above this would gain only an additional 265 BTU and require more line insulation and higher temperature materials in the auxiliary heater construction. Therefore, saturated steam is generally desirable for these services, except for a slight degree of superheat in the air ejectors. As a general rule, then, it can be said that all auxiliary components which utilize steam as a *heating* substance operate so that the steam is condensed in the process; and those auxiliary units which use steam as a *working* substance operate so that the steam is not directly condensed, but is delivered into an *auxiliary exhaust* line. The inevitable exception occurs in the air ejectors, where the steam is used as a working fluid, and is condensed immediately afterward.

The auxiliary condensate formed in the heating services is piped back to the D.F.T. either directly or indirectly, where it becomes a part of the boiler feed water. Most of it is drained from the various units at atmospheric pressure and flows into a low-pressure drain collecting tank, mentioned previously, from which it is pumped into the D.F.T. Some of it, especially from fuel oil heating (service and tank heating coils) is condensed at a relatively high pressure (150 psig) and is led directly to the D.F.T. Figure 7-17 is a schematic of some of the units, showing steam supply and the condensate piping.

708. THE AUXILIARY EXHAUST STEAM LINE.

All of the auxiliaries which operate non-condensing are designed to exhaust into a 15 psig auxiliary exhaust steam line. This line collects the auxiliary exhaust steam and distributes it to three points of utilization: (A) a portion of it is led to a reducing valve, where its pressure is reduced to 2 psig and then it is utilized as gland seal steam for the main turbines and ship's service generator turbines; (B) a portion is led to the evaporators, where its pressure is reduced to 5 psig, to provide the heat supply required for conversion of sea water into fresh water, and (C) the remainder is led to the D.F.T. where, along with high pressure drains, it is utilized to deaerate and heat the boiler feed water.

Normally, the machinery plant operation can be balanced, by proper selection of turbine and motor-driven units so that the auxiliary exhaust steam available is closely equal to that required for the above three services. This condition is the most conducive to economical operation.

Should the auxiliary exhaust steam supply be insufficient, its pressure will tend to drop. This will cause a reducing valve from the 150 psig auxiliary steam line to open and supply it as

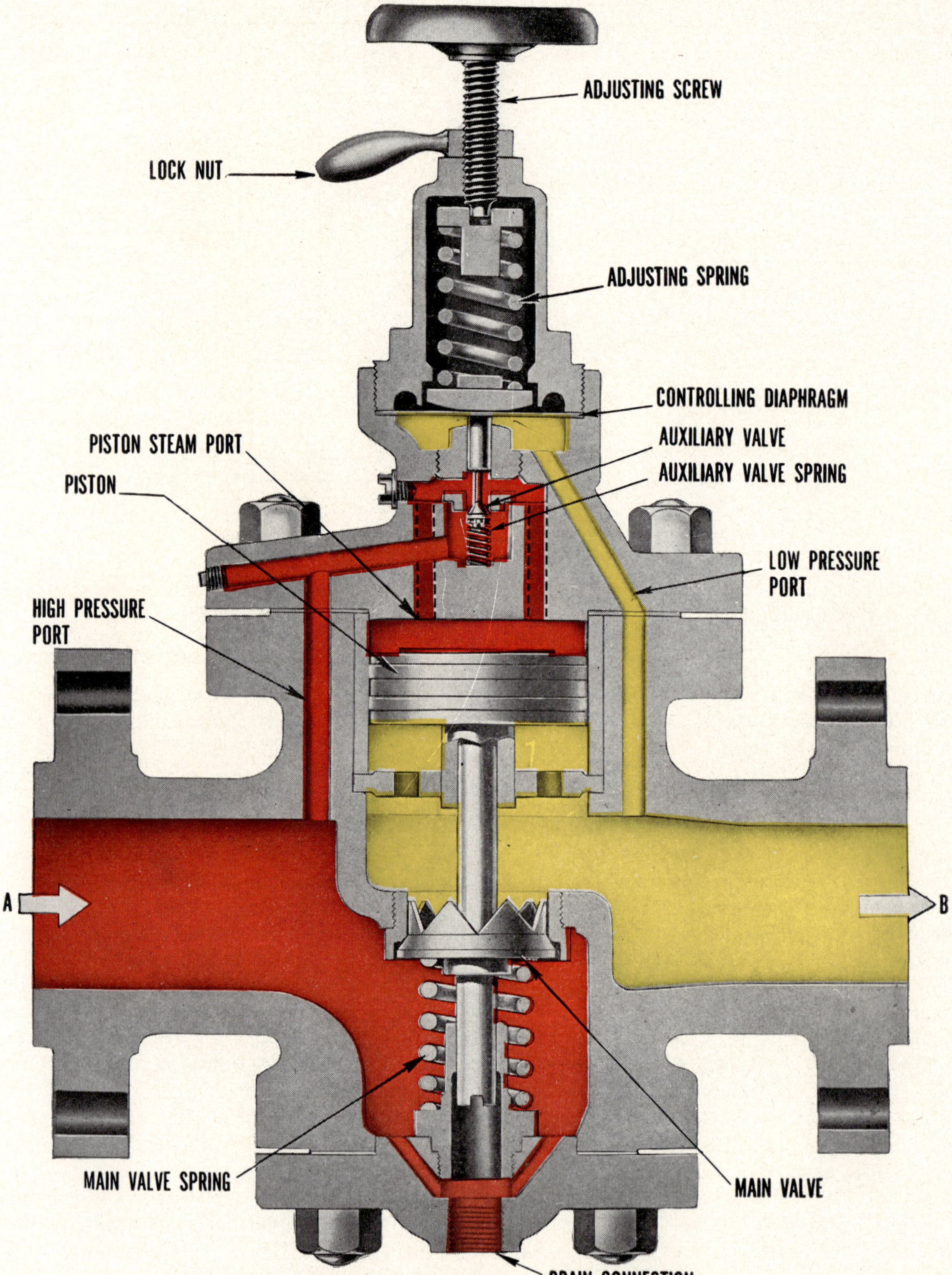

FIG. 7-14. Spring Loaded Reducing Valve.

necessary to maintain the pressure. This is not economical, since the throttling process in the reducing valve is wasteful.

If the supply of auxiliary exhaust steam is too great, its pressure will tend to rise. In this event a pressure relief, or unloading, valve will open allowing the excess to bleed off to the main condenser. This process is also not economical, since the thermal energy contained in the exhaust steam is transferred to the circulating sea water and lost overboard.

One type of automatic unloading valve is shown in Figure 7-18. This provides an accurate and sensitive control of the auxiliary exhaust line pressure. Suppose that the valve is set to "dump" steam into the condenser when the auxiliary exhaust line pressure rises above 17 psig. When this pressure is reached, it is sufficient to force the diaphragm in the right hand unit down against the spring pressure. This allows the double-seated actuating valve to drop, closing the lower seat and opening the upper seat. Pressure previously existing below the main diaphragm in the left hand assembly is now allowed to vent, through open valves "B" and "C," to the condenser. This releases pressure on the lower side of the main diaphragm and allows the steam pressure on top of the unloading valve disc to push it open against spring pressure, allowing flow of auxiliary exhaust steam to the condenser. When the line pressure has dropped to, say, 15 psig, it is no longer sufficient to hold the actuating valve diaphragm down against the spring pressure. The spring then forces the double-seated valve to rise, sealing off the connection from bottom to top of the main valve diaphragm, and admitting line pressure to the lower side of the main diaphragm. This pressure, combined with the unloading valve spring force, is sufficient to overcome steam pressure on top of the valve disc and force it to close. The valve can be hand operated if necessary by opening valve "D" and closing valves "B" and "C". It can be locked shut, if desired, by the manual control valve stem.

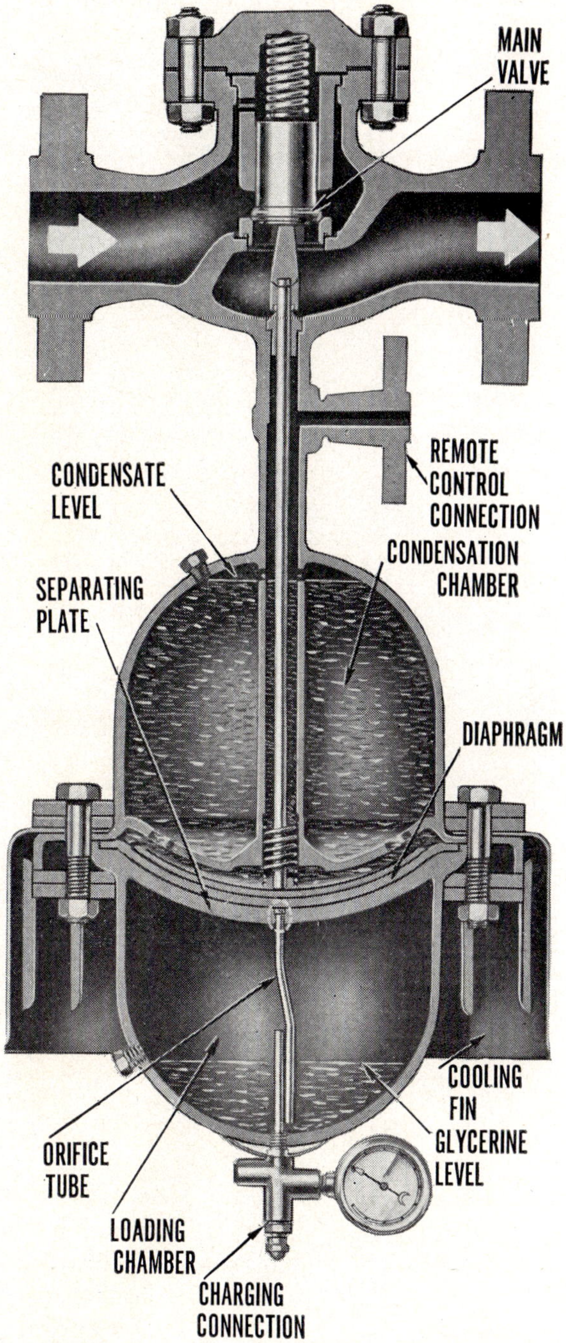

FIG. 7-15. Gas Loaded Reducing Valve for High Temperature Fluids.

709. PLANT ECONOMY AND SPEED CONSIDERATIONS.

The power required of the main turbines varies approximately as the cube of the speed, in ranges of 15 knots and above. The consumption of main steam by the turbines follows this relationship, if changes in turbine efficiency with speed are ignored. Below about 15 knots the power (and main steam consumption) varies roughly as the square of the speed. The steam consumption *per mile* then would vary directly as the speed up to about 15 knots, and then as the square of the

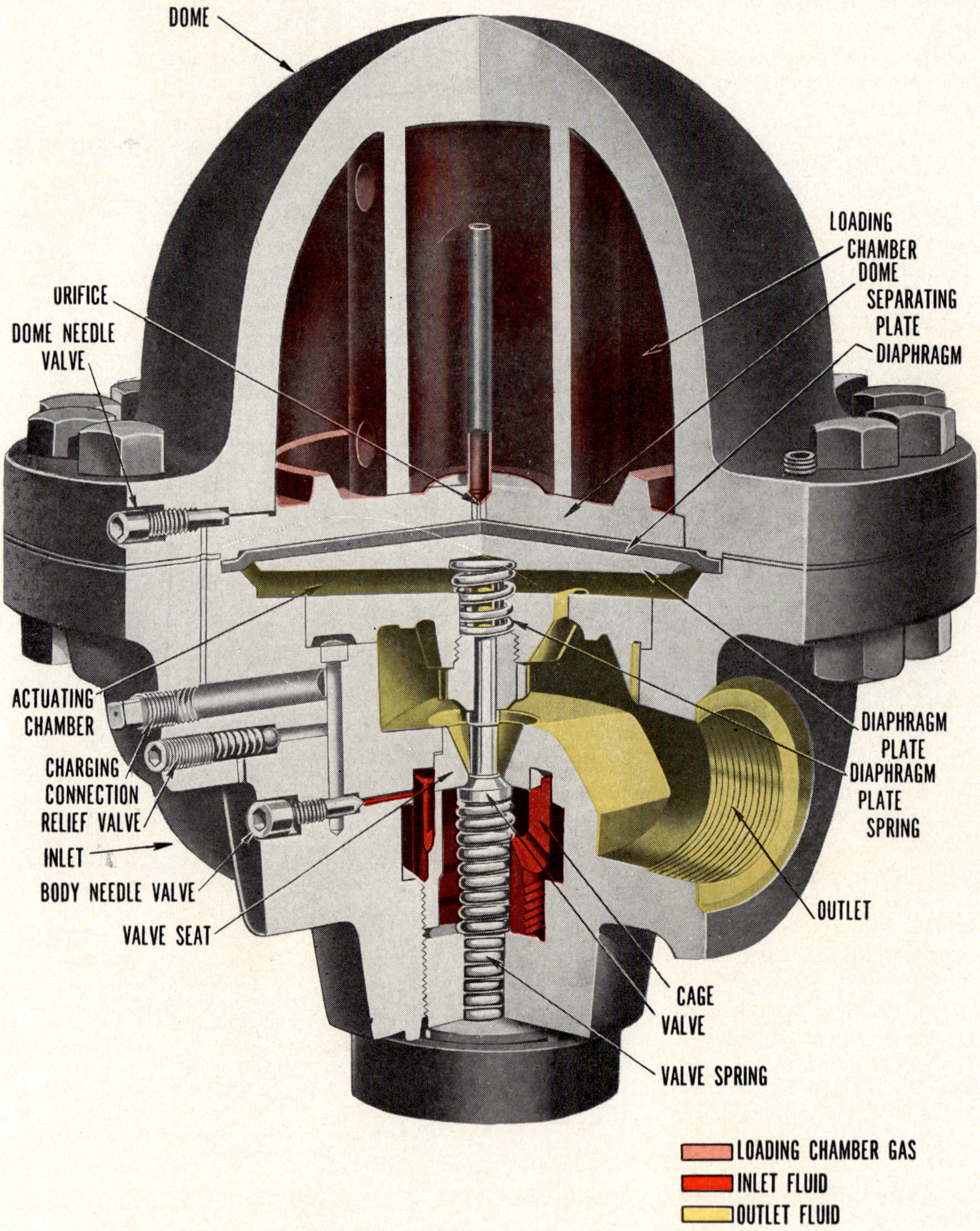

FIG. 7-16. Gas Loaded Reducing Valve.

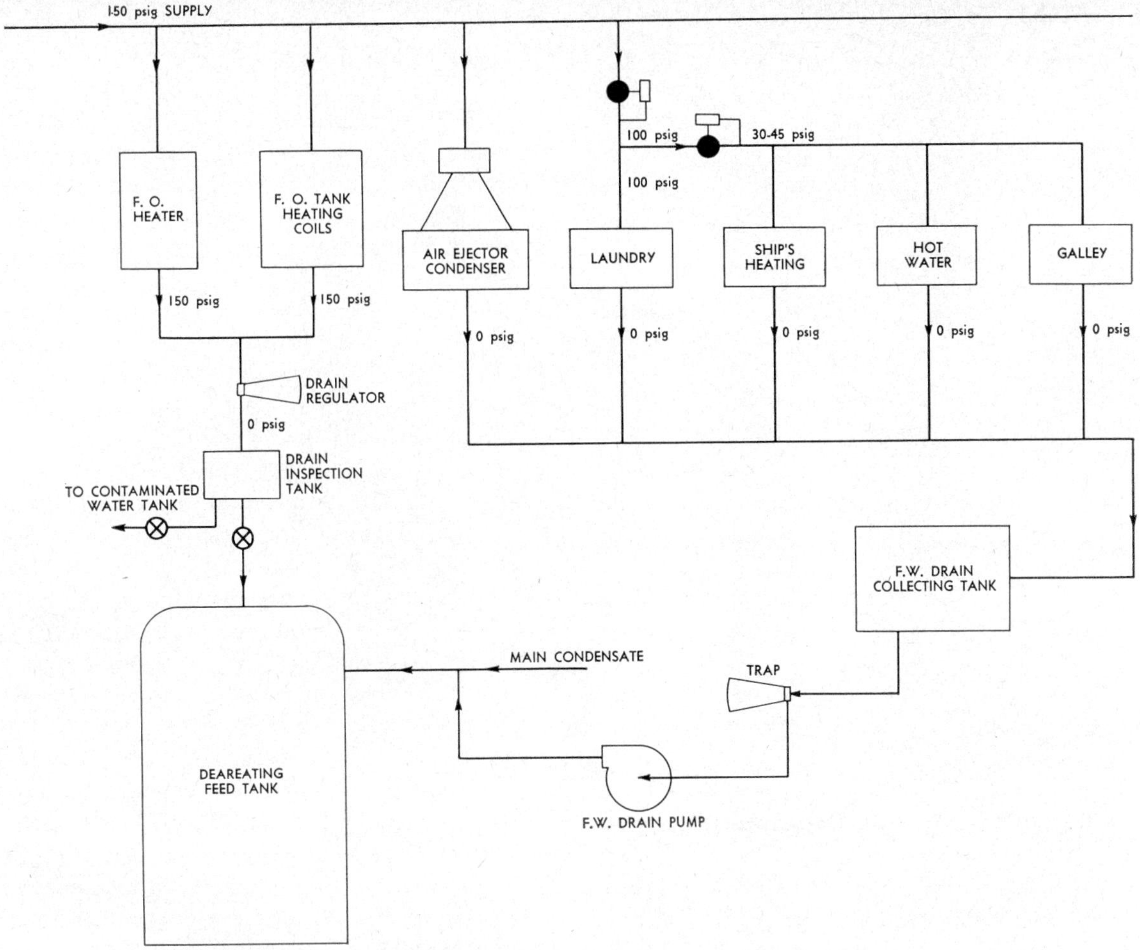

FIG. 7-17. Schematic of Auxiliary Condensate Flow.

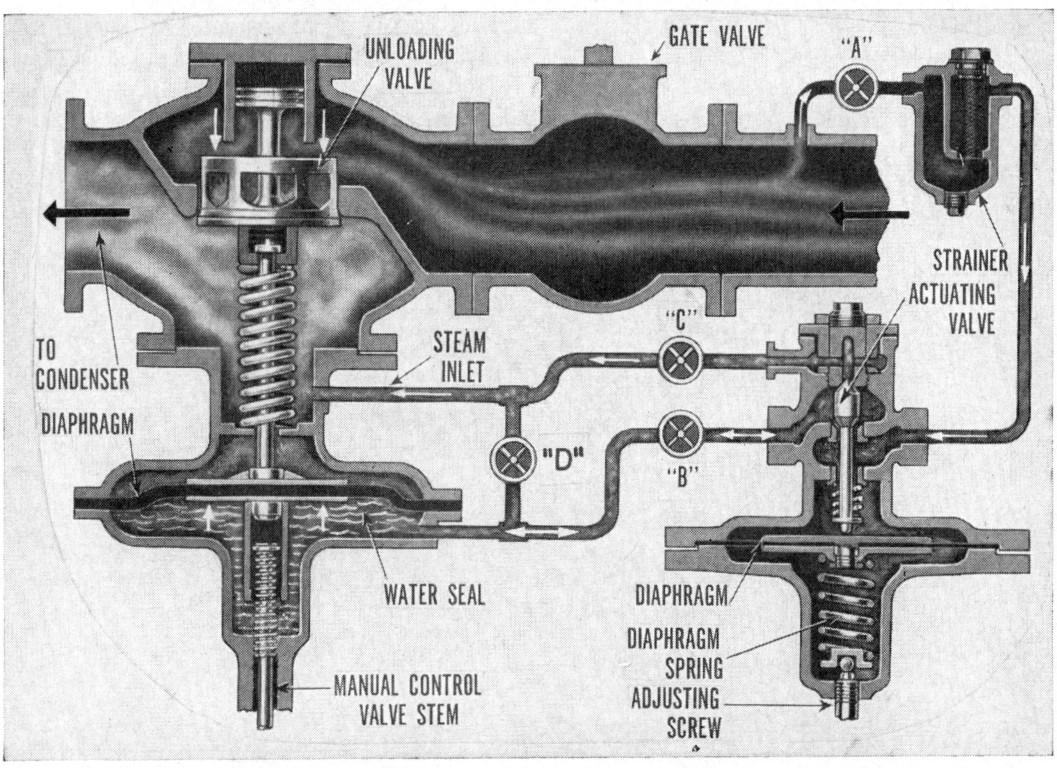

FIG. 7-18. Automatic Unloading Valve.

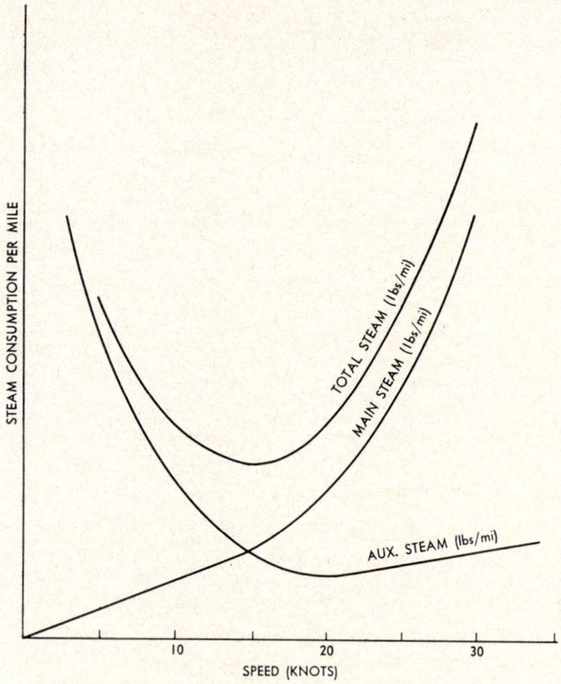

FIG. 7-19. Typical Steam Consumption Curves.

speed. Plotting the steam consumption per mile, on this basis, versus speed results in a curve similar to that in Figure 7-19.

On the other hand, the auxiliary steam requirement consists of many demands which do not vary with speed. For instance, the air ejectors take a constant steam supply regardless of speed. Steam required for the constant electric load, the laundry, and the heating system, do not vary with speed. Furthermore, in low speed ranges, up to about 20 knots, the efficiencies of pumps, blowers and many other components, are low. Thus, although their load may increase markedly during this range, the quantity of steam required to drive them varies only a small amount. Above 20 knots, however, the main feed pump and forced draft blower loads increase very rapidly (the forced draft blower steam consumption at 30 knots may be five times that at 20 knots), so that the total auxiliary steam requirement increases about as the square of the speed. However, considered on a steam-consumption-per-mile basis, the above relations must be divided by the speed. This gives an auxiliary steam consumption per mile about proportional to the reciprocal of the speed below 20 knots, and about directly proportional to the speed above 20 knots. Furthermore, at 20 knots, the total auxiliary steam used is about 40% of the main steam required at this speed. Figure 7-19 indicates the auxiliary steam consumption per mile plotted against speed.

The total steam consumption per mile is the sum of these two curves, and is represented by the upper curve in the figure. The lowest point on this curve represents the most economical speed for the ship. It will be noted that the curve is fairly flat in the 10–15 knot range, but tends to rise rather steeply above 20 knots. This curve is typical of nearly all combatant types, and graphically demonstrates the price that must be paid in fuel consumption, and range before refueling, for high speed operation.

CHAPTER VIII
THE ELECTRIC PLANT

800. GENERAL.

The electric plant has been reserved for discussion as an entity since it is the largest independent auxiliary plant in the ship. A modern aircraft carrier, for instance, has a 60 cycle electric plant of 12,500 KW capacity, large enough for a sizeable city. In addition, it contains 1200 KW capacity for 400 cycle alternating current, and 3000 KW capacity in emergency (diesel-driven generator) 60 cycle power. The standard generator output for combatant ships at present is 60 cycle, 3-phase, 450 volt alternating current. Aircraft carriers are provided in addition with a 400 cycle, 3-phase, 450 volt alternating current system for aviation service.

801. TURBINES.

The ship's service generator turbines are designed to operate condensing, that is, they exhaust into a vacuum maintained by a condenser. The condenser is a separate unit for each generator turbine, supplied with its own circulating water system, including auxiliary circulating pumps, for cooling, and with its own two-stage air ejectors for maintaining the vacuum.

The turbine is normally a five to seven stage impulse type with a velocity compounded first stage. It is designed to use superheated steam at the boiler superheater outlet pressure and temperature. In installations provided with 600 psig saturated auxiliary steam, it can also operate on this supply when the superheaters are secured. It exhausts into about a 26″ Hg vacuum maintained in its condenser. The condensate formed in the auxiliary condenser is pumped out by an electric driven auxiliary condensate pump, through the auxiliary air ejector inter- and after-condensers. It is then piped to the D.F.T. where it joins the main condensate just prior to entry into the tank.

Steam is admitted to the turbine through a throttle valve (fully opened or fully closed) and a set of nozzle control valves which are automatically operated. The nozzle control valves are mounted on a lifting beam. The arrangement is shown schematically in Figure 8-1. The beam, by moving vertically, opens the valves sequentially to the different sets of nozzles. The motion of the lifting beam is controlled by a speed-governor mechanism. In an AC generator, it is important to obtain a constant speed in order to maintain a constant generator frequency. Efficiency of electrical equipment designed for a given frequency normally falls off rapidly with a departure from this frequency, and in some instances, electrical casualty results. The control mechanism is discussed in Article 802.

Figure 8-2 is a cross section of a typical generator turbine. It is readily identifiable as an impulse turbine by the presence of the interstage diaphragms, the balancing holes in the bucket wheels, and the ample tip clearance of the bucket wheels. Many features common to propulsion turbines will be noted: the labyrinth packing at each end of the shaft, 48 and 49; the oil seal ring, 35; and the thrust bearing, 5, are examples. This turbine turns at 11,921 RPM. It is connected to a double reduction gear, shown in part at 26 and 29. The gear reduces the speed to 1800 RPM at the generator shaft. The turbine is fitted with its own lube oil system, including a lube oil pump and cooler (not shown). An air cooler for cooling the air circulating through the generator coils is also provided in conjunction with the auxiliary condenser.

802. CONTROL SYSTEM.

An integral control system is provided for the generator turbine. In addition to constant speed control, the system provides the following safety features: (a) overspeed trip, which shuts the turbine down in the event of failure of the speed control mechanism, (b) a back-pressure trip, which closes the throttle if excessive back pressure develops due to failure of circulating water, or for any other reason, (c) low-oil pressure trip, which closes the throttle in the event of oil pressure failure, and (d) a manual trip by means of which the operator can shut the turbine down by simply pushing a lever.

A typical system is shown in Figure 8-3. Before a detailed consideration of the system's opera-

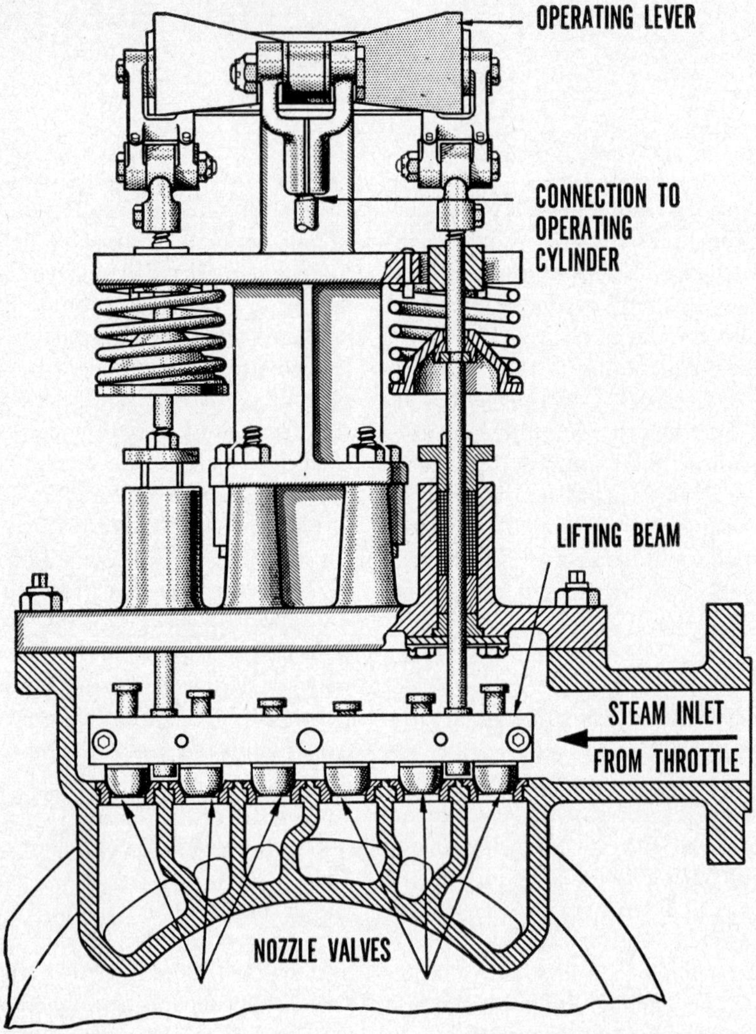

FIG. 8-1. Details of Lifting Beam and Nozzle Control Valves.

tion, it will be fruitful to examine the main system. Note the throttle valve assembly on the right side of the figure. This is the "on-off" valve admitting steam to the nozzle chest through the control valves, shown in right center of the figure. Immediately under this is the trip-reset mechanism. On the left side of the figure, the oil sump, the oil pump, the governor, and the pilot valve assembly are shown. The pilot valve controls oil flow to the operating cylinder which, in turn, controls the position of the lifting beam and control valves, shown in end view in the top center of the figure. Note that the nozzle chest and control valve assembly is shown in two separate views.

Consider first the throttle valve assembly. Figure 8-4 is an enlarged view of this and the trip-reset assembly. The throttle valve is so arranged that oil pressure is required for it to remain in the open position. Turning the handwheel rotates the miter gear *A*, which is splined to the lower part of the valve stem, and thus causes the lower stem to rotate. The spline arrangement (not shown) allows the stem to move freely up or down within the bore of gear *A*, which is fixed, but causes any rotation of the gear to be transmitted to the stem. Suppose the handwheel is turned counter-clockwise. The lower valve stem forces the control piston down, to the position shown. When oil pressure is admitted to the lower side of the piston, it forces the piston and the lower and upper valve stems upward, opening the valve at the upper

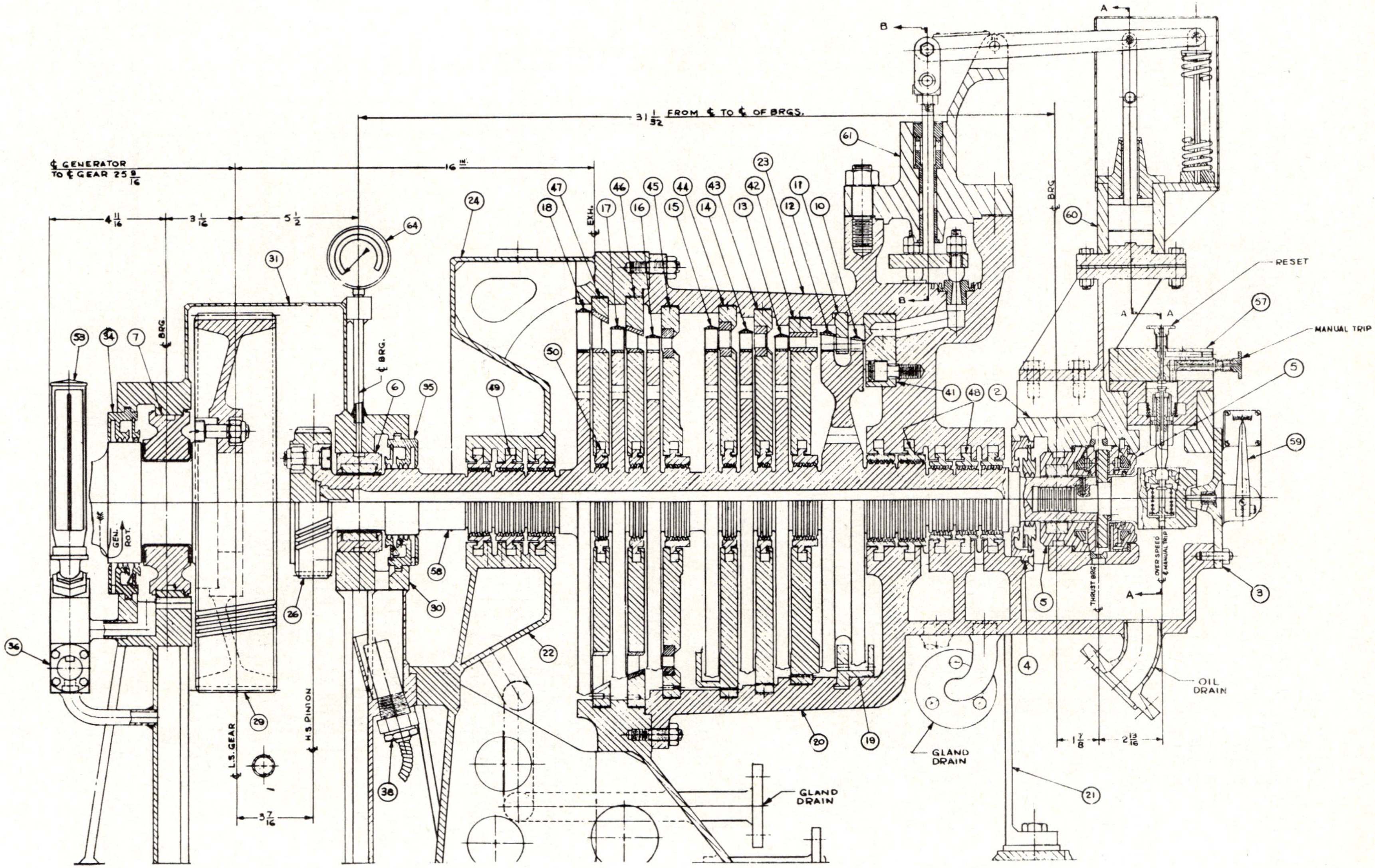

FIG. 8-2. Ship's Service Generator Turbine—Cross Section.

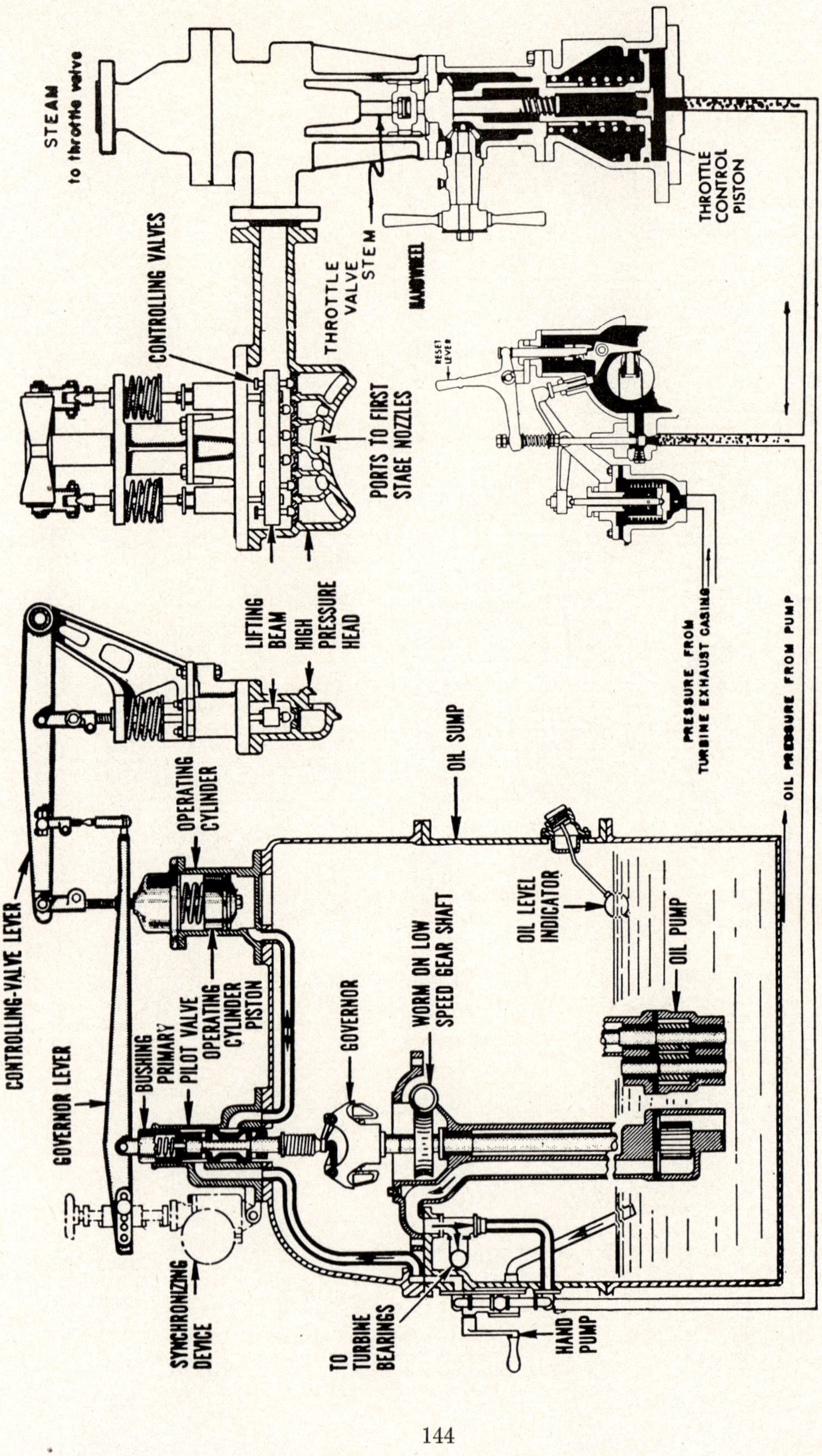

FIG. 8-3. Schematic of Ship's Service Generator Control System. (Note: Threads on throttle valve stem should be left handed.)

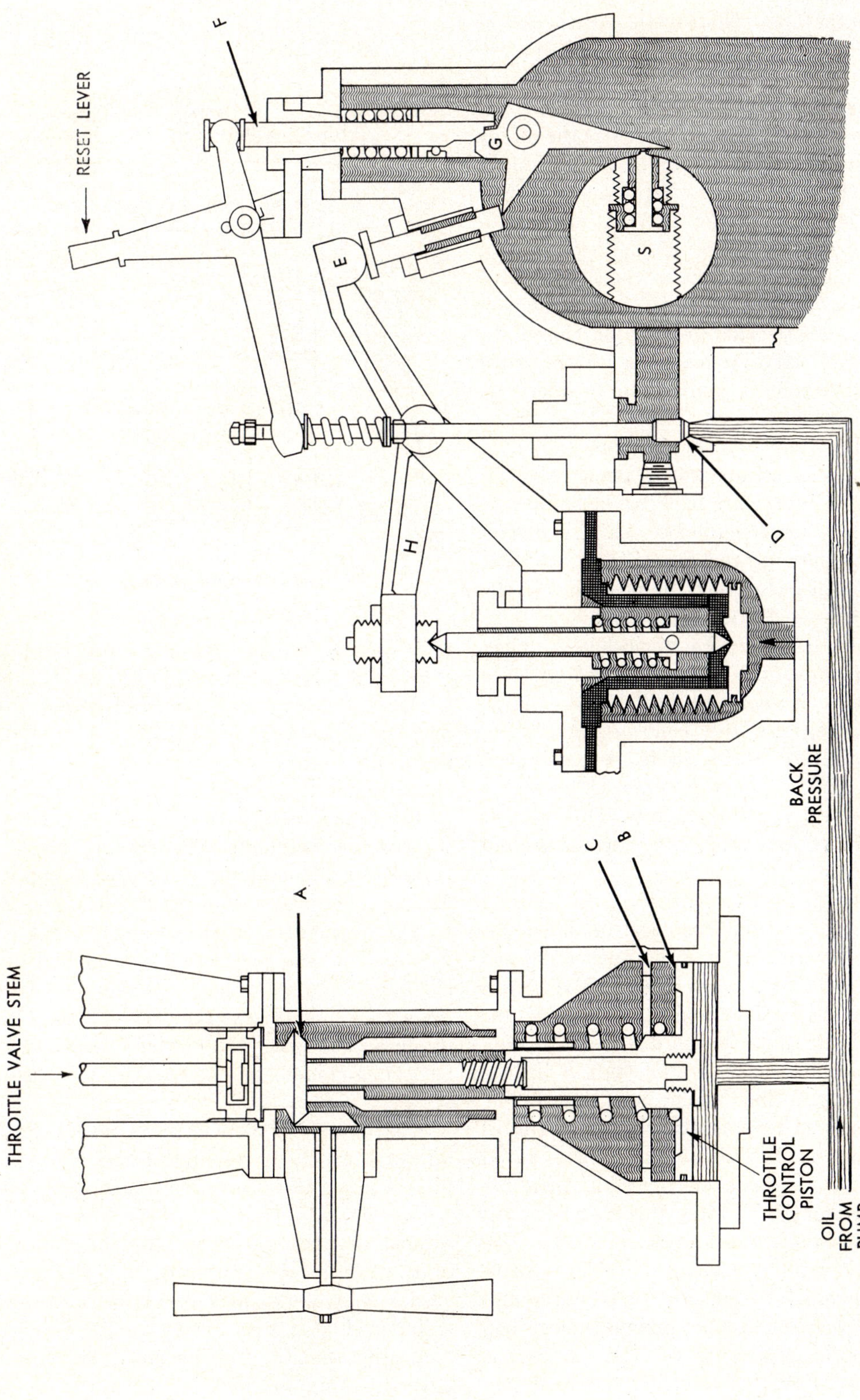

FIG. 8-4. Details of Throttle Valve and Trip Assemblies. (Note: Threads on throttle valve stem should be left handed.)

end of the throttle valve stem. When oil pressure fails, the spring forces the throttle control piston down, causing the valve to close. On starting up, the initial pressure is raised by a hand pump shown in Figure 8-3. After the turbine is turning, it drives the gear pump which supplies the oil for both lubrication and control. This arrangement constitutes the low oil pressure protection previously mentioned. When the throttle valve handwheel is completely closed, the throttle control piston is raised with respect to the lower valve stem until its shoulder, *B*, rests against the surface, *C*, the throttle valve remaining closed. In this position the valve stem is locked in the down position and cannot be moved up to open the valve.

Three other protective devices are shown in Figure 8-4. These all operate to open valve *D*, which relieves oil pressure back to the sump and causes the throttle valve to close. Valve *D* is held on its seat by the push rod *F*, acting on the reset lever, and itself being held in position on the small detent in the bell crank, *G*.

The first of the protective devices is the back pressure trip. When the turbine exhaust pressure rises to a value of about 5 psig it becomes large enough to compress the bellows and spring shown. This rotates the lever *H*, depressing its right end *E*, and tripping the bell crank, releasing the push rod and opening valve *D*. This relieves the oil pressure and stops the turbine as described above.

The second protective feature is the manual emergency trip, accomplished by manually depressing the lever at point *E*, actuating the bell crank, push rod and valve *D* as described above.

The third protective device consists of the actuation of this same system due to overspeed. A plug is held in the generator shaft *S* by the screw insert and spring arrangement shown. The spring is designed to hold the plug seated until the generator reaches about 10% above rated speed. Centrifugal force on the plug then becomes great enough to force the plug outward causing it to strike the lower arm of the bell crank *G* as it rotates, causing the bell crank to turn around its pivot to release the push rod and initiate the chain of events previously described which results in closing the throttle valve. The reset lever provides for resetting the push rod

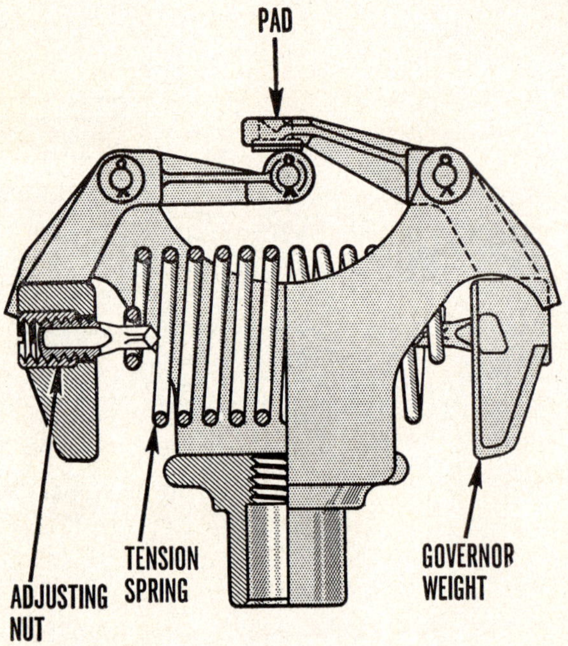

FIG. 8-5. Details of Governor.

after the trip has operated due to any of the three causes above.

Details of the governor and speed control mechanism are shown in Figures 8-5 and 8-6. The governor weights in Figure 8-5 are responsive to the speed of the governor shaft which is geared to a turbine reduction gear shaft. They move outward as speed increases, or inward as it decreases, translating the motion to a vertical movement of the pad. The pad is connected to the pilot valve by the spring and pivot rod assembly shown in Figure 8-6. The pivot rod seat on the pad is slightly eccentric, causing the lower end of the pivot rod to revolve as the pad revolves with the weights. This constant motion eliminates static friction, and increases the sensitivity of the mechanism.

The pilot valve fits within a sleeve, containing oil ports which correspond to the lands on the pilot valve. The sleeve, or bushing, is free to move vertically within the housing. The sleeve is attached to the governor lever, as shown. The pilot valve, its sleeve, and the housing taken together control the flow of oil to the operating cylinder (Figure 8-3).

Assume that the turbine speed decreases. The governor weights move in, forcing the pad up.

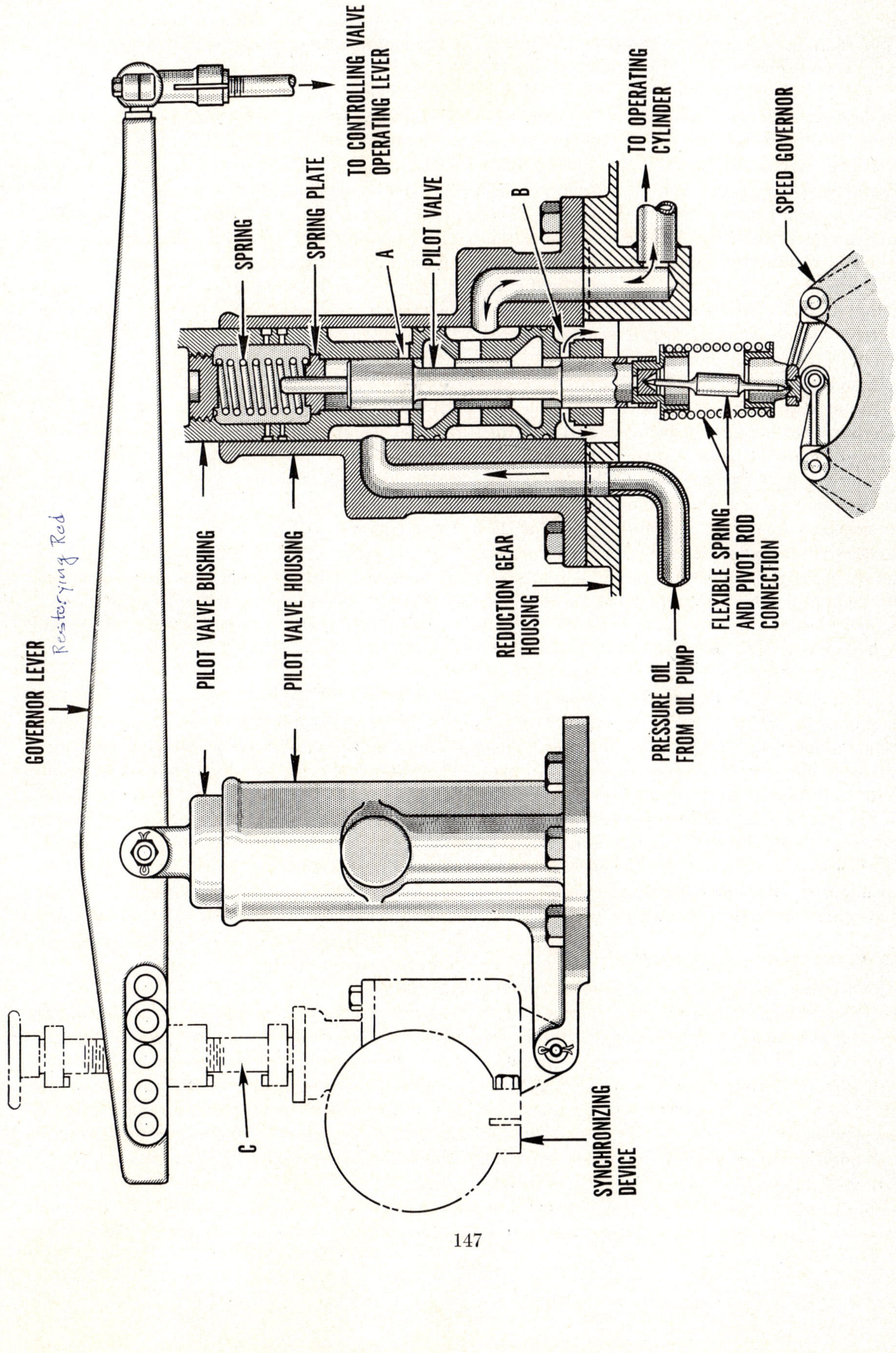

FIG. 8-6. Details of Pilot Valve Assembly.

This motion is transmitted to the pilot valve. As it moves up with respect to the sleeve, the upper land uncovers port A, admitting high pressure oil around the central portion of the valve and out the discharge line to the operating cylinder. Referring to Figure 8-3, the operating cylinder is forced up, raising the control valve lever, which raises the lifting beam and its nozzle control valves, to admit more steam to the turbine. The same motion of the control valve lever raises the right end of the governor lever, which lifts the pilot valve bushing to close off port A and stop further oil flow. This "follow-up" feature acts to prevent "over-controlling," and gives a precise control action which prevents hunting.

Should the turbine overspeed, the same sequence of action takes place in reverse. Now the lowering of the pilot valve uncovers port B in the sleeve allowing oil pressure under the operating cylinder to spill to the sump. The operating cylinder spring forces the piston down, lowering the control valve lever and closing some of the nozzle control valves to decrease steam flow. Simultaneously the governor lever lowers, moving the pilot valve bushing down to shut off further release of oil.

The synchronizing device is a small motor, controlled from the switchboard. (See Figure 8-6.) Through a gear train this device moves the shaft C up or down. Shaft C carries the pivot point of the governor lever, and thus its motion is transmitted to the sleeve. This permits fine adjustment of the speed by the switchboard operator for (1) setting the speed at which the governor system will maintain the generator and (2) varying the speed slightly to bring the generator into synchronism when paralleling. Handwheel D permits control of these features locally.

803. ELECTRICAL DISTRIBUTION.

Each ship's service generator supplies a main switchboard from which electrical power is distributed. The boards may be paralleled. but are designed to provide separate portions of the ship with primary supply. Essential equipment is so arranged as to receive power from at least two primary sources.

In addition, the emergency power system, consisting of one or more diesel driven generators, supplies power, each to its own emergency distribution board. These distribution boards are kept "hot" from the primary boards normally, but should primary power fail, the diesel starts automatically and feeds power to the emergency board, which meanwhile has been separated from the primary system by a tripped circuit breaker when the primary power failed. Each emergency board supplies a third source of power to each essential component.

The emergency boards are never paralleled with each other, or with the main boards, when supplied by the emergency generators.

An additional source of power is provided by the casualty power system. This consists of small diesel-driven generators located in widely separated parts of the ship and normally disconnected from any circuits. If their use is required, casualty power cables, kept available in damage control lockers, are strung by hand. Casualty power connections are permanently installed through watertight bulkheads and decks, so that power runs may be led as required without violating watertight integrity.

To summarize, all essential electrical services are provided with a normal and alternate source of primary power, a source of emergency power, and, as a final resort, can be provided with casualty power.

804. ENERGY RELATIONS.

It is common practice to use the term *water rate* in connection with turbo electric machinery. The water rate is the number of pounds of steam delivered to the turbine throttle per unit of output power. If this is known, along with the turbine inlet and exit conditions, the mechanical efficiency of the unit can be calculated. The entrance and exit conditions determine energy release per pound of steam at the turbine buckets; this multiplied by the water rate gives the power delivered to the buckets per unit of output power.

Alternatively, if a reasonable mechanical efficiency is used, with the water rate and turbine inlet conditions known, a close estimate of the exit conditions can be obtained. The latter information is useful in computing the overall quantity of thermal energy rejected from the machinery plant. The following examples illustrate the above.

Example 81: Steam is delivered to a turbogenerator throttle at 1100 psig and 950°F. The water rate is 11.7 lbs. per kilowatt-hour. The turbine exhausts saturated steam into its condenser at a 27″ Hg vacuum. Find the mechanical efficiency of the turbo-generator plant.

Solution: Given the inlet pressure and temperature, the enthalpy of the entering steam is

$$h_1 = 1474 \text{ BTU/lb.}$$

The exit enthalpy is determined for saturated steam at 27″ Hg as

$$h_2 = 1112 \text{ BTU/lb.}$$

The energy released per pound is

$$h_1 - h_2 = 1474 - 1112 = 362 \text{ BTU/lb.}$$

The total energy released per kilowatt-hour output is found by multiplying this value by the water rate, or

$$E_1 = 362 \times 11.7 = 4240 \text{ BTU/KWH.}$$

But

$$1 \text{ kilowatt-hour} = 3413 \text{ BTU.}$$

Therefore the efficiency is

$$\eta = \frac{3413}{4240} = \underline{\underline{80.5\%}}.$$

Example 82: With the conditions given in Example 81, except that the condenser pressure is 3″ Hg, the state of exit steam is unknown and the mechanical efficiency is assumed at 85 per cent, find the state of the exhaust steam.

Solution: The total energy released at the buckets per kilowatt-hour is

$$\frac{3413}{.85} = 4030 \text{ BTU/KWH.}$$

Each pound of steam must deliver

$$\frac{4030}{11.7} = 344 \text{ BTU.}$$

The enthalpy of the entering steam, as in Example 81, is

$$h_1 = 1474 \text{ BTU/lb.}$$

The exit enthalpy is

$$1474 - 344 = 1130 \text{ BTU/lb.}$$

Entering the Mollier Chart with an exit pressure of 3″ Hg abs. and an enthalpy of 1130 BTU, the exhaust steam is found to have 40° superheat.

In the two examples above, sufficient data is available to determine the internal efficiency of the turbine (refer to Article 437). From the inlet conditions and the outlet pressure, the isentropic enthalpy drop can be computed. The actual enthalpy drop divided by this value gives the internal efficiency, or the efficiency with which the turbine converts thermal energy into mechanical energy at the *turbine wheels.*

Example 83: Find the internal turbine efficiency of the turbo-generator operating under the conditions of (a) Example 81, and (b) Example 82.

Solution:

(a) $h_1 = 1474$ BTU/lb (as before)

$S_1 = S_2 = 1.622$ BTU/lb-°F

$\therefore \quad h_2' = 936$ BTU/lb (at 27″ Hg)

$$\eta_i = \frac{362}{1474 - 936} = \frac{362}{538} = 67.2\%$$

(b) $$\eta_i = \frac{344}{538} = 64.0\%$$

It is not uncommon for ship's service turbines to exhaust steam in a slightly superheated condition except at very low powers, since the size and cost would become excessive if enough stages were provided to avoid this. Consequently, the internal efficiency is normally less than that of main propulsion turbines.

CHAPTER IX
THE DISTILLING PLANT

900. GENERAL.

In order to supply the large quantities of fresh water needed aboard ship for the machinery plant, and for personnel requirements, a distilling plant is provided on every naval ship. The distillation process is based on the fact that the heating of sea water produces pure water vapor. The plant consists of a means for accomplishing this heating, and condensing the vapor so formed. Under proper conditions, the distilling plant is extremely effective in producing fresh water free from chemical salts, impurities, animal or vegetable growths, and bacteria. The process is not effective in removing volatile gases or liquids which have a boiling point below that of water, and which may be present in polluted harbors.

Two basic types of distilling plants are used aboard ship: (1) the steam distilling plant, which receives its source of operating energy from a steam supply, and (2) the vapor compression plant, which uses electricity as its source of operating energy. The former is always found in steam driven ships. The latter is used aboard submarines and small diesel driven surface craft with daily requirements not exceeding 4,000 gallons per day. Steam distilling plants in combatant ships are also of two basic types: *the low-pressure submerged tube type* and *the flash type*. Standard steam plants are built in nominal sizes of 4,000; 8,000; 10,000; 12,000; 20,000; 30,000; 40,000 and 50,000 gallons per day. Vapor compression plants are built in 750; 1,000 and 2,000 gallon per day units.

All Navy designed steam units are classified as *low pressure distilling plants* and use steam supply at 5 psig. Some commercial types using steam pressures of 20 to 100 psig are found on auxiliary type ships; these will not be discussed here.

901. PRINCIPLES OF THE SUBMERGED TUBE EVAPORATOR.

In discussion of steam distilling plants, the word *steam* will be used to indicate the steam supplied to the plant for heating purposes. The term *vapor* will be used to indicate the product resulting from heating the brine. The word *condensate* is reserved for describing the condensed steam; *fresh water* or *distillate* is the product of the plant and will be used in describing the condensed *vapor*. That portion of the sea water coming into the plant destined to provide the vapor will be called *feed* before any evaporation takes place. After any portion of it has evaporated, concentrating the remaining salts, it is called *brine*. Sea water used for cooling purposes only will be called *circulating water*.

Basically, the distillation process need consist only of (1) a heat exchanger, where the steam boils the feed water, and which is designed to collect the vapor formed and send it to (2) another heat exchanger, where the vapor is condensed by circulating water. See Figure 9-1.

Such a system has several disadvantages. First, a provision is needed to insure that the steam condenses in the evaporator, giving up all its latent heat of vaporization, in order to make maximum use of its thermal energy. Second, the pressure in the evaporator is uncontrolled as shown. Assuming it to be atmospheric, the feed must be heated up to 212°F before it will boil. This process is wasteful of heat and furthermore, the boiling of sea water at so high a temperature causes hard scale to form on the steam coil in the evaporator which is very difficult to remove, and which greatly impairs the transfer of heat across the tube surfaces. Boiling at a lower pressure, and a lower temperature, can be achieved by placing the evaporator shell under a vacuum, as will be described. Third, considerable heat is given up by the vapor to the circulating water and thus dumped overboard. The heat of condensation is considerable, and this is entirely lost in the system shown in Figure 9-1. This energy could be utilized to heat the incoming feed.

To get an idea of overall performance, assume that saturated steam is supplied at 15 psig to the unit in Figure 9-1; that all of it condenses and is cooled to 212°F; and that the sea temperature is 75°F. Assume further that for each pound of vapor formed, three pounds of feed are supplied, in order to insure a continual disposal of the concentrated salts via the brine going overboard. Then for each pound of vapor formed, three

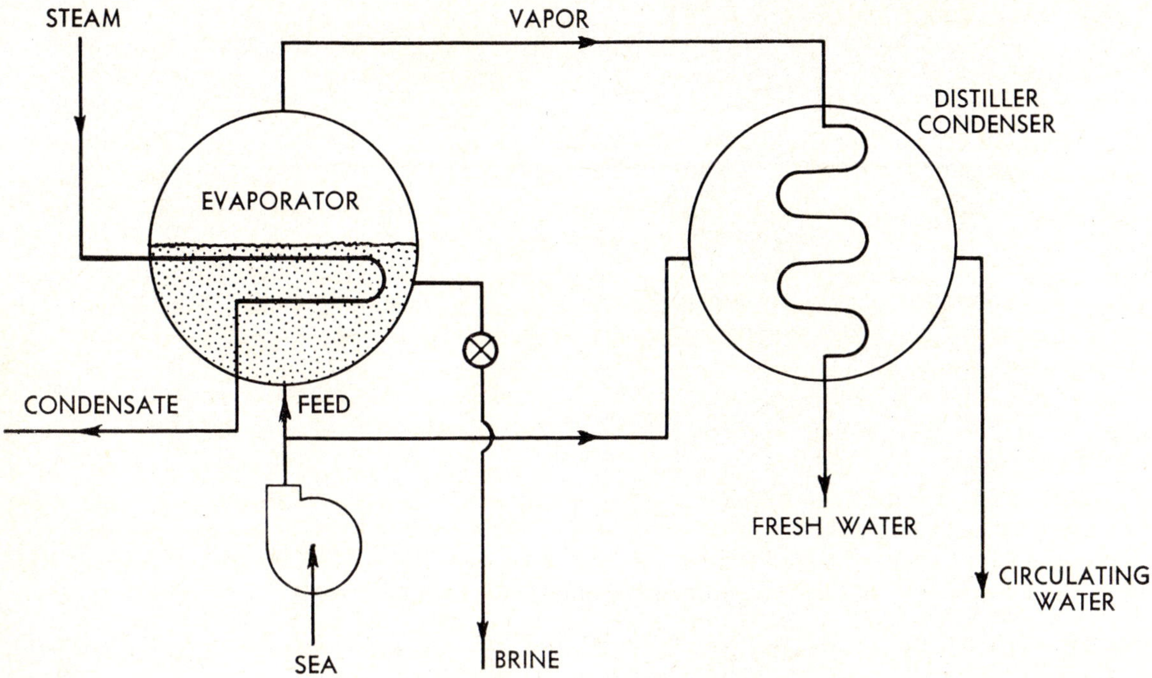

FIG. 9-1. Basic Distillation System.

pounds of sea water must be heated from 75°F to 212°F. If the specific heat of sea water is 0.94, this requires 387 BTU. An additional 970 BTU is required to evaporate one pound of the feed, making a total of 1099 BTU required for each pound of vapor produced. From steam tables, it is seen that the steam gives up 945 BTU per pound in condensing, and that its condensate gives up approximately 1 BTU per pound per degree of cooling, or a total of 38 BTU per pound in cooling from 250°F (its saturation temperature at 15 psig) to 212°F. The total heat supplied by each pound of steam is thus 983 BTU. Therefore, the production of one pound of vapor requires, in this case, the supply of 1099÷983 or 1.12 pounds of steam. Modern plants which recover much of the heat in the vapor can produce 1 pound of fresh water with about 0.4 pounds of steam.

902. THE LOW PRESSURE, TRIPLE EFFECT DISTILLING PLANT.

The utilization of a vacuum makes possible the use of two or more evaporators in a distilling plant, each operating at a different temperature level. Consider the system shown in Figure 9-2. The three evaporator shells are connected by the brine lines. The air ejector takes a suction on the air cooling section of the distiller condenser, creating a vacuum of about 26″ Hg therein. Since this condenser is directly connected to the third effect evaporator shell, the same pressure is maintained therein. After the air has been initially exhausted from the first effect shell and the second effect shell, valves *A* and *B* are closed. Thereafter the pressure in *A* and *B* can be regulated by the brine valves *C* and *D*, if the brine is flowing. Proper regulation of the flow, for instance, can create a vacuum of, say, 22″ Hg in the 2nd effect shell and 15″ Hg in the first effect shell. This permits boiling to occur at a different temperature in each shell, i.e., at the saturation temperature corresponding to the shell pressure. Thus the first effect water boils at 179°F, the second effect boils at 152°F and the third effect at 125°F.

The Steam Circuit. Steam is the source of heat in the first effect. The steam is taken from the 15 psig auxiliary exhaust line, its pressure is reduced to 5 psig (20 psia) through a reducing valve; it then passes through an orifice into the first effect coils where it gives up heat to the brine and is condensed. It then flows through a drain regulator, to the drain pump, which pumps it back to the ship's main condensate system.

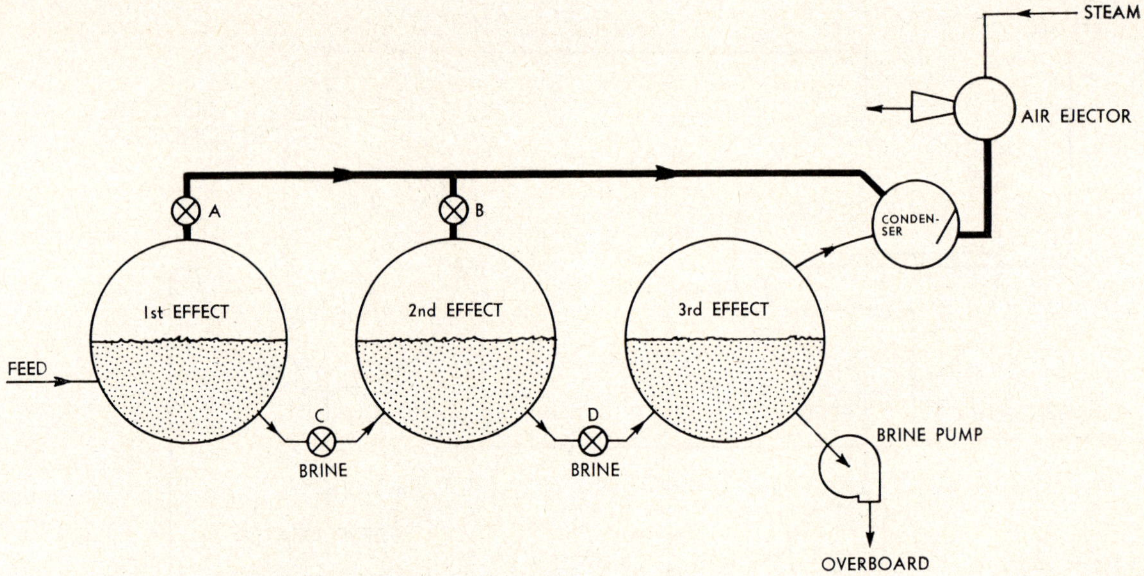

FIG. 9-2. Triple Effect Distilling Plant, Vacuum System.

This system is illustrated schematically in Figure 9-3.

The drain regulator controls the valve connecting the coil with the line to the ship's main condensate system. It opens at a given level of water in the coils, and closes if only steam is present. Thus it insures that all the steam supplied to the first effect shell is condensed. The drain pump runs continuously, keeping pressure on the drain valve. It also recirculates a part of the condensate through a desuperheating connection and nozzle, to keep the supply steam saturated as explained below.

In order to understand the function of the orifice, consider a heat balance expression for the first effect. The heat transferred, Q, through the tube surfaces is

$$Q = UA\Delta T_m \tag{1}$$

where

U = overall heat transfer coefficient, BTU/hr-ft^2-°F
A = area of tubes, ft^2
ΔT_m = mean temperature difference between the two fluids, °F.

This heat must be supplied by the incoming steam due to its change in enthalpy between points 1 and 2, shown in Figure 9-3, or

$$Q = w_s(h_1 - h_2) \tag{2}$$

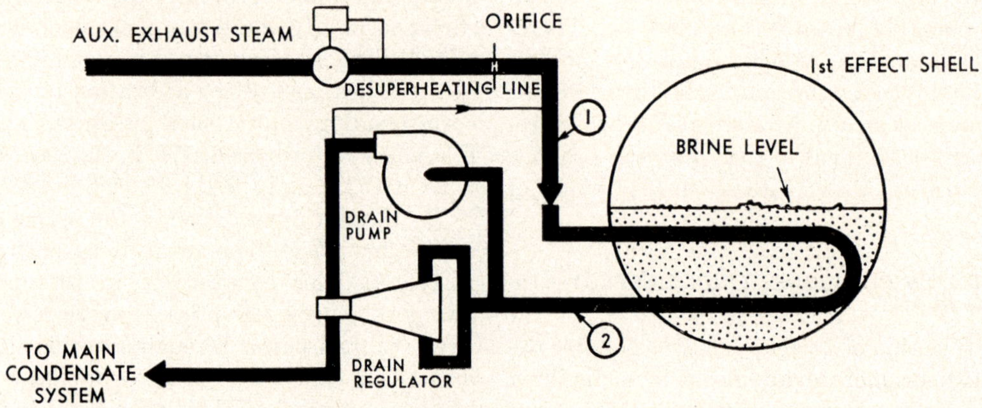

FIG. 9-3. Distilling Plant Steam Circuit.

where

w_s = weight rate of steam flow, lbs/hr
h_1 = enthalpy at section 1, BTU/lb
h_2 = enthalpy at section 2, BTU/lb.

Also, this heat is supplied to the brine, causing it to change from a liquid at enthalpy h_a to a vapor at enthalpy h_b. Therefore

$$Q = w_v(h_a - h_b) \tag{3}$$

where

w_v = weight rate of vapor formation, lbs/hr
h_a = enthalpy of brine, BTU/lb
h_b = enthalpy of vapor, BTU/lb.

Equating (1), (2), and (3),

$$w_s(h_1 - h_2) = UA\Delta T_m = w_v(h_a - h_b). \tag{4}$$

Since steam is forced to condense in the coils by the drain regulator, the quantity $(h_1 - h_2)$ represents the latent heat of condensation. Examination of the steam tables shows that this remains nearly constant (within 3 per cent) at pressures varying from 7 psia to 20 psia. Therefore, if w_s can be kept constant, the three groups of terms in equation (4) all remain constant. Since $(h_a - h_b)$ is a function of the first effect shell pressure, and remains constant, then a constant rate of evaporation is achieved by supplying a constant rate of steam. This is done by installing the orifice which passes a constant rate of steam as long as its discharge pressure is less than 58 per cent of its inlet pressure, 5 psig (20 psia). The coil is designed so that U, A, and ΔT_m provide the necessary heat transfer for the rated w_v when the tubes are clean, at pressures well below this 58 per cent value. As the tubes become scaled, decreasing U, the value of ΔT_m increases, raising the pressure in the coils. Thus, scaling does not decrease the output until such time as scaling causes the coil pressure to rise above about 12 psia.

Expanding the steam through the orifice causes it to become superheated. Since the heat utilized is the latent heat, the desuperheater connection reduces the sensible temperature without causing a loss of latent heat. This permits the coils to operate at a somewhat lower temperature and inhibits the formation of scale.

The Vapor Circuit. The vapor formed in the first effect shell at, say 179°F, is hot enough to cause boiling in the second effect shell, the boiling point of which is 152°, in our example. In fact, it can give up some of its sensible heat to the incoming feed water and still cause boiling in the second effect. It is passed through a series of baffles in the first effect shell, to separate out any droplets of brine which it may tend to carry in suspension, then through a heat exchanger called the first effect vapor feed heater, where it gives up some of its heat to the incoming feed, before flowing into heating coils submerged in the brine in the second effect shell. The brine is at its boiling point, having come from the first effect shell at a higher pressure. The vapor condenses to fresh water in the second effect coils, being forced to do so by a drain regulator, similar to that already described. This causes the second effect brine to boil and produce more vapor. The schematic in Figure 9-4 shows the flow. In exactly the same manner, the vapor from the 2nd effect at 152°F can give up some heat to the incoming feed in the second effect vapor feed heater and still have enough heat to boil the brine in the third stage, the boiling temperature of which is 125°F in this example.

The vapor in the third stage flows through a baffle to the distilling condenser where it gives up heat to incoming feed water *and* circulating water and is condensed. The arrangement is shown in Figure 9-5.

Fresh water system. The above description completes the vapor circuit. We now have distillate, or fresh water, coming from the second and third effect coils (at different pressures) and fresh water coming from the distiller condenser at still a third pressure. Since a saturated liquid at high pressure will partially flash to vapor when its pressure is lowered, these three fresh water drains must be combined in some manner which provides for disposal of the vapor formed. This is done by combining the drains in *flash chambers* from which the vapor is led back to some point suitable for recondensation. For instance, the second effect drains pass into a flash chamber into which the third effect drains flow. This chamber is at the third effect coil pressure. As a portion of the second effect drains flash here, the vapor rejoins the vapor going to the third effect. In a similar manner the drains from the second and third effects are combined with the drains from the distiller condenser. Figure 9-6 shows this part of the circuit. The fresh water passes to a distil-

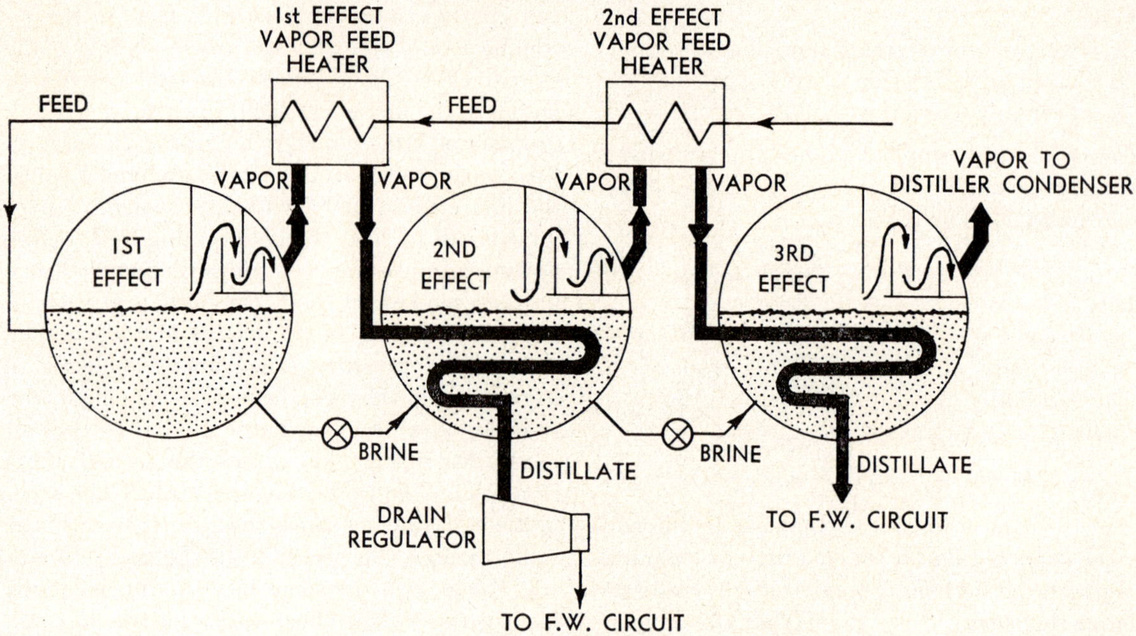

FIG. 9-4. Triple Effect Distilling Plant, Vapor System.

late cooler, giving up heat to the circulating water, and passing to a test tank for a salinity check before finally going to the ship's tanks.

Circulating and Feed Circuit. Salt water originally drawn in from the sea by the circulating water pump first passes through the distillate cooler, where it cools the final fresh water output from the plant. It then passes through the distiller condenser where it condenses the vapor from the third effect. At this point, more circulating water is required for this service than is eventually needed for feed, so a portion of it branches off

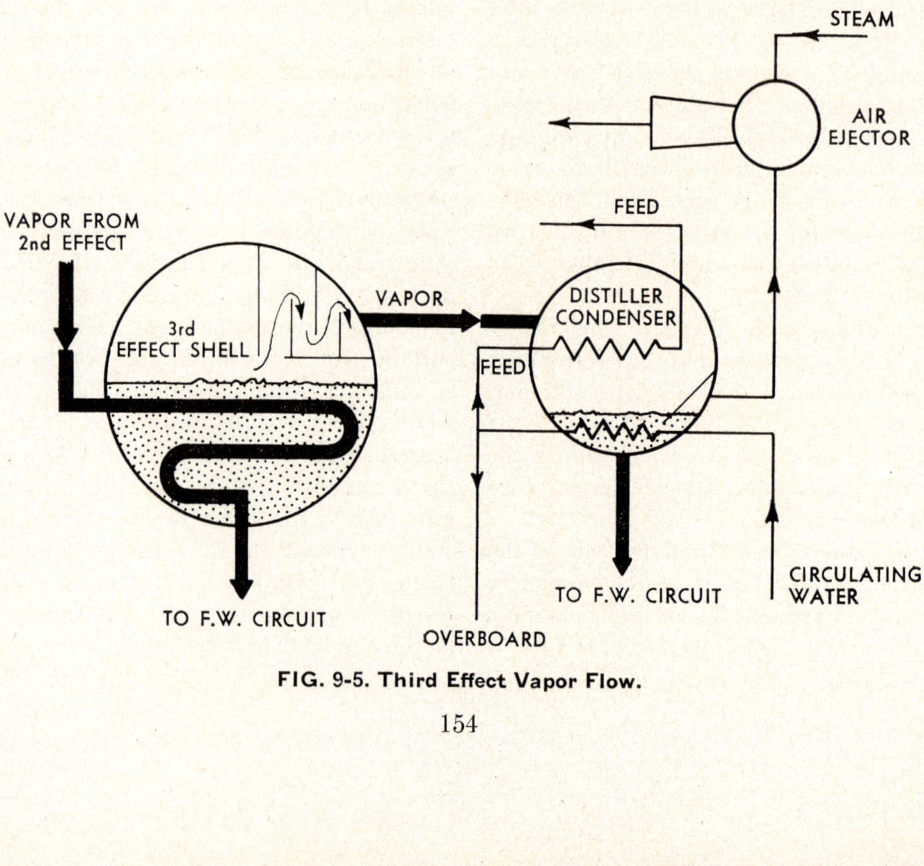

FIG. 9-5. Third Effect Vapor Flow.

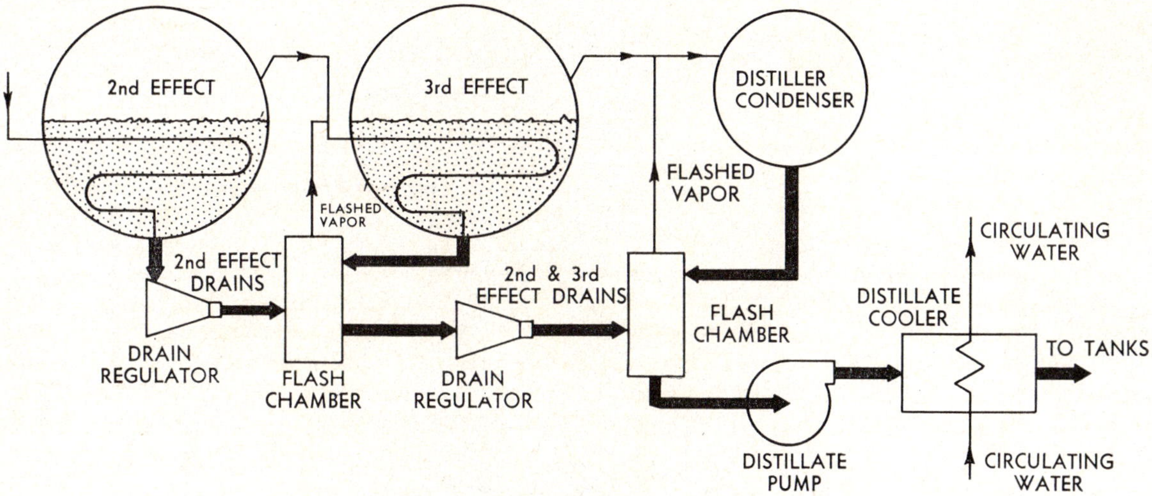

FIG. 9-6. Fresh Water System.

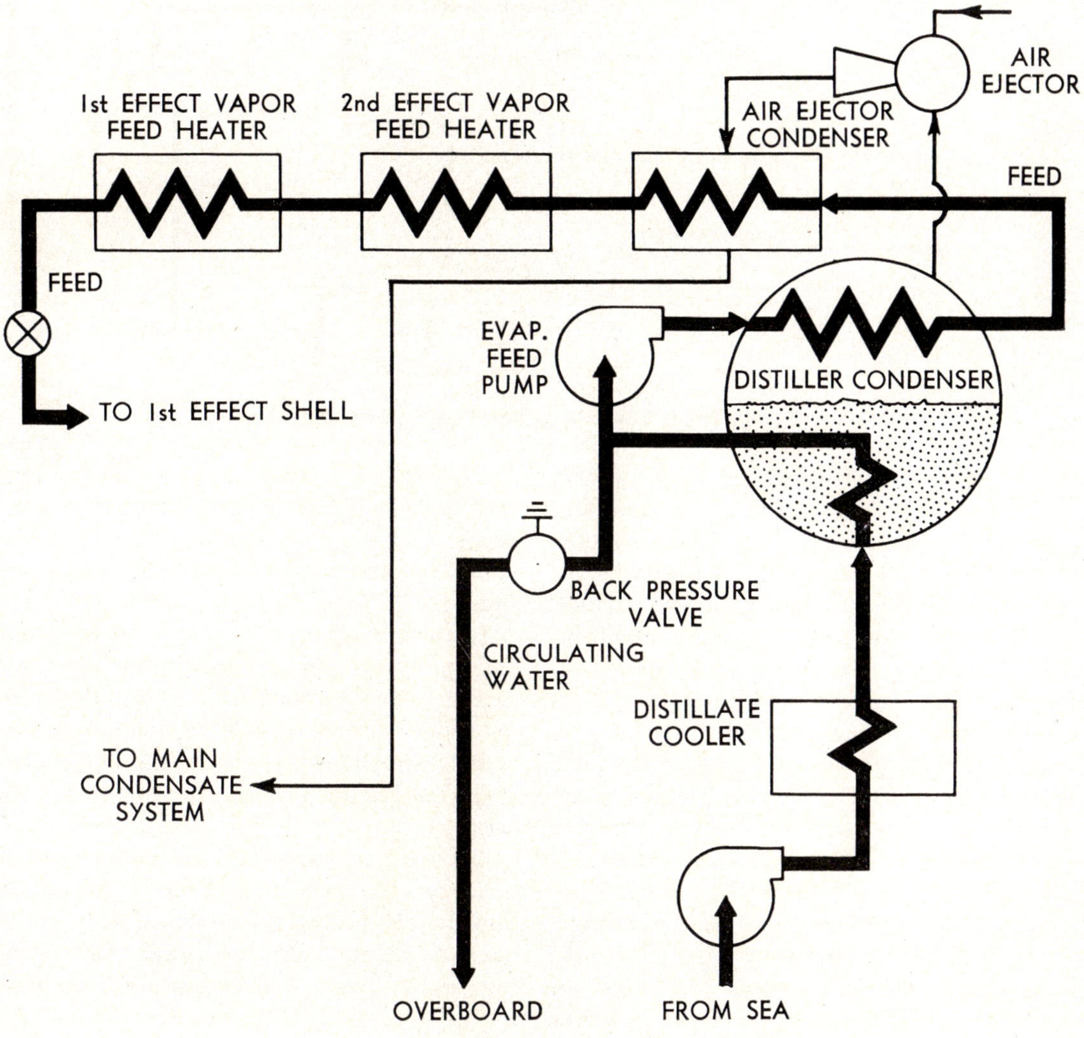

FIG. 9-7. Circulating and Feed Water System.

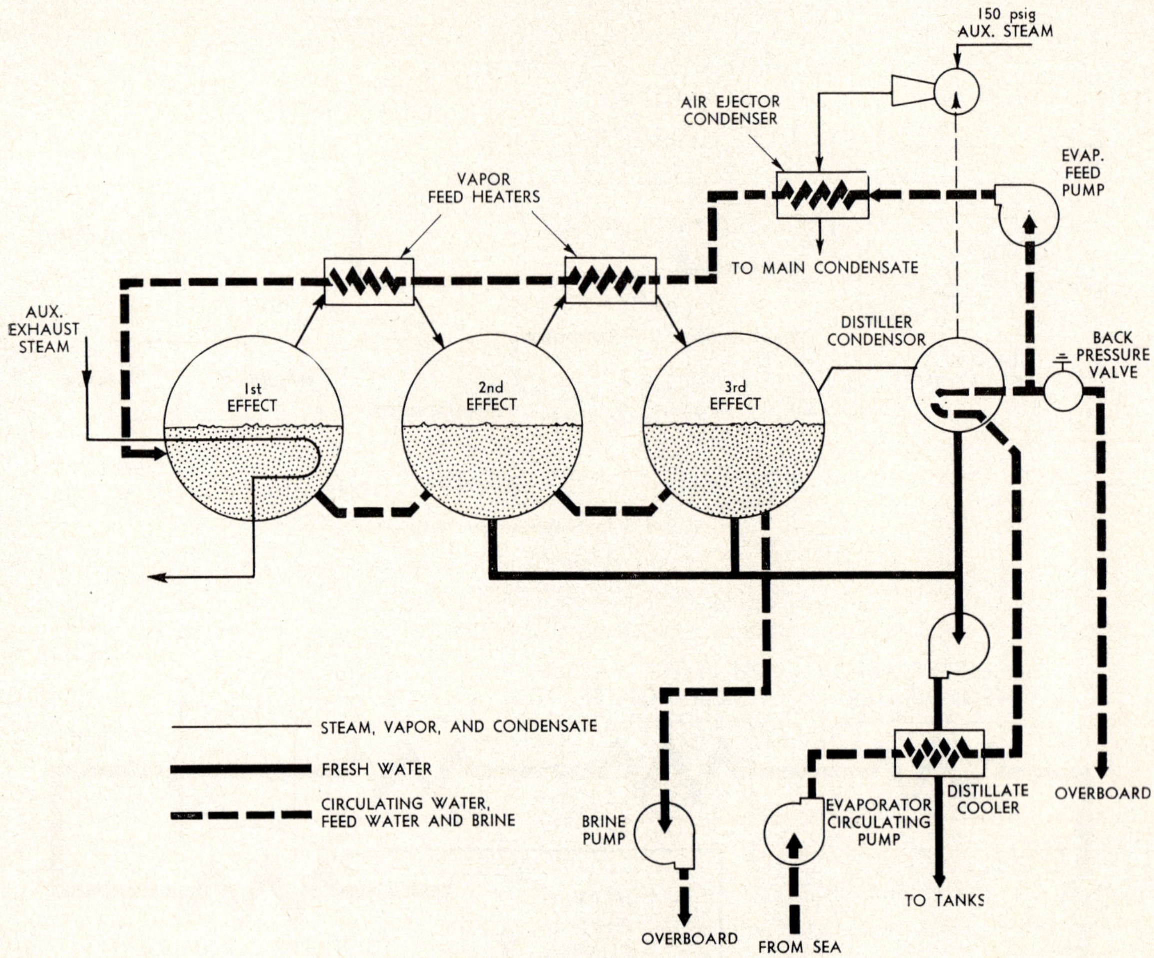

FIG. 9-8. Triple Effect Distilling Plant—Flow Diagram.

and is discharged overboard. See Figure 9-7. A back pressure valve is provided which maintains a constant pressure on the feed pump suction, assuring a constant rate of flow of feed to the plant. The feed, branching off from the circulating water, now makes one more pass through the distiller condenser, then flows to the air ejector condenser, where it picks up heat from exhaust actuating steam of the air ejector, condensing it. (The condensate formed flows back to the main condensate system.) The feed then picks up additional heat in the second effect vapor feed heater and finally in the first effect vapor feed heater before passing into the first effect shell. The heat picked up in these five heat exchangers is sufficient to bring its temperature nearly to the boiling point corresponding to the pressure in the first effect shell, so that the steam supply to the first shell need supply little more than the heat of vaporization.

The Brine Circuit. The brine circuit consists only of connections between the three evaporator shells, and the brine pump and the overboard connection from the last shell, as shown in Figure 9-2.

In the foregoing description, considerable detail was covered to indicate exactly how the various inter-related portions of the plant function. In order to give a better integrated view, a greatly simplified sketch of the whole plant is given in Figure 9-8; several components have been omitted, but the various circuits and major components are shown. Note the two sources of energy to the plant: the auxiliary exhaust steam, and the 150 psig auxiliary steam which actuates the air ejector. At rated capacity, about 13 pounds of auxiliary exhaust steam is supplied for each pound of 150 psig auxiliary steam, so that the major portion of the energy is obtained from the former. Note also that the major sources of heat lost from the plant are via the brine overboard, and the circulating water overboard. Some heat is also carried away to the fresh water tanks, but this is normally not much in excess of that brought into the plant from the sea.

903. DOUBLE EFFECT SUBMERGED TUBE PLANTS.

In the four smaller standard plant sizes (4000; 8000; 10,000; and 12,000 gallon per day) only two effects are used, the lowered efficiency and output being accepted to conserve weight and space in the smaller ships. To provide the maximum space economy, both effects are combined in one shell, in a highly complex and formidable appearing arrangement. The flow is exactly the same as in plant described, except that the second effect is omitted. Figures 9-9 and 9-10 show two views of this arrangement, called a "Soloshell" plant, since it is assembled in a single unit. Occasional 20,000 gallon per day sizes will be found having two separate effects. All larger sizes of the submerged tube type are triple effect.

904. BRINE CONTROL AND SCALE FORMATION.

As indicated before, the formation of scale on the tubes in the evaporators reduces the heat transfer coefficient. The rate of scale deposit is a function of two factors: the temperature of the surface and the density of the brine. Under normal operation, the brine density is controlled to minimize its effect. The scaling, then, is more serious in the first stage, where the tube nest surfaces approach the temperature of the supply steam.

Two provisions are made to keep the tube surfaces free of scale. The first is a tank and the necessary piping to supply a water mixture containing boiler compound and cornstarch to the first effect shell. This mixture is supplied continuously at the rate of about 0.6 per cent of the rated output of the evaporator, i.e., about 0.6 gallons per hour for every 100 gallons per hour of fresh water produced. This mixture causes the solids from the sea water to deposit in a soft sludge rather than a hard scale.

The second provision is a spray head in each shell to permit spraying fresh, cold water on the outside of the tubes, a process called chill shocking. If feed treatment has not been used, the differential expansion, caused by the rapid temperature change, results in cracking and flaking of the scale. If feed treatment has been used, the spray washes sludge off the tube surfaces.

Sea water normally contains about one part in thirty-two (by weight) of solid material. The concentration of solids in the brine from the last shell is controlled so that it does not rise above one and one-half parts in thirty-two. This means that one gallon of fresh water is obtained from each three gallons of feed. A considerably greater quantity of sea water in excess of this is required in order to condense the last effect vapor in the distiller condenser, the excess being discharged overboard after the quantity necessary for feed is taken off.

905. THE FLASH TYPE DISTILLING PLANT.

A recent development in distiller design is the flash type. In this type, *all* of the heat supplied to feed from external sources is supplied while the feed is under pump pressure. No heat is supplied directly to the evaporator shell. The heated feed then passes through successively lower pressure shells, where a portion of it flashes to vapor in each shell. The vapor is condensed and collected in a manner similar to that in the submerged tube type.

The advantage of this type is that no boiling occurs at the tube surfaces, so that very much less scale is deposited in the evaporator. A schematic diagram of a flash type, two stage, distiller is shown in Figure 9-11.

Note that all of the salt water taken in passes through the plant. The brine density at discharge is very much less than in the submerged tube type, about 1.05/32 is a reasonable value. This type plant has been built with up to five effects.

906. THE VAPOR COMPRESSION DISTILLING PLANT.

In smaller ships which do not have an ample supply of steam, the vapor compression plant is used. In this cycle, the energy necessary to evaporate the sea water is initially supplied in the form of mechanical work by compression of the vapor formed. The schematic of Figure 9-12 illustrates the principle of operation. Assuming the salt water to be boiling, the compressor adds energy to the vapor sufficient to cause the boiling to continue from the heat given off by the compressed vapor coils. This condenses the vapor in the coils which is then sent to the feed heater in the form of distillate, or fresh water, where it is further

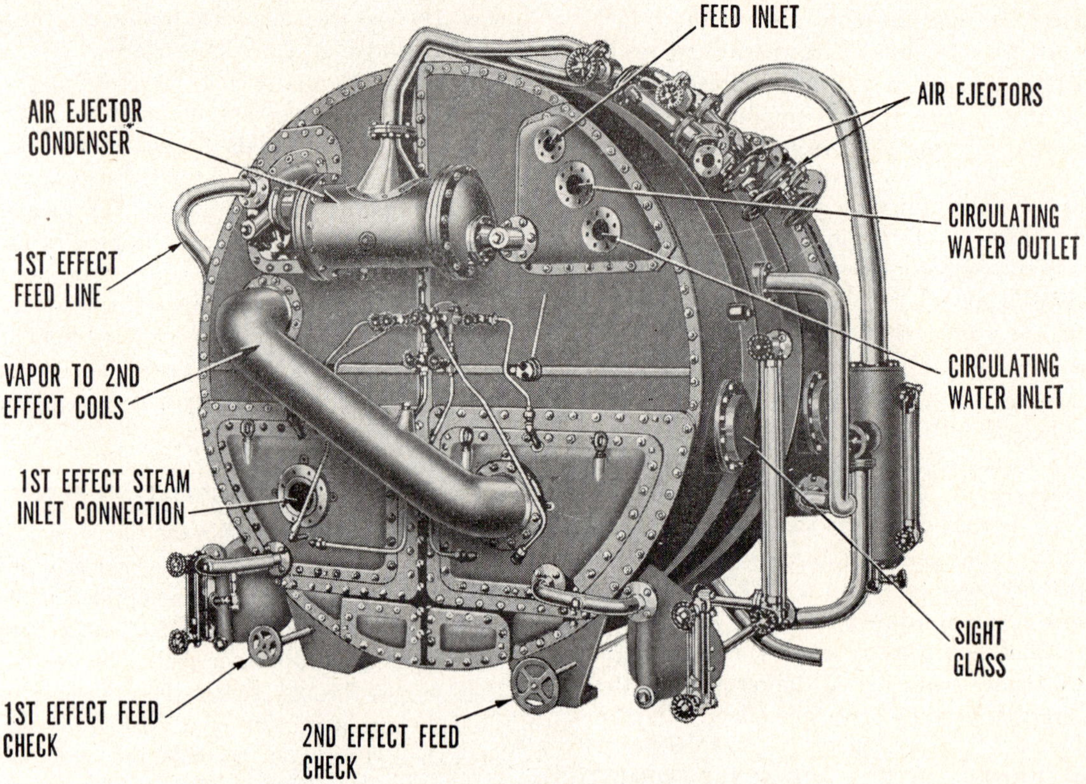

FIG. 9-9. Double Effect "Soloshell" Distilling Plant.

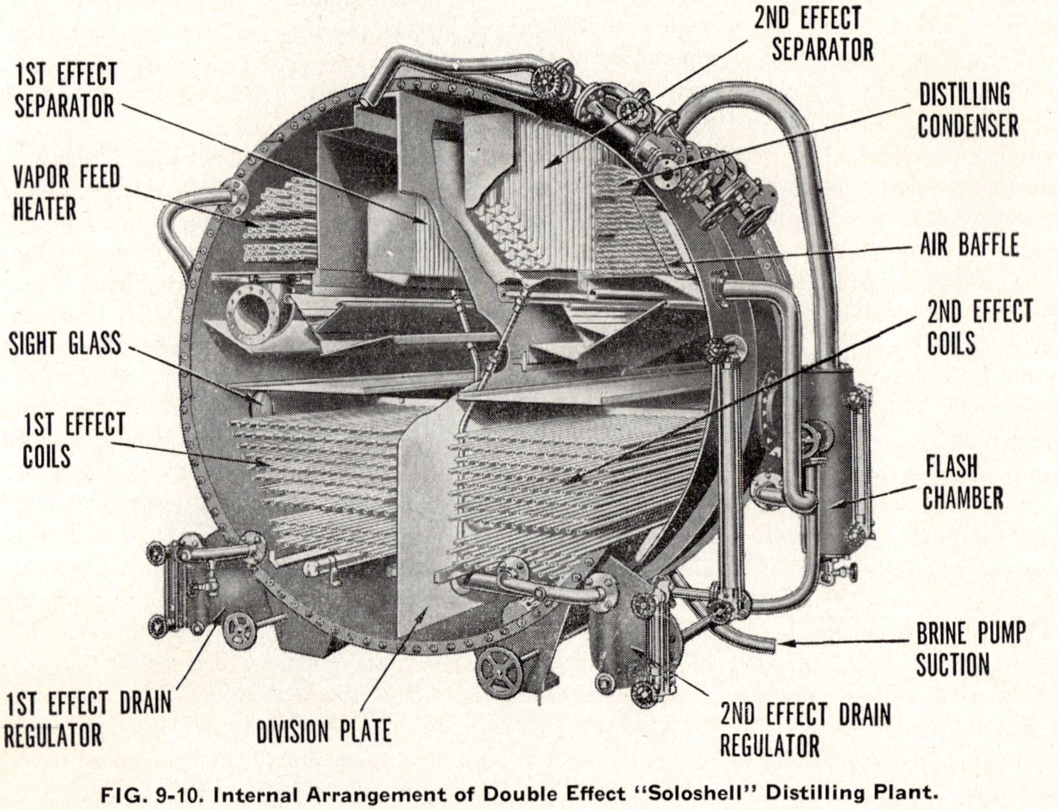

FIG. 9-10. Internal Arrangement of Double Effect "Soloshell" Distilling Plant.

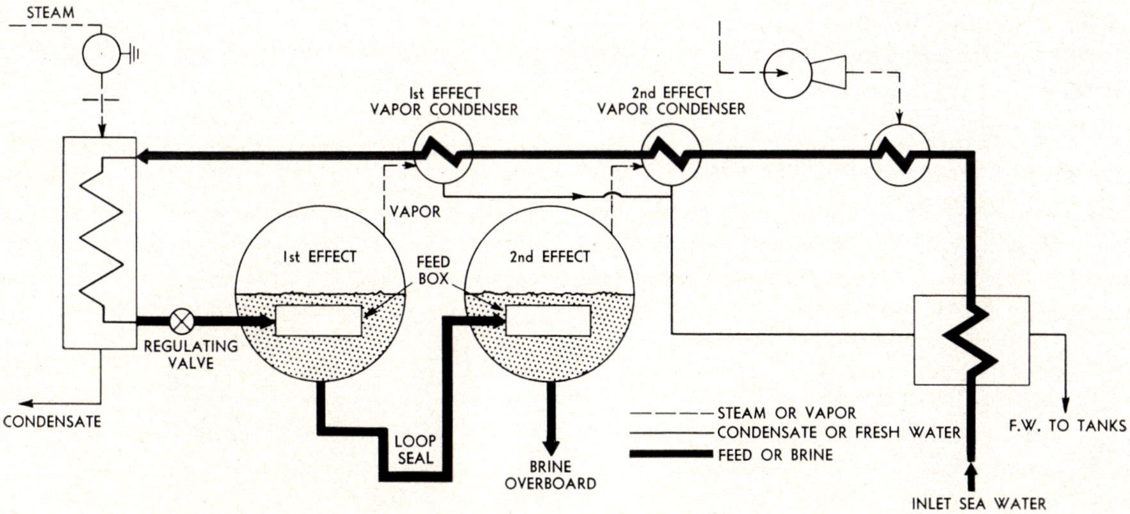

FIG. 9-11. Flash Type Distilling Plant, Flow Diagram.

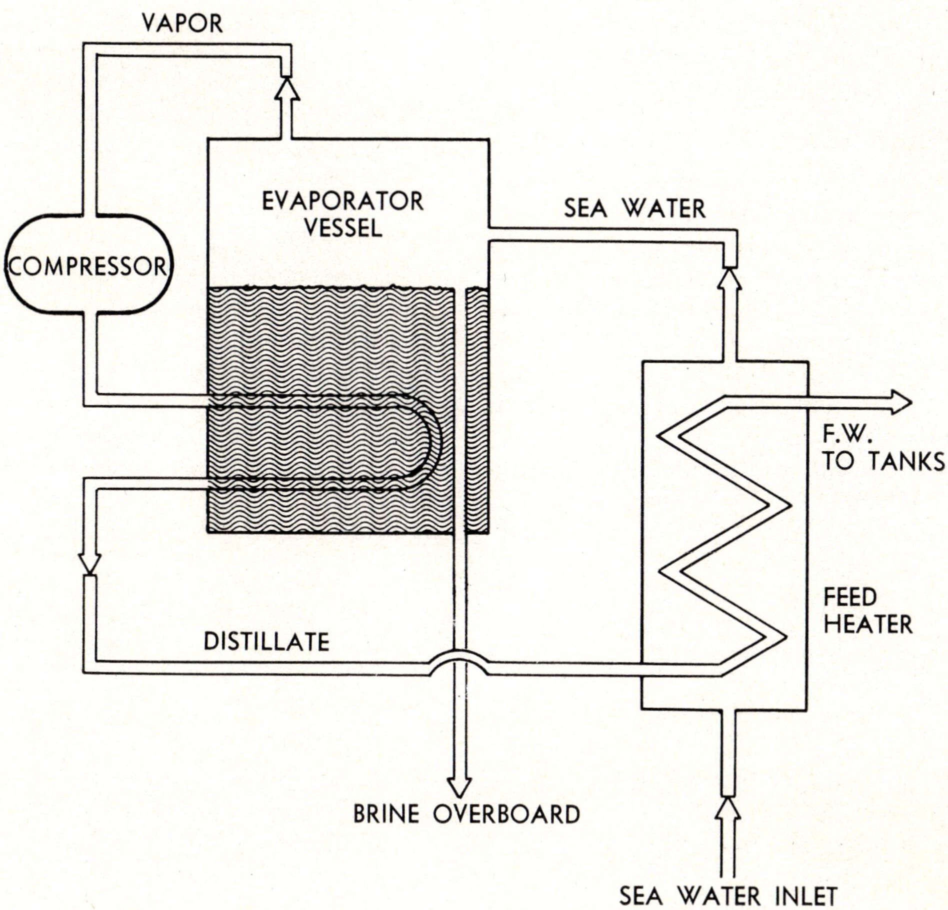

FIG. 9-12. Schematic of Vapor Compression Distilling Plant.

cooled in heating the incoming feed. The unit is started by initially boiling the brine with electric resistance heaters, which are secured after the compressor can be started.

907. SALINITY INDICATING SYSTEM.

A salinity indicating system is used to determine the purity of the fresh water produced in the distilling plant, and to monitor the steam condensate system in the distilling plant to guard against salt contamination of the drains which are returned directly to the main feed system.

The system is based on the fact that electrical conductivity of water varies in proportion to its salt content. Two electrodes immersed in the water with a voltage impressed across them will pass a current proportional to the salt content. The electrodes are manufactured in cells with a threaded connection which are inserted into the various lines or units at the desired points. These cells are connected to a meter panel where the meters are scaled to read the salinity directly in equivalent parts of chloride ion per million parts of water. (One gram equivalent weight of chlorine equal approximately 35 grams.)

908. BACTERIOLOGICAL PURITY OF WATER.

Pure water vapor is incapable of carrying bacteria or solid matter. However, the vapor can carry entrained water, and even very small particles of entrained water can carry bacteria and dissolved salts. If the separation of minute liquid particles from the vapor were complete in the evaporator, the distilled water would be sterile and chemically pure. Complete separation is nearly impossible to achieve.

From a standpoint of salt content, a concentration of a few hundred equivalent parts per million (e.p.m.) can be acceptable for human consumption. However, such a salt concentration would indicate a water carry-over which is dangerous from a bacteriological standpoint. Furthermore, boiler feed water should contain a low chloride ion content, 15 e.p.m. being the maximum allowable concentration. The United States Public Health Service and the Bureau of Medicine and Surgery have conducted tests which indicate that disease carrying bacteria cannot be found in water from distilling plants having a chloride ion concentration of less than 0.065 e.p.m., even when the feed water was purposely contaminated. This represents, then, the upper limit in salinity concentration allowable when operating the distilling plant with sea water. However, when operating with brackish or fresh water in coastal areas and harbors, this low salinity reading will not assure bacteriological purity, since the salt concentration in the feed is low, and a low salinity output reading is not necessarily indicative of low carry-over. If necessary to operate the distilling plant in such an environment, the temperature in the first effect shell must be maintained at not less than 165°F at all times to insure sterilization of the feed. This is done by either choking the brine flow, or raising the inlet steam pressure above 5 psig by a change in the setting of the weight loaded inlet steam valve.

909. SUMMARY.

The submerged tube low pressure type, the flash type and the vapor compressor type evaporators are the principal types in use on naval ships. The submerged tube was discussed in some detail to illustrate how each necessary process is carried out. Only broad principles were discussed in the latter two. Details of their operation are readily available in the manufacturer's instruction books or in the Bureau of Ships Manual.

CHAPTER X
NUCLEAR POWER

1000. GENERAL.

In the generation system of the main steam cycle, energy is normally introduced into the cycle by the controlled chemical combustion of conventional fuels. (See *Naval Boilers* by Latham.) In the past few years, the controlled release of energy from nuclear fission has been attained. The fuel quantity necessary for this process is minute compared to that necessary from conventional fuels for equal quantities of power delivered. Thus a large energy source is contained in a comparatively small volume. One of the main problems associated with nuclear power is the effective removal of large quantities of thermal energy from its source while maintaining temperatures which are not beyond the physical limits of available engineering materials, when these materials are subjected to the peculiar environment which exists in the presence of nuclear reactions.

The release of nuclear energy and its transformation to thermal energy takes place in a *nuclear reactor*. The design and construction of reactors is in a stage of rapid development at present. Many different types of reactors have been proposed and designed, and a large number are under construction. Of the several different types presently being invested with heavy research effort, doubtless a few will emerge with a clear-cut superiority, and others will prove to be burdened with limitations and undesirable characteristics impracticable or impossible of solution. Therefore in considering this field which is changing so rapidly, and which is leading in so many different directions at present, only certain basic principles will be discussed, together with a few of the presently more promising arrangements.

1001. ADVANTAGES.

Among the many advantages of using a nuclear reactor as an energy source aboard ships or submarines are:

a. A small amount of fuel is needed for operation of the reactor, which can release a tremendous amount of energy.

b. A nuclear plant is relatively independent of any material external to the system, i.e., it does not require a continuous supply of fuel or air.

c. The rate of consumption of nuclear fuel is so small that a change in displacement of the ship due to a change in fuel weight is negligible.

d. Hull arrangements can be simplified by elimination of large fuel storage spaces, smoke pipes and uptakes.

Many problems are also created by the use of nuclear fuel. These will be pointed out in the following discussion, and summarized later, when their causes can be better understood.

1002. UNITS.

In the discussion which follows, use will be made of certain units which have become conventional in nuclear engineering. Since the science of nucleonics was born in the physicist's laboratory, it is natural that scientific units became conventional rather than the units more familiar to the engineer.

Force: Dyne—The force acting on one gram necessary to give it an acceleration of 1 cm./sec.^{2}.

Energy: Erg—The energy necessary to exert a force of one dyne through a distance of 1 cm.

Electron volt (ev.)—The energy acquired by a particle carrying a unit charge (as possessed by one electron) as it moves through a potential difference of one volt.

Since these two units are both units of energy, they are mutually convertible. The ev. is so small that it is conventional to use the Mev. (million electron-volts). The relationship is

$$1 \text{ ev.} = 1.6 \times 10^{-12} \text{ ergs}$$

or

$$1 \text{ Mev.} = 1.6 \times 10^{-6} \text{ ergs.}$$

In conventional engineering units,

$$1 \text{ Mev.} = 1.52 \times 10^{-16} \text{ BTU.}$$

Mass: Atomic Mass Unit (a.m.u.)—1/16 of the mass of one oxygen atom.

$$1 \text{ a.m.u.} = 1.66 \times 10^{-24} \text{ gms.}$$

Conversion of mass to energy: According to

Einstein's Law, when mass is converted to energy, the following relation holds:

$$E = mc^2$$

where

E = energy in ergs, when
m = mass in grams,
c = velocity of light in cm./sec. (3×10^{16}).

From this relationship, and the above conversions, it can be shown that 1 a.m.u. is equivalent to 931 Mev. or 1.412×10^{-13} BTU.

1003. THE SOURCE OF NUCLEAR ENERGY.

The nuclei of all elements are made up of two relatively stable kinds of basic particles of matter, the *neutron* and the *proton.* The neutron bears no electrical charge, and alone, has a mass of 1.00898 a.m.u. The proton carries a positive unit electrical charge and, alone, has a mass of 1.00759 a.m.u. These particles are often called *nucleons.* When neutrons and protons are formed into an atomic nucleus, the total nuclear mass is found to be somewhat less than the simple addition of the masses of the individual nucleons would indicate. For simplicity, it may be assumed that somehow when the nucleus was formed, a quantity of mass was converted into energy according to Einstein's relationship. Since the same amount of energy would be needed to break the atom into its component parts, this loss of mass, or *mass defect,* is equivalent to the *binding energy.*

In nuclear physics, it is conventional to write the symbol of an element with a preceding subscript and a following superscript. The subscript represents the *atomic number* which is equal to the number of orbital electrons in the atom, and also to the number of protons in the nucleus. The superscript represents the *mass number,* or the total number of nucleons in the atomic nucleus. Thus ${}_2He^4$ is the symbol for helium, denoting a nucleus made up of four nucleons, of which two are protons (and therefore, two neutrons). In a similar manner, the symbol for iron, ${}_{26}Fe^{56}$ reveals that its nucleus has 56 nucleons, of which 26 are protons and 30 are neutrons. It will be borne in mind that the atomic mass of an atom includes also the mass of the orbital electrons, equal in number to the nuclear protons and having a mass of 0.00055 a.m.u. each.

Now, let us investigate the binding energy of three elements, lithium, iron and natural uranium, representing light, intermediate and heavy nuclei.

Lithium. ${}_3Li^7$ is made up of 3 protons, 4 neutrons, and 3 orbital electrons. The sum of the masses of its components is

$$\begin{aligned} 3 \times 1.00759 &= 3.02271 \\ 4 \times 1.00898 &= 4.03592 \\ 3 \times 0.00055 &= 0.00165 \\ \hline &\quad 7.06028 \text{ a.m.u.} \end{aligned}$$

The actual mass of the lithium atom, by experiment, is 7.01818 a.m.u. The mass defect is

$$7.06028 - 7.01818 = 0.04210 \text{ a.m.u.}$$

Converting to the equivalent energy, the binding energy is obtained. This is

$$0.04210 \times 931 \text{ Mev./a.m.u.} = 39.3 \text{ Mev.}$$

Since there are seven nucleons, the average binding energy per nucleon is

$$39.3 \div 7 = 5.61 \text{ Mev./nucleon}$$

Iron. ${}_{26}Fe^{56}$ has 26 protons and 30 neutrons in the nucleus, and 26 orbital electrons. The sum of the individual parts is

$$\begin{aligned} 26 \times 1.00759 &= 26.19734 \\ 30 \times 1.00898 &= 30.26940 \\ 26 \times 0.00055 &= 0.01430 \\ \hline &\quad 56.48104 \text{ a.m.u.} \end{aligned}$$

The actual mass of the atom by measurement is 55.95272 a.m.u. The mass defect is

$$56.48104 - 55.95272 = 0.52832 \text{ a.m.u.}$$

The binding energy is

$$0.52832 \times 931 \text{ Mev./a.m.u.} = 492.0 \text{ Mev.}$$

And the binding energy per nucleon is

$$492.0 \div 56 = 8.79 \text{ Mev./nucleon}$$

Uranium. ${}_{92}U^{232}$ has 92 protons, 146 neutrons and 92 electrons. The sum of its component masses is

$$\begin{aligned} 92 \times 1.00759 &= 92.69828 \\ 146 \times 1.00898 &= 147.31108 \\ 92 \times 0.00055 &= 0.05060 \\ \hline &\quad 240.05996 \text{ a.m.u.} \end{aligned}$$

Actual mass = 238.12340

Mass defect = 1.93656 a.m.u.

Binding energy = 1.93656 × 931 = 1803 Mev.

Binding energy/nucleon = 1803 ÷ 238 = 7.58 Mev.

From the above, it might be inferred that it is necessary to *supply* energy to break an atom into its ultimate component parts. This is the fact. However, a little further thought shows that the variation of the *binding energy per nucleon* holds the key to *obtaining* energy by nuclear processes.

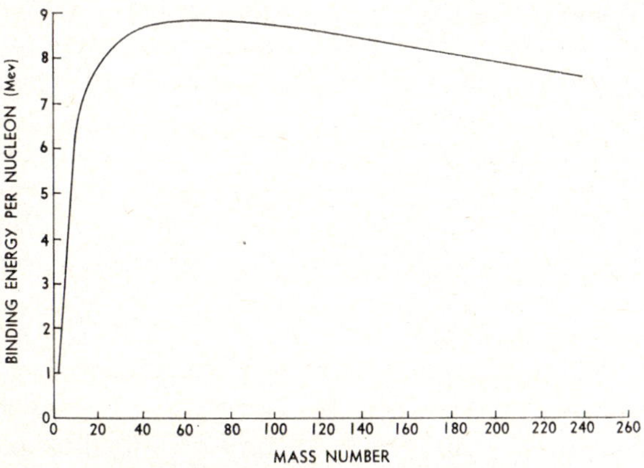

FIG. 10-1. Variation of Binding Energy per Nucleon with Mass Number.

In the three examples given above for lithium, iron and uranium, it is noted that the binding energy per nucleon is respectively 5.61, 8.79 and 7.58 Mev. This means that more energy per nucleon accompanies the formation of nucleons into an iron atom than into either a lithium (light) atom or a uranium (heavy) atom. If the binding energies of all naturally occurring nuclei are plotted against mass number (number of nucleons per nucleus), the graph shows a steady increase in energy with mass number until the intermediate elements are reached, and then a gradual falling off. See Figure 10-1.

The variation in the energy per nucleon suggests that energy will result when two light nuclei fuse into an intermediate nucleus, or when a heavy nucleus splits into two or more light nuclei. Practical processes have been found for the latter case which can be so controlled as to produce useful energy. The only processes presently known for the former case cannot be controlled for production of useful industrial energy; however, they can be used in weapons.

1004. THE PRODUCTION OF ENERGY IN THE FISSION PROCESS.

The highly publicized and well known uranium fission process consists of the peculiar ability of the uranium-235 atom, consisting of protons and neutrons, to absorb another neutron and become an unstable compound nucleus which splits or fissions into two smaller parts. The sum of the mass of the parts is less than that of the original nucleus plus the captured neutron, the difference, or mass-defect, being converted into energy in accordance with the Einstein relationship. A detailed analysis of the compound nucleus shows that it may fission in any one of 40 different ways; in other words, there are 80 primary fission products. These range from an isotope of zinc (mass number 72) up to probably an isotope of gadolinium (mass number 160). Most of the fission products fall into two broad groups: a so-called light group with mass numbers from 80 to 110, and a heavy group with mass numbers from 125 to 155. The most probable type of fission, occurring 6.4 times in every 100 fissions, is that in which the products are elements with mass numbers 95 and 139 respectively. Two other possible fuels, uranium-233 and plutonium-239, form about the same fission products.

In the case of fission of U-235 by a neutron, the following process is typical.

$$_{92}U^{235} + _0n^1 \rightarrow (_{92}U^{236})^* \rightarrow _{53}I^{139} + _{39}T^{95} + 2_0n^1$$

The immediate fission products, iodine (I) and yttrium (Y), are radioactive, and decay further as indicated in the two following equations. An electron (beta particle) and radiation in the form of gamma rays (highly penetrating wave energy with no mass and no charge) accompanies each step of the decay. The numbers below the arrow indicate half-lives in seconds, minutes or days.

$$_{53}I^{139} \xrightarrow[2.7\text{ s}]{} {_{54}Xe^{139}} \xrightarrow[41\text{ s}]{} {_{55}Cs^{139}} \xrightarrow[9.5\text{ m}]{} {_{56}Ba^{139}}$$

$$\xrightarrow[85\text{ m}]{} {_{57}La^{139}} \text{ (stable)}$$

$$_{39}Y^{95} \xrightarrow[10\text{ m}]{} {_{40}Zr^{95}} \xrightarrow[65\text{ d}]{} {_{41}Nb^{95}} \xrightarrow[35\text{ d}]{} {_{42}Mo^{95}} \text{ (stable)}$$

The final result, then, of the fission of the compound nucleus, in this example, is

$$(_{92}U^{236})^* \rightarrow {_{42}Mo^{95}} + {_{57}La^{139}} + 2_0n^1$$
$$+7 \text{ Beta particles} + \text{Energy.}$$

The two neutrons produced are available, under favorable conditions, to produce fission in other uranium atoms and so to keep the process sustained. The energy appears as kinetic energy of the large fission fragments and of the neutrons; and as radiation energy, mostly in the form of gamma rays.

The total amount of energy released can be computed by a mass balance of the last equation, as follows:

Mass Before Fission		Mass After Fission	
${}_{92}U^{235}$	235.124	${}_{42}Mo^{95}$	94.945
${}_{0}n^{1}$	1.009	${}_{57}La^{139}$	138.955
		$2{}_{0}n^{1}$	2.018
	236.133 a.m.u.		235.918 a.m.u.

The mass defect is the difference between the totals, or 0.215 a.m.u. The energy is then

$$E = (0.215 \text{ a.m.u.})(931 \text{ Mev./a.m.u.})$$
$$= 198 \text{ Mev./fission.}$$

It is recalled that Avagadro's number, 6.02×10^{23}, is the number of atoms per gram molecular weight, or in this case, per 235 grams (approximately) of U^{235}. Since there are 453.6 grams per pound, one pound of uranium contains approximately 11.6×10^{23} atoms. If every atom in one pound of uranium fissioned, the energy produced would be

$$11.6 \times 10^{23} \times 198 \text{ Mev.} = 2.3 \times 10^{26} \text{ Mev.}$$

or

$$3.48 \times 10^{10} \text{ BTU.}$$

At 17,400 BTU per pound of fuel oil, this is the equivalent of 2,000,000 pounds, or 1000 tons, of fuel oil.

Over 80% of the energy indicated above appears immediately as kinetic energy of the fission fragments. As these high velocity fragments collide with adjacent atoms and molecules, they are slowed down, while increasing the molecular velocities of the particles surrounding them, and the energy is thus manifested as sensible heat. Some of the energy remaining appears instantly in the form of gamma rays and some of it is in the form of kinetic energy of the neutrons and other small fission fragments born in the fission process. The final portion is released gradually as kinetic energy of the beta particles produced by the decaying fission products, and as gamma radiation accompanying this decay.

1005. NEUTRON PROCESSES AND CROSS SECTIONS.

As a moving neutron travels about in interatomic space, it may interact in certain modes with proximate atoms. First, it may collide with the atom and rebound. This action is an elastic collision, named descriptively by the term "scattering," since the result is a "scatter" of the neutron. Second, it may enter the nucleus of the atom. If it enters the nucleus, one of three things may happen: it may be re-ejected at a different energy level, this process being one form of inelastic scattering; it may remain in the nucleus or be captured; or it may cause the nucleus to break into fragments (or fission). Any process in which the neutron remains in the target nucleus is called absorption, whether it results in capture or in fission. For a given energy, or speed, of a neutron there is a definite, measurable probability that the above processes will occur which varies with the nature of the target atom. The neutron may be thought of as a projectile of a finite size; and a measure of the probability of one of the above processes occurring for a given target atom may be thought of as target area presented to the projectile. Thus, if the target area is large for a given process, there is a good chance that it will occur; if small, there is a slight chance that it will occur. This area is called a "cross-section" whose symbol is σ, and is measured in a unit called a "barn," one barn being an area of 10^{-24} cm^2. An element may have four different cross sections for neutron processes, as follows:

Scattering cross section—measure of the probability of rebound of neutron after collision.

Capture cross section—measure of the probability of neutron capture, without causing fission.

Fission cross section—measure of the probability of fission of the atom after capture of a neutron.

Absorption cross section—sum of the capture cross section and the fission cross section.

It is important to note that these cross sections are not constant for the atoms of any given element, but may vary with the speed of the approaching neutron. For instance, Figure 10-2 shows a plot of the variation of the absorption cross section of U^{235} with the energy of the incident neutron. This gives rise to the general classification of neutrons according to their speed, or kinetic energy, as follows:

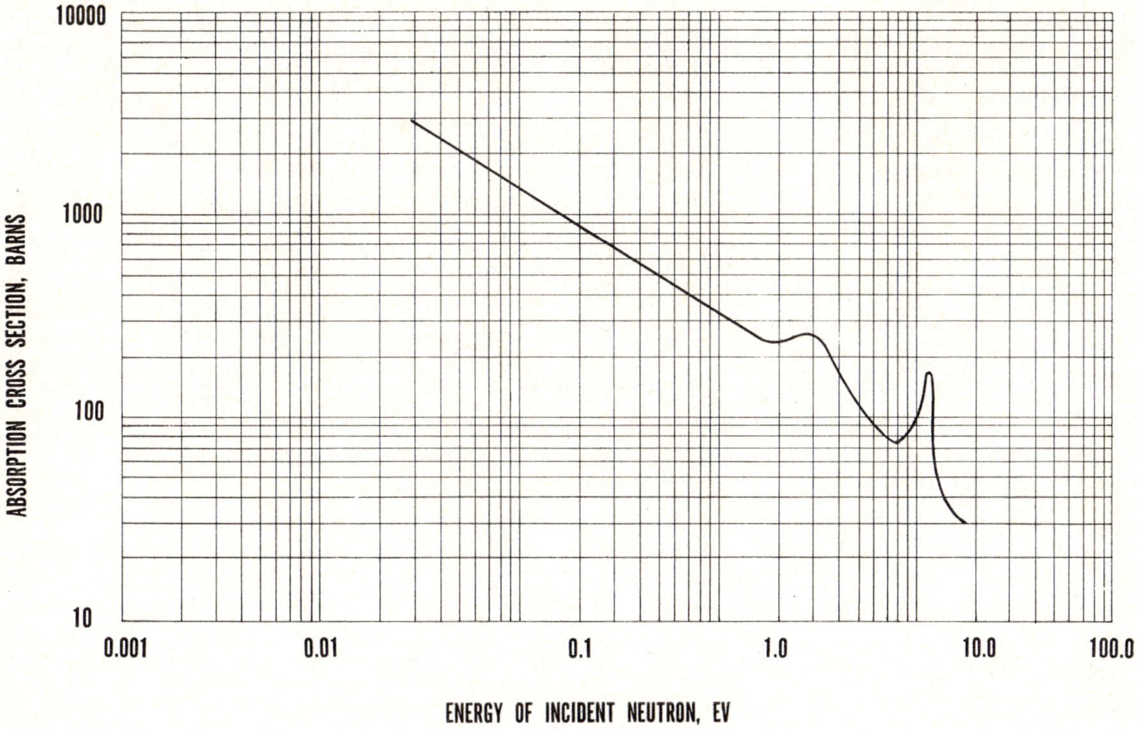

FIG. 10-2. Variation of Absorption Cross Section of U-235 with Neutron Energy.

Thermal neutron—possesses an energy of below 1 ev., comparable to that of thermally agitated atoms or molecules.

Intermediate neutron—possesses an energy between 1 ev. and 10^5 ev.

Fast neutron—possesses an energy above 0.1 Mev. or 10^5 ev. The probability of neutron absorption occurring for fast neutrons is usually very slight—they do not remain in the vicinity of a target atom for a long enough time to be affected by the nuclear forces involved. Thus, to increase the probability of absorption and subsequent fission, the neutrons must be slowed down.

It has been stated that the cross section measured in barns is a measure of the probability of a process. A barn is a measure of area. This area is called the *microscopic* cross section. However, it is used to compute the true probability of the process by multiplying it by the number of atoms per unit volume. This product is called the *macroscopic* cross section and is designated by the symbol Σ. Thus

$$\Sigma = N\sigma = (N \text{ atoms/cm}^3)(\sigma \text{ cm}^2\text{/atom})$$
$$= N\sigma \text{ cm}^{-1}.$$

This is interpreted as the probability that a reaction will occur in 1 cm. of neutron travel, or its reciprocal is the average distance the neutron will travel in the target mass before the reaction will occur.

Example: Find the probability of capture of a thermal neutron by water. For thermal neutrons, hydrogen has a microscopic capture cross section of 0.33 barns and oxygen has a microscopic capture cross section of 0.0002 barns.

Solution: The capture cross section for water is the sum of the cross sections of each atom in one water molecule. Thus

$$\sigma_c = [(2)(0.33) + (0.0002)] \times 10^{-24}\text{cm}^2$$
$$= .6602 \times 10^{-24}.$$

One mole of any substance contains 6.02×10^{23} atoms, or molecules. One mole of water is 18 grams, and each gram occupies 1 cm^3. Therefore the number of molecules of water in one cubic centimeter is

$$\frac{6.02 \times 10^{23}}{18} = .334 \times 10^{23}.$$

Therefore,

$$\Sigma_c = .6602 \times .0334 = 0.022 \text{ cm}^{-1}.$$

Or, the average distance a neutron will travel prior to capture is 1/.022 or about 50 cm.

1006. NUCLEAR FUELS.

There are good theoretical reasons which indicate the nuclei which most readily undergo fission by thermal neutrons are those of high mass number containing an odd number of neutrons and an even number of protons. Natural uranium consists of 0.714% of U-235 and 99.286% of U-238. Uranium-235 has 92 protons and 143 neutrons. The symbol ${}_{92}U^{235}$ indicates this, 92 being the atomic number and 235 being the mass number (neutrons plus protons). U-235 is then called an even-odd nucleus and is the only naturally occurring fuel with such a nucleus to warrant any practical interest. However, there are two other artificially prepared nuclear fuels, which have even-odd nuclei, and which are of practical interest. These are ${}_{94}Pu^{239}$ and ${}_{92}U^{233}$. They are prepared from naturally occurring ${}_{92}U^{238}$ and ${}_{90}Th^{232}$ respectively, by bombarding with neutrons as follows:

$${}_{92}U^{238} + {}_{0}n^{1} \rightarrow {}_{92}U^{239} \xrightarrow[23.5\text{ m}]{} {}_{93}Np^{239} \xrightarrow[2.33\text{ d}]{} {}_{94}Pu^{239}$$

$${}_{90}Th^{232} + {}_{0}n^{1} \rightarrow {}_{90}Th^{233} \xrightarrow[23.3\text{ m}]{} {}_{91}Pa^{233} \xrightarrow[27.4\text{ d}]{} {}_{92}U^{233}$$

It appears, from consideration of the above reactions, that placing natural uranium-238 or thorium-232 in an operating reactor, where neutrons are available, will give a continuous generation of fuel. This is done in a breeder reactor.

It is possible to fission pure U-238 and Th-232 directly. However, their cross-sections are such that neutrons of 1 Mev. or more are necessary to cause fission. A large proportion of the neutrons produced as a result of fission have such energy, but as a result of non-capture collisions with other nuclei, their energies are quickly reduced to values below the fission threshold values. For this reason, a self-perpetuating chain reaction using these fuels alone *will not occur*. However, the other three fuels mentioned, U-235, U-233 and Pu-239, have very large fission cross sections for slow neutrons—this cross section continuously increases as neutron energy decreases.

1007. MODERATORS.

From the previous discussion, it is seen that it is necessary to slow down or "thermalize" high energy neutrons to increase the probability of absorption and fission by the fuel nuclei, before the escape of the neutrons from the fuel mass can occur. The slowing down of these high energy neutrons to a thermal energy of about 0.025 ev. is due almost entirely to scattering by elastic collisions of the neutrons with nuclei of a substance added to the fuel called a "moderator." In the elastic collision process, the amount of energy a neutron loses in each collision depends on the mass of the target nucleus. Just as a billiard ball will rebound from a very heavy object, say a wall, with little loss of energy, so will a neutron rebound from a heavy nucleus with little energy loss. On the other hand, a billiard ball striking a group of other balls will quickly lose its kinetic energy and come to rest. In a similar manner, a neutron striking light atomic nuclei, each having a mass comparable to that of the neutron, will transfer its energy to the target nuclei and quickly slow down. Thus the lighter elements, in general, make good moderators *provided* that they do not have large capture cross sections themselves.

A moderating ratio may be defined as being a constant for a material indicating its effectiveness as a moderator, taking into account its mass, its scattering cross section and its absorption cross section. It will vary as some inverse function of its mass (the lighter elements being more effective in slowing down); directly as its scattering cross section (the chances of collision being larger for a larger cross section); and inversely with its capture cross section (a large capture cross section increasing the chance that the neutron will be "taken out of circulation"). The moderating ratios for several substances are given in the table below.

Material	Moderating Ratio
Hydrogen	66
Water	67
Heavy Water	12,000
Helium	94
Beryllium	160
Carbon	169
Uranium	negligible

It is seen that heavy water is a superior moderator. Its cost has been high but improved production methods promise increasing future utilization. The choice of a moderator in a particular situation will depend on type of fuel, design and structural problems.

1008. POWER.

For a continuous production of power in a reactor, it is necessary that some of the neutrons produced by the fission reactions be available to initiate further fissions. The fission process produces, on the average, about 2.5 neutrons, but all such neutrons are not available for further fission. Leakage of neutrons from the reactor, capture by "inert" materials present, and non-fission absorption by the fuel itself result in a non-useful dissipation of the neutrons. To sustain a reaction at a given power level, exactly one neutron born in each fissioning nucleus must initiate a fission process in another nucleus. The chart in Figure 10-3 shows the typical life history of a generation of 1000 neutrons in a reactor, produced by the fission of approximately 400 fuel nuclei.

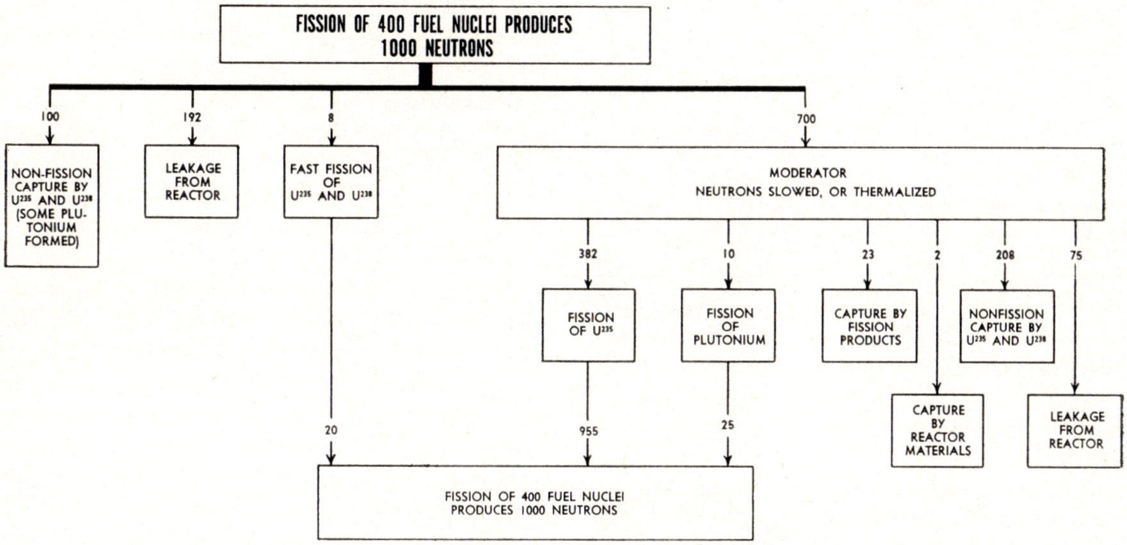

FIG. 10-3. Typical History of One Neutron Generation.

The achievement of the balance shown in Figure 10-3 will be discussed more fully under *Control.* It is apparent, however, that the total number of fissions per unit of time determines the power level of the reactor. The number of fissions per unit of time in turn depends on the number of neutrons available per unit of time for fission, the number of fuel nuclei present, and the probability that the neutrons will cause fission of these fuel nuclei (which is the macroscopic fission cross section, Σ_f.

The number of neutrons available per unit of time is measured by the flux density of the reactor, ϕ, which is the average number of fission neutrons passing through each square centimeter in the reactor per unit of time. The power, then, can be expressed as follows:

$$P = \frac{V_f \Sigma_f \phi}{3 \times 10^{10}}$$

where

V_f = volume of fissionable material, cm^3

Σ_f = fission cross section of fuel nucleus, cm^{-1}

ϕ = flux density, no. neutrons/cm^2-sec

3×10^{10} = no. of fissions/sec for 1 watt.

The volume of the fissionable material is fixed by the reactor design; Σ_f is a property of the fuel nucleus and concentration of fuel atoms; therefore, the control of power consists of control of the flux, and theoretically the power level is nearly unlimited.

The flux density of the fission neutrons depends on two categories of factors: those arising from the reactor design and those which are associated with its operation. The latter will be discussed under *Control.* The design of the reactor includes the following elements which affect the flux (assuming a fixed volume of fissionable material):

(a) *Geometry*—The geometry of the reactor determines the ratio of fission neutrons which will escape from the fuel mass to the total number born in the reactor. For configurations having a high ratio of surface to volume, the proportion of escaping neutrons will be high. From this point of view, a spherical geometry of the fuel would minimize the proportion of neutrons escaping, since it gives the lowest surface to volume ratio of any geometry.

(b) *Reflector*—A substance surrounding the fuel which will "reflect" escaped neutrons back into the fuel mass will increase the flux density of neutrons in the fuel. This permits use of a smaller volume of the fuel than for an unreflected reactor. A good reflector material is one which possesses a large scattering cross section and a small absorption cross section. The term "albedo" is used to describe the reflecting properties. It is defined as follows:

$$\text{Albedo} = \frac{\text{no. reflected neutrons}}{\text{no. incident neutrons}}.$$

The thickness of a reflector material beyond a certain value does not significantly increase its albedo. Examples of reflector materials are given below.

Material	Albedo (*infinite thickness*)	Albedo (*40 cm. thick*)
Water	0.821	0.821
Heavy Water	0.968	0.919
Beryllium	0.889	0.881
Carbon	0.930	0.892

It is noted that the requirements for a moderating material are almost the same as those for a reflector material.

1009. COOLANT.

Since most of the energy liberated in fission appears as heat, some means of removing this heat must be provided to transport it to the location where it will be used and, equally important, to remove it from the reactor to prevent the attainment of destructive temperatures. It is the latter consideration which is one of the most vexing problems in reactor design, and which dictates a definite limitation on the allowable flux, and thus the power output of a given reactor. If heat could be removed rapidly enough to prevent the attainment of destructive temperatures within the reactor, a very small reactor could be designed to produce almost any power desired. The temperature limits in the reactor are fixed by the physical properties of the fuel, the moderator, the reflector, structural materials present in the reactor, and of the coolant itself.

Desirable properties of a coolant are:

(1) High specific heat and thermal conductivity.
(2) Low power requirement for pumping.
(3) High boiling point and low melting point.
(4) Chemical stability in the presence of heat and radiation.
(5) Suitable corrosion characteristics in the system.
(6) Small neutron capture cross section.
(7) Freedom from hazard in operating, including low induced radioactivity.

No known substance can meet all these conflicting requirements to the degree desired. Some more common ones in use, or now under experimentation are described in the following.

Water has a high specific heat and a fair thermal conductivity. It has a relatively low viscosity and is easily pumped, the pumping power required being about 1/10 of that required for a gaseous coolant at a pressure of 10 atmospheres. One of the greatest advantages of water is that it can act simultaneously as a moderator and a coolant as it passes through the reactor. Its disadvantages are that it has a fairly large capture cross section; it must be highly purified; its atomic structure is altered by radiation, (radioactive nitrogen-16 being formed from oxygen-16); its corrosive action on metals is unfavorable; and it must be placed under high pressures in order to remain liquid at the reactor temperature (necessitating placing the reactor in a pressure vessel). Means have been found, however, to largely circumvent or overcome these disadvantages, and water is a tried and proven coolant substance.

The excellent thermal properties of liquid metals bring such substances as sodium and NaK into prominence. For a given rate of heat removal, smaller heat transfer areas are required for a liquid metal coolant than for water coolant. Liquid metals have a low vapor pressure, making possible the attainment of high temperatures at pressures considerably lower than required for water. They are physically and chemically stable at high temperatures and in intense radiation fields. Their corrosive properties at high temperatures are relatively favorable. Pumping power for most liquid metals is not significantly different from water. On the debit side, particularly for sodium, is the formation in the reactor radiation field of sodium-24, a radioactive substance, necessitating shielding of the coolant circuit. The hazard existing due to contact with air or water must be carefully guarded against, necessitating extremely careful system sealing and leak prevention. Sodium solidifies at about 210°F, thus some external means must be provided to liquify it after extended shut-down.

Gaseous coolants are promising for gas turbine and possibly turbo-jet applications. Air, helium, and carbon dioxide are of interest as gaseous coolants.

Organic coolants—hydrocarbons—have certain qualities which make them of interest. Like water, they would serve well as a moderator. Higher boiling points and lower vapor pressures than possessed by water are distinct possibilities and advantages. Most hydrocarbons are less corrosive than water. It appears that for organic coolants under consideration, only slight radioactivity is induced, making the entire heat transfer system relatively accessible. The main drawback is the relative instability of known substances in the presence of radiation. However, considerable effort is being made to solve this problem, and it appears that definite progress is being made.

1010. CONTROL.

For a reactor to be able to commence operation, or go critical, it must innately possess the property of increasing the number of fissions with each generation of neutrons. This property is measured by the effective multiplication factor, k_{eff}, which must be capable of being made greater than 1.0 for a reactor of finite size. If k_{eff} is only slightly greater than 1.0, the reactor will have to be large; if k_{eff} is considerably greater than 1.0, the reactor can be made smaller. It is the product of the total average number of neutrons produced per fission, multiplied by factors which account for all losses from further fission processes. This produces a ratio which is the number of neutrons present in the reactor at a given time to the number present at one finite lifetime earlier. If k_{eff} for the reactor were to remain greater than 1.0, the number of fissions per unit time and thus the power would build up continuously until temperatures attained would destroy the reactor. Therefore, some means of controlling the multiplication factor must be provided.

The primary means for accomplishing this is the insertion of material in the reactor which has a large neutron capture cross-section. Boron, cadmium, and hafnium are such substances. Control rods made of these materials absorb neutrons at a rate approximately proportional to the neutron flux in the reactor at the point where they are inserted. By having sufficient control rod volume in the reactor, neutrons can be absorbed at a greater rate than they are produced, reducing the multiplication factor to below 1.0, and causing the reaction to come to a halt. When the control rods are sufficiently withdrawn, k_{eff} becomes greater than 1.0, the reaction builds up, and when the desired power is reached the rods are moved in until k_{eff} becomes exactly 1.0. This maintains the reactor in equilibrium at the desired power level. Since the rate of neutron absorption is closely proportional to the neutron flux density, the position of the rods is the same for all equilibrium power levels, regardless of their magnitude, as long as the reactor characteristics such as fuel concentration, amount of "poisons" etc., remain constant. As the amount of fuel decreases with time, and as fission products build up in the reactor, less control rod material is required within the reactor to reduce the multiplication factor to 1.0. Thus the control rod position for equilibrium power varies with the age of the reactor, and is one indication of the amount of power remaining. The two graphs in Figure 10-4 show how control rod position is used to vary power. The reactor is shut down by full insertion of the control rods. The rods are arranged to be automatically tripped and inserted when dangerous operating conditions obtain. Such emergency shut-down is called "scramming" the reactor.

In a reactor using enriched uranium as a fuel (i.e. fuel in which the proportion of U^{235} to U^{238} has been increased considerably more than that found in natural uranium) the mean lifetime of one generation is about 0.0001 second. This is the length of time elapsing from the birth of a neutron in a fission process until its escape or absorption. If k_{eff} is greater than 1.0, then $k-1$ is the factor by which the number of neutrons will increase in one lifetime, l. If n neutrons are present at any given time, then $n(k-1)$ will be pres-

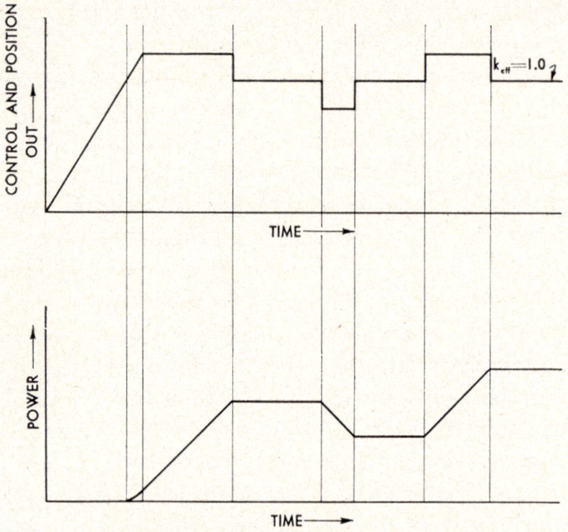

FIG. 10-4. Variation of Power with Control Rod Position.

ent one generation later. Therefore, the rate of change of the number of neutrons is $n(k-1)/l$. Then

$$\frac{dn}{dt}=\frac{n(k-1)}{l} \tag{1}$$

Integrating, and assuming n_0 neutrons at $t=0$, the number of neutrons present after a time, t, is

$$n=n_0e^{(k-1)t/l} \tag{2}$$

Assuming $n_0=1$, and that the neutron lifetime in an enriched reactor is 0.0001, and that $k=1.002$; at the end of 1 second

$$n=e^{.002/.0001}=e^{20}=4.9\times10^8$$

The above example illustrates the very rapid build up in activity possible in the reactor, and the necessity for quick, accurate control to prevent a dangerous condition from arising.

Fortunately, two conditions present in the reactor serve to make it less sensitive than indicated above. They are the action of the delayed neutrons and the negative temperature coefficient, to be described in Article 1013.

1011. FUEL ELEMENTS.

In most power reactor designs, the fuel is in the form of long rods, clad with a protective metal. These rods, or fuel elements, are arranged in a matrix forming the core of the reactor. Fuel elements have been made in a variety of sectional shapes, their surface forming the heat transfer boundary between the fuel and coolant. Figure 10-5 is an example.

It is necessary to clad the fuel (1) to prevent chemical attack on it by the coolant; and (2) to prevent contamination of the coolant with radioactive fuel particles and fission products. The material used for cladding must be adequately resistant to corrosion at the reactor temperatures encountered in the presence of the fuel and coolant used. It must possess a low neutron capture cross section to prevent diminishing the neutron flux density. It appears that the high temperature corrosion characteristics of presently available cladding materials is one of the chief factors responsible for the limitation of reactor temperatures. For instance, zirconium is one of the best materials for fuel element cladding with a water coolant, and, according to Glasstone, its maximum permissible surface temperature in water is about 660°F. Stainless steel is another material of considerable interest and possibility for cladding. It is resistant to attack by sodium and can sustain a surface temperature of about 1200°F.

1012. REACTOR ARRANGEMENTS.

1. *General.* The objective of all power reactors is to provide thermal energy in a usable form to an energy conversion device which can change it

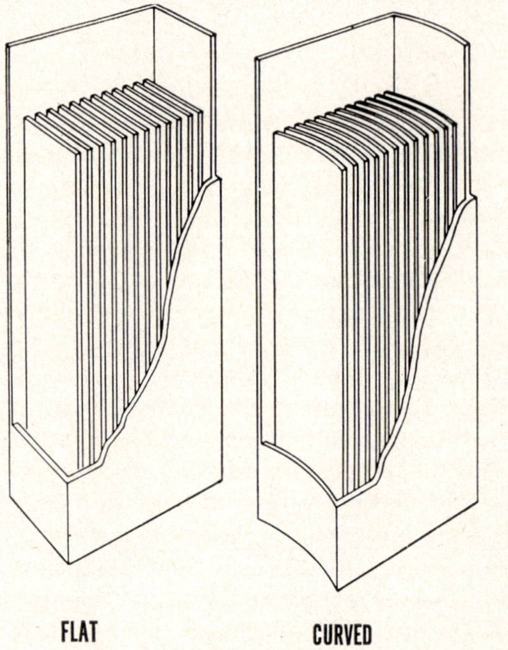

Courtesy of Assn. of Iron and Steel Engineers

FIG. 10-5. Typical Fuel Elements.

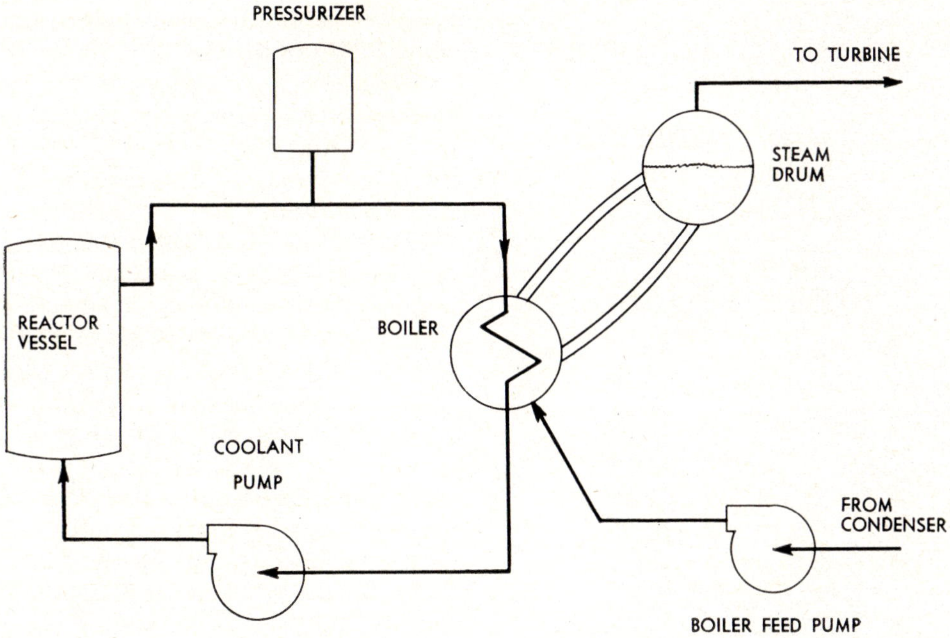

FIG. 10-6. Schematic of PWR Cycle.

into useful mechanical work. The work then may be used directly as in the shipboard plant or converted into electrical energy as in the shore based plants.

The reactor coolant is the agent which carries the thermal energy out of the reactor. Its properties and physical state for optimum use as a coolant are not necessarily those which make it readily adaptable as the working fluid in a steam turbine. Also, as a coolant, it may become radioactive, and it is desirable for operation and maintenance of the mechanical plant that such "induced" radioactivity be suppressed to a safe level. For the above reasons, it is usually necessary for the coolant to transfer its heat to a working fluid. Thus two or three fluid circuits are provided in most cases. (See paragraph 3 below for the exception to this.) The coolant circuit is called the *primary* loop, and the working fluid is called the *secondary* loop. Where three loops are necessary, an *intermediate* loop is inserted between the primary and the secondary loop.

2. *Pressurized Water Reactor—PWR.* Figure 10-6 is a schematic of a typical pressurized water reactor cycle. In this category, the Shippingport reactor is of some interest because its size and power output are comparable to that required on one shaft of a large ship. Figure 10-7 shows some of the features of the reactor itself. It is enclosed in a pressure vessel 33 feet high with an inside diameter of 9 feet. The tank wall is $8\frac{1}{2}$ inches thick with stainless steel cladding on the internal surfaces.

The coolant is high purity water, entering the vessel at the bottom and flowing upward along the passages between and around the fuel elements, then out from the mixing chamber. The coolant flow is 45,000 gal/min with a temperature rise in the reactor from 508°F at inlet to 542°F at the outlet. This gives a heat flow out of the reactor of about 11,000,000 BTU/min. The coolant and the pressure vessel are maintained at a pressure of 2000 psi to prevent vaporization of the coolant. This is accomplished by the pressurizer indicated in the simplified cycle diagram.

Most of the primary fuel is contained in highly enriched U-235 "seed assemblies" arranged in the form of a square in the center of the core. They are surrounded by U-238 fuel elements constituting a "blanket" where plutonium is produced from the U-238 and subsequently is used as fuel for the reactor. The seed assemblies contain about 110 lbs. of enriched U-235 and the blanket contains about 12 tons of natural uranium.

The coolant is pumped by a canned rotor pump to a boiler where it transfers its heat to the secondary loop, producing dry steam at 600 psi for the turbine. Note that the reactor tempera-

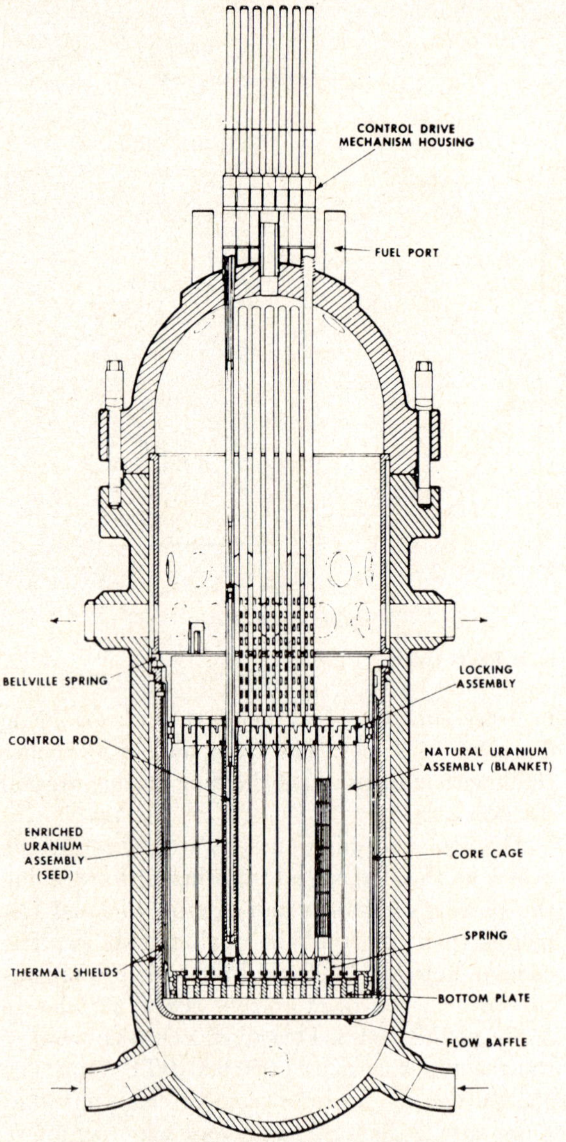

Courtesy of Westinghouse Electric Corporation

FIG. 10-7. Shippingport Reactor Vessel.

tures do not permit production of steam superheated to the degree currently used with conventional boilers. In a similar reactor under construction for another shore power station, an oil-fired superheater is included in the secondary loop to give the desired superheat.

3. *Boiling Water Reactor*. Originally it was thought that it was impracticable to permit water to change to the vapor phase within the reactor. This has now been proven possible, and permits simplification of the reactor. The pressure vessel can be designed to withstand the required steam pressure much more readily than it can be built to withstand the 2000 psi required in the PWR to maintain the water in a liquid state. Furthermore, no separate pressurizer and boiler are required as in the case of the PWR and the pumping power required to circulate the coolant is significantly less.

The Argonne National Laboratory has designed and built a boiling reactor (EBWR) which produces steam at 600 psi, with a fuel element temperature limit of about 500°F. One difficulty of this cycle is the presence of some radioactivity in the propulsion machinery, which is not a problem in the PWR where the secondary loop is used.

Another boiling water reactor of interest is the Dresden power plant reactor, now in the design stage, scheduled to be in operation in 1960. This is called a dual cycle reactor. A schematic diagram is shown in Figure 10-8. The coolant boils in the reactor which provides steam for the turbine as in the EBWR. This steam will be produced at 1000 psi (at about 545°F). In addition, liquid coolant is pumped into a steam generator where it heats some of the turbine condensate to produce steam at 500 psi which is led into the turbine at a suitable stage. The advantage claimed for this scheme is that it permits a greater rate of energy extraction from the reactor than is possible by using primary steam alone.

4. *Homogeneous Reactor*. The homogeneous reactor uses a liquid substance as a fuel, which also serves to remove the heat from the reactor core. The fuel solution, or "soup," is uranyl sulphate dissolved in water. The principal mass of the fuel is in the reactor core which is the only part of the system in which fission occurs, since a critical mass is not accumulated at other points in the system. The fuel solution is circulated through a heat exchanger in which steam is generated in the secondary system. The advantages of this system are: the elimination of the heat transfer problem in the reactor vessel; the possibility of continuously bleeding off and purifying a portion of the fuel and returning it to the system without shut-down; the inherent operational stability of the system which makes control extremely simple; the possibility of continuous Xenon removal (see Article 1013); the absence of metallurgical damage to fuel elements which limits fuel element life in other types. Disadvantages are: the

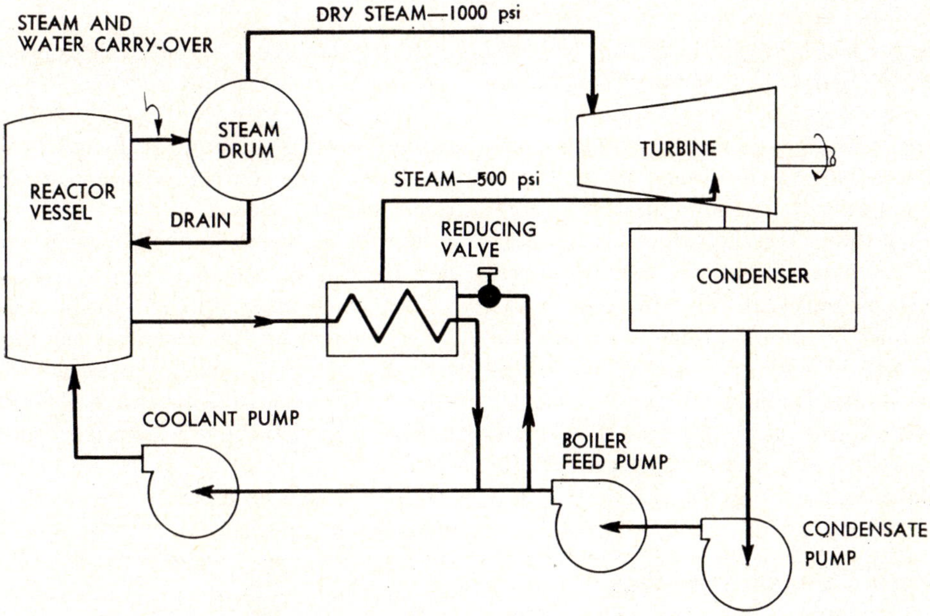

FIG. 10-8. Schematic of Dual Cycle BWR.

high radioactivity in the primary loop outside the reactor, and the corrosive nature of the liquid fuel. A small power reactor of this type has been built and operated at Oak Ridge National Laboratory, and a more advanced type is being designed by Westinghouse for shore power installation, scheduled for operation in 1962.

5. *Liquid Metal Reactor*. A scheme for a liquid metal reactor cycle is shown in Figure 10-9.

One design calls for use of liquid sodium as a coolant with a reactor exit temperature of 960°F without pressurization of the primary system. This temperature is high enough to bring the intermediate cycle to a sufficiently high temperature to both generate and superheat steam, as indicated in the cycle diagram. The intermediate loop is necessary to separate the steam cycle from the intensely radioactive primary sodium. The intermediate fluid may be liquid sodium, or NaK. To guard against the serious effects of leakage in

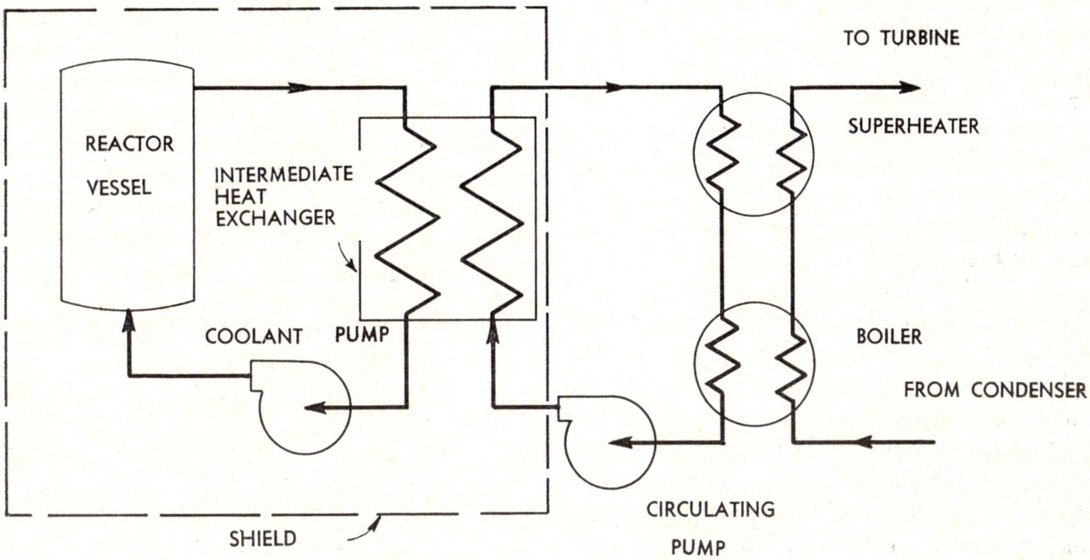

FIG. 10-9. Schematic of Liquid Metal Reactor.

the sodium-water heat exchangers, a film type heat exchanger has been proposed in which the liquid film is mercury. The mercury may be circulated and monitored to detect sodium or water leakage before the consequences become serious.

6. *Summary.* The above cycles, plus other proposed types using gas or organic liquids as coolants are being fully investigated. Although each has certain obvious advantages and disadvantages, only experience can tell which types can survive to fulfill future requirements. Doubtless other types will be conceived and many refinements and perhaps simplification of presently known types will take place. Even at the present state of knowledge and technology, adoption for naval propulsion on an ever increasing scale is a certainty.

1013. OPERATING CHARACTERISTICS.

Three particular features of operating reactors are worthy of mention. They are (1) the negative temperature coefficient, (2) the delayed neutron action, and (3) the poisoning of the fuel.

In order to understand the negative temperature coefficient, refer to the PWR cycle diagram, Figure 10-6, in connection with the following discussion. Suppose that the cycle is operating at a fixed power rate, steam being delivered to the turbine at a constant rate. Now the throttle is opened up to speed up the turbine. This allows steam to flow in greater quantity to the turbine and through the boiler. The immediate result is to lower somewhat the temperature of the reactor coolant fluid leaving the boiler. As its temperature drops, its density increases slightly but significantly. The increase in density increases the magnitude of its macroscopic neutron scattering cross section. Consequently, in the reactor, the coolant in its capacity as a moderator begins to thermalize high speed neutrons at a faster rate, increasing the number of thermal neutrons, and thus the number of fissions. The consequent increase in the reactor power output raises the temperature of the coolant leaving the reactor and entering the boiler. The heat transfer rate in the boiler is thus increased to accommodate the additional power demanded by the turbine. The boiler exit temperature of the coolant is re-established at its former value, the macroscopic scattering cross section is re-established and the reactor settles down at the new power level. Thus the only control in the system used to change power was the turbine throttle. A similar sequence takes place when the throttle is closed, causing the reactor power to decrease to that required. This property of the coolant also operates to help prevent a runaway fission build up and serves with the control rods as a double measure of control and safety.

The above discussion is somewhat over-simplified for clarity. Temperature affects other constants which make up the multiplication factor, but the effect on the coolant is the most important one. However, when the net effect of a temperature increase causes a reactivity decrease, the reactor is said to have a *negative temperature coefficient.*

When a nucleus fissions, most of the new generation neutrons appear immediately. However, the appearance of a small percentage (about 0.73 per cent for U-235 fission) is delayed for a finite time. About one-half of them appear within 3 seconds and some of the remainder are delayed up to a minute or more. By weighting the values of the various delayed groups of neutrons appropriately, it is found that the mean lifetime of these neutrons is about 12 seconds. These neutrons, although small in number, have a very significant effect in increasing the average neutron lifetime.

Although rigorous analysis is somewhat more complex, the following simplified example will show the effect of these delayed neutrons.

Example 101: Assume, as in Article 1010, that the average prompt neutron lifetime time is 0.0001 second. Using 12 seconds as the average delayed neutron lifetime, find the weighted average lifetime of all neutrons in one generation, if .73 per cent of the neutrons are delayed. Determine the number of fissions at the end of one second which result from the original fission of one neutron, if $k=1.002$. Find the time required for 4.9×10^8 fissions to occur (as in Art. 1010).

Solution: The weighted average lifetime is

$$.9927\times.0001=.0001 \text{ (approx.)}$$
$$.0073\times12 \quad =.0876$$
$$l=.0877 \text{ secs.}$$

At the end of one second,

$$n=n_0e^{(k-1)t/l}=e^{.002/.0877}=e^{.0228}$$
$$=1.023 \text{ fissions.}$$

For 4.9×10^8 fissions to occur

$$4.9\times10^8=e^{.0228t}=e^{20}$$

$$\therefore\quad t=\frac{20}{.0228}=875 \text{ seconds}$$

The above example shows that with the delayed neutron effect, the reactor assumed in Article 1010 requires 875 seconds to reach the same power level which was reached in one second when no delayed neutron effect was included. Of course, this power level could be reached more rapidly than 875 seconds if the control rods are pulled out further, increasing the value of k_{eff}. The important point is that the effect of the delayed neutrons is to decrease the sensitivity of the reactor so that accurate and effective control by the rods is made feasible.

The third particular feature to be discussed is xenon poisoning. As operation of the reactor proceeds, an ever increasing quantity of fission fragments is produced. Most of these fragments or fission products have small neutron absorption cross sections. A few, the most important of which is xenon-135, are strong neutron absorbers and act to "poison" the chain reaction. Most of the xenon is produced, according to the following reaction, from tellurium-135, which itself is a product of fission.

$$Te^{135}\xrightarrow{2\text{ min}}I^{135}\xrightarrow{6.7\text{ hrs}}Xe^{135}\xrightarrow{9.2\text{ hrs}}Cs^{135}$$

$$\xrightarrow{2\times10^6\text{ yrs}}Ba^{135}$$

During operation of the reactor, Xe^{135} is removed by the above decay, and also by a process of neutron capture, in which it becomes a non-poisonous material. Thus, during operation, an equilibrium amount of xenon is reached, after which it is removed by the two processes as fast as it is absorbed. However, if the reactor is suddenly shut down, the xenon concentration will build up to a higher level, since its decay is slower than its formation by I^{135} remaining at shut-down. (Note the half-lives indicated in the above equation.) The xenon concentration reaches its maximum about 11 hours after shut-down. Sufficient excess fuel must always be available to overcome this poisoning effect after shut-down, else the reactor either must be started up shortly after shut-down, or wait a day or two for the I^{135} and Xe^{135} to decay to small values.

1014. PUMPS.

In the design of the different types of pumps for use in Nuclear Engineering applications, the design characteristics with reference to a purely hydraulic point of view are relatively conventional. There are, however, general requirements which have made special design and development necessary for the high temperatures and pressures encountered. This applies to both pressurized water and liquid metal applications.

Among the general requirements are:

(1) Ability to pump high temperature liquids. This is not a particularly unique problem as it has been encountered in the pumping of high-temperature salts and oils in other applications. However, it calls for special features in the design of windings and drive motors which are more troublesome than those encountered in conventional pump design.

(2) Ability to operate for extended periods without maintenance. Since maintenance may be difficult because of radiation hazard, an attempt should be made to minimize or eliminate parts subject to wear and fatigue, and components requiring adjustment.

(3) The necessity of leakless, or nearly leakless operation. This is usually the governing factor in the design and selection of pumps and has led to the development of unusual means for sealing mechanical pumps as well as to the development of electromagnetic pumps which can be made inherently leakless.

(4) Method of speed control . . . whether the pump operates at constant flow or whether the flow is variable, and by what means the variation is obtained.

(5) Satisfactory mounting of the pump to operate under conditions of pitch, roll, and mechanical shock expected in a mobile plant.

Sufficient development has been done on certain types of pumps to allow some general conclusions which are of interest. These types of pumps can be divided into two general groups, *mechanical* and *electromagnetic*.

Mechanical Pumps. The hermetically sealed (canned) motor pump is the type of pump generally employed in the primary circuit of pressurized water reactors. This type can also be used for plants using a liquid metal coolant. System pressures up to 2800 psi, temperatures up to 680°F, and pumping capacities as high as 20,000 gpm have been designed.

Some of the advantages are:

(a) Zero leakage. System may be hermetically sealed by welding.

(b) Low maintenance. No mechanical seal or stuffing boxes are used—no auxiliary equipment is required. System fluid lubricates the bearings.

(c) Built in heat exchanger. Wrap around type heat exchanger maintains safe winding temperatures.

(d) Totally enclosed. All parts in contact with the pumped fluid (primary water) are made of corrosion-resistant alloys. The pumped fluid is circulated in the motor cavity, (being called auxiliary water in this connection) but is excluded from the rotor and the stator windings by sealing jackets in the magnetic gap. There are no external shaft seals, and suction and discharge nozzles are designed to be welded into the pipe line.

(e) The only auxiliary requirement is an external supply of ordinary clean cooling water (secondary cooling water) which flows through the jacket over coils wrapped around the stator. These coils contain the auxiliary water.

Secondary cooling water flows over the wrap-around coils. An auxiliary impeller provides circulation of auxiliary water around the rotor, and through the wrap-around coils which transmit the motor heat to the secondary cooling water flowing around the coils. The motor rotor is hermetically enclosed by a welded stainless steel or Inconel sealing "can." The motor stator winding is completely isolated from the pumped fluid by a stainless steel or Inconel sealing "can." Heat from electrical losses is conducted radially outward and removed directly by the cooling fluids. A self-equalizing pivoted-pad bearing absorbs the axial thrust. The labyrinth seal together with the thermal barrier minimize fluid flow and heat transfer between the primary water in the pump casing and the auxiliary water in the motor cavity and thus insulates the motor from high system temperatures. The main impeller is keyed and locked to the motor shaft. Figure 10-10 is a schematic half-section of a hermetically sealed pump, which illustrates many of the features discussed above.

Electromagnetic Pumps. There are several types of electromagnetic pumps which have been designed for nuclear engineering purposes but the two principal types are the D.C. Faraday and the A.C. types. These may be used to pump any liquid metal which will wet the pumping section and which has a good electrical conductivity. Oxide concentration in the liquid metal has no effect on the pump performance.

The D.C. Faraday type pump operates on the basic principle of the "left-hand rule." The liquid metal in the pumping section is a conductor of electricity. When a current is passed through it, perpendicular to a magnetic field, a force is produced on the liquid metal which results in motion within the pumping section. This motion is at right angles to the current and the flux. The liquid metal is the conductor which passes through a duct located between the poles of an electromagnet. A current from an external source is then passed across the liquid metal in a direction perpendicular to both the magnetic field and the desired direction of flow of the liquid metal. The force exerted on the conductor by the magnetic field then causes flow to take place. One advantage of the electromagnetic pump is that it has no moving parts. Maintenance is thus greatly simplified. This is important because the pump may be contaminated with radioactive material.

Figure 10-11 is a sketch of the Faraday pump, and the Figure 10-12 shows typical head-capacity performance curves for a pump of this type.

The A.C. electromagnetic pump operates on the same "left-hand rule" principle. The magnetizing coils are so connected that the magnetic field changes sense simultaneously with the flow of current through the conductor, with the result that the induced force on the liquid metal conductor is always in the same direction.

Both types of electromagnetic pumps have full reversible flow and full flow rate control from zero to maximum.

1015. PRESSURIZING TANK.

In order to keep the primary coolant under sufficient pressure in the PWR type plant to avoid boiling, and to provide a surge tank to accommodate sudden changes in pumping rates when the load is changed, a pressurizing tank is installed in the loop. Figure 10-13 is a cut-away view of the Shippingport tank.

Heat supplied to the tank by the electrical heating units shown at the bottom causes the formation of a steam blanket on the liquid surface. By controlling the temperature, the pressure

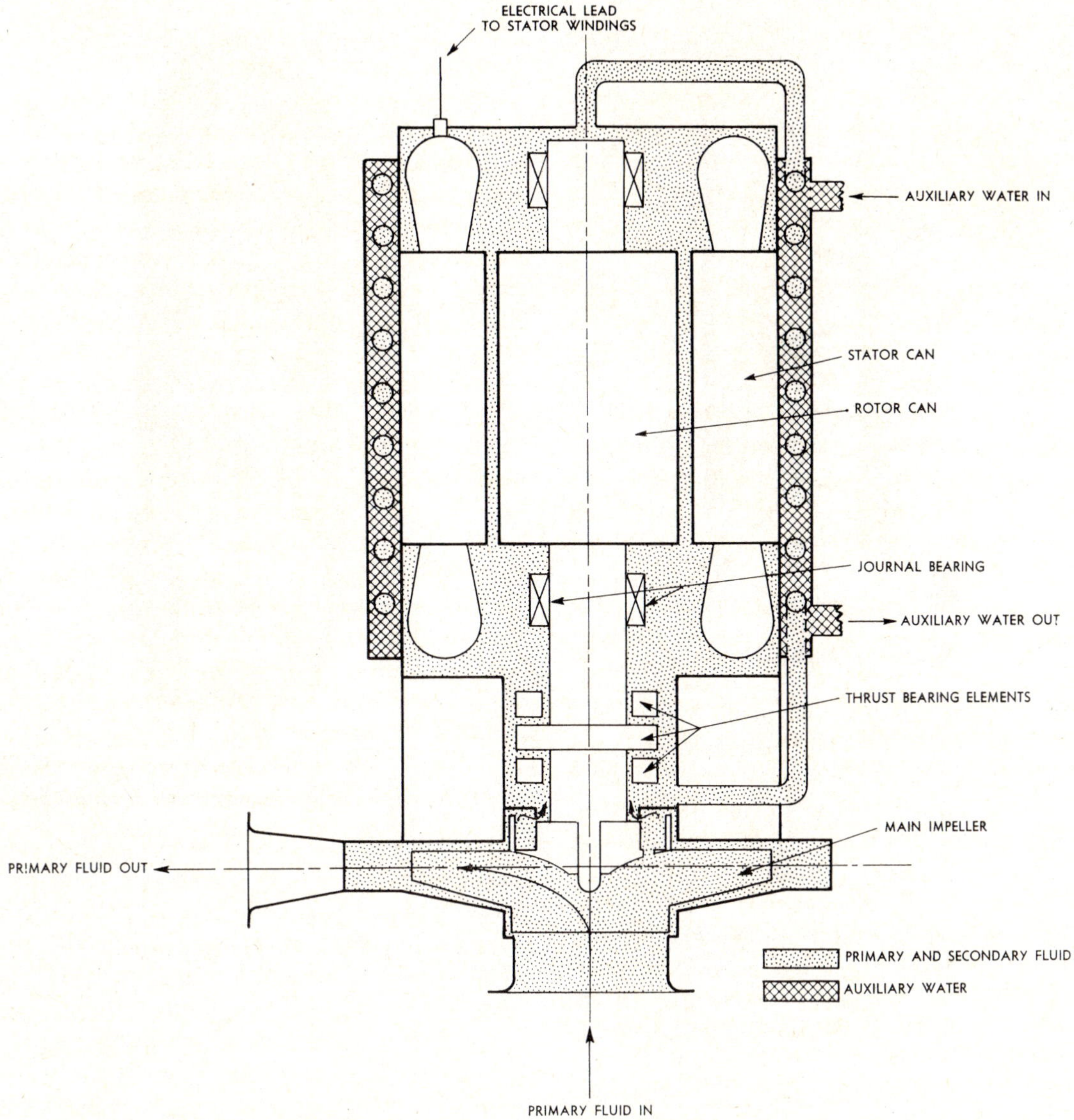

FIG. 10-10. Schematic of Hermetically Sealed Pump.

of the steam bubble, or blanket, can be maintained at the desired level. This pressure is transmitted to the primary loop through the line provided at the bottom of the tank. A spray nozzle is located at the top of the tank to limit the pressure of the steam by spraying colder water into the bubble during a positive surge in the water level when the load changes. Gauges are provided to inform the operator of the water level and pressure conditions, and a low water level alarm serves to warn the operator when the level drops to a point where negative surges could uncover the heater units.

1016. OTHER COMPONENTS.

The steam generators are large heat exchangers in which the primary coolant transfers its thermal energy to the secondary working fluid. Since the temperature difference in the primary inlet and outlet is relatively small, particularly in the PWR type, and since no latent heat accompanying a phase change in the primary fluid is involved, the steam generator must be designed to accommodate large flow rates on the primary side.

A schematic of one type is shown in Figure 10-14. This is called a natural circulation steam

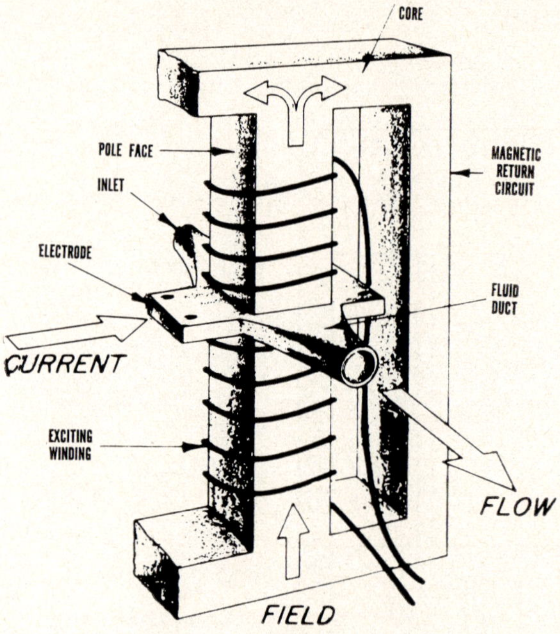

FIG. 10-11. Faraday Pump.

generator. The primary coolant flows through U tubes in the lower unit. The turbine loop condensate is introduced to the shell of this exchanger; the steam formed passes to the steam drum through the risers and accumulates in the steam drum, where it passes through separators on its way out to remove all moisture. Water level is maintained in the steam drum in a manner similar to a conventional natural circulation boiler, but all the heat exchange takes place in the lower unit, called the feed-heating and evaporating section.

Other steam generator units, including separate superheaters are under development.

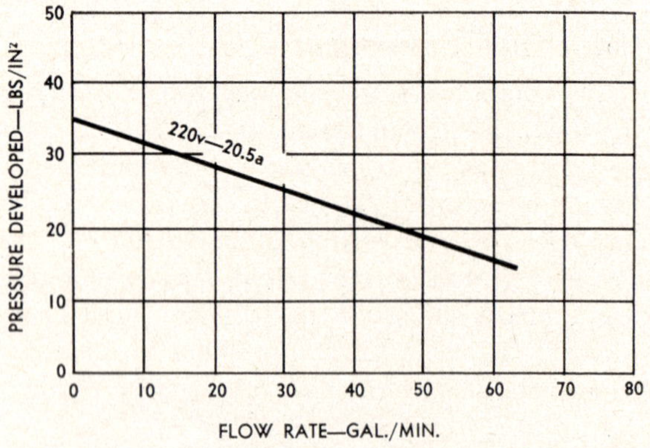

FIG. 10-12. Typical Head-Capacity Curve for Faraday Pump.

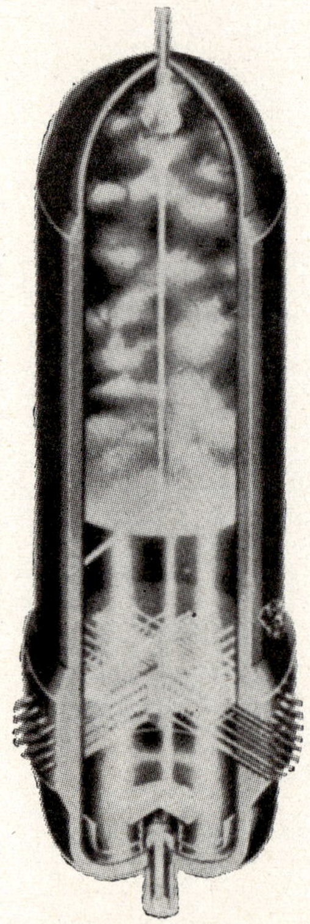

Courtesy of Westinghouse Electric Corporation

FIG. 10-13. Shippingport Pressurizing Tank.

Leak proof valves are necessary in the primary loop. Several types have been designed and tested. Two typical arrangements are the "frozen-stem" type and the bellows type. The frozen stem valve has been tested in some liquid metal loops. In this type the liquid metal is allowed to accumulate in a "pocket" along the stem, far enough removed from the main flow, so that it will freeze and form a leak proof seal. Sufficient torque to shear this seal must be available to change the valve position.

1017. PROBLEMS.

Having briefly examined some salient features of the process of obtaining power from nuclear fission, we conclude with a list of problems. Some of these have been stated and others have been suggested. It is not intended to end on a pessimistic note, for the outlook for the future of nuclear power is extremely bright. It is necessary however, to know what the problems are before they can be solved, and the below list will suggest areas in which progress can bring fruitful rewards.

(1) Some form of shielding from the hazardous

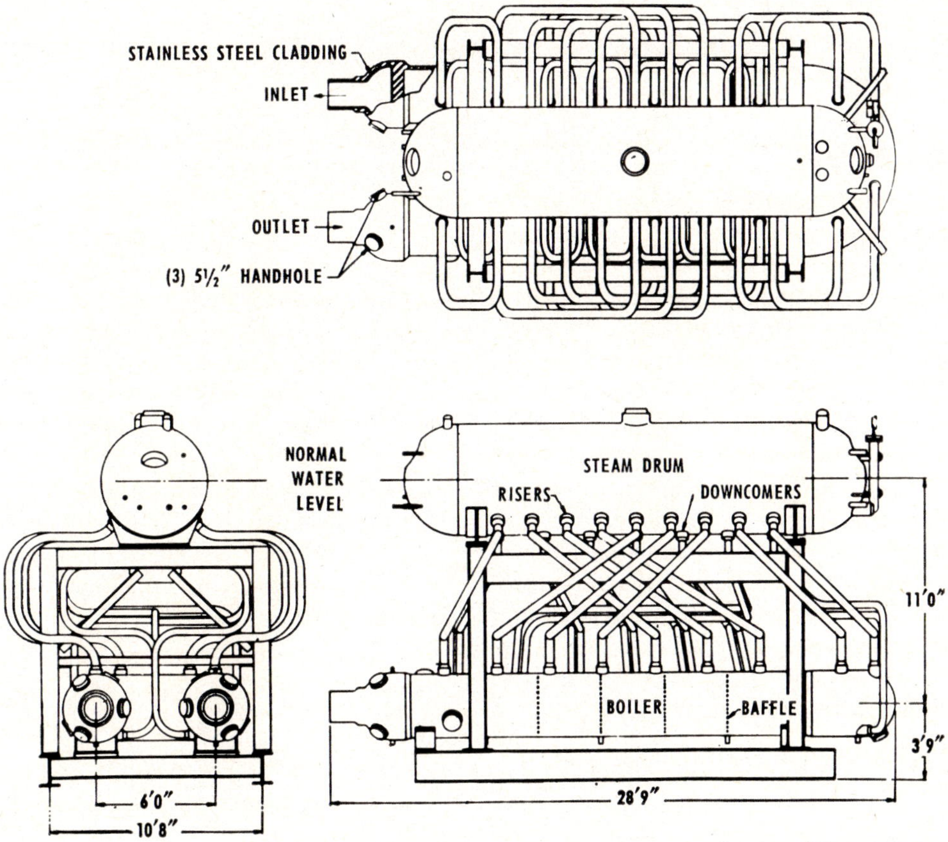

Courtesy of Foster Wheeler Corporation

FIG. 10-14. Natural Circulation Steam Generator.

neutrons and gamma radiations is necessary to protect the operating personnel. The weight of such a shield is an important item.

(2) Hot spots within the reactor core must be avoided. This means that the coolant must flow fast enough and carry away enough heat with each pass through the core, to avoid any excessive temperature buildup in any local section of the core.

(3) Many reactor power systems operate under pressures as high as 2000 psia. in their primary coolant circuit. These high pressures require a special leak proof type of centrifugal pump as well as specially designed bearings which are lubricated with coolant water inside the "canned motor" pump. All connections must be extra leak tight.

(4) The corrosive properties of both water and liquid sodium become a problem at higher temperatures.

(5) If sodium is used as a coolant, extra care must be taken to avoid leaks because of the fire hazard and the rapid attack on the sodium by oxygen and water.

REFERENCES

Principles of Nuclear Reactor Engineering—S. Glasstone, D. Van Nostrand Co., 1955.

Introduction to Nuclear Engineering—R. Stephenson, McGraw-Hill Book Co., 1954.

Introduction to Nuclear Engineering—R. L. Murray, Prentice-Hall, Inc., 1954.

Reactor Handbook Series, Vol. 2—U. S. Atomic Energy Commission, 1955.

"Nuclear Reactors for Power"—G. Leppert, *Journal of the Society of Naval Engineers*, Vol. 68, No. 4, November, 1956.

"The Physics of Nuclear Power"—W. E. Shoupp, one of a selection of papers published by Westinghouse Electric Corporation.

CHAPTER XI
HEAT BALANCE

1100. INTRODUCTION.

A complete heat balance of a ship's machinery plant is an accounting of the source and use of all the thermal energy involved. Used in design, it is invaluable in computing the required quantity of energy to be provided and in interrelating the characteristics and size of the components to provide the most efficient plant. Used in trials, it provides a determination of the adequacy of the plant as a whole to meet specified conditions. In operation, it provides a check of proper and most economical combinations of the various units.

No attempt will be made to completely develop a heat balance here, but a brief investigation will serve to acquaint the student with its character and scope and will provide a final overview of the plant as previously discussed. It will be used to calculate fuel consumption in this case.

1101. THERMAL ENERGY REQUIREMENTS.

For the purposes of this description, the thermal energy used by the plant will be divided into six categories. Each will be briefly investigated and quantitative estimates will be made, based on assumptions given. One propulsion unit, and its associated equipment will be considered. While the actual figures used are not taken from any specific ship, they are representative, for the most part, of modern destroyer performance.

The six categories of thermal energy requirements which will be considered are:

(a) Main turbines;
(b) Main steam cycle pumps;
(c) Boiler auxiliaries;
(d) Ship's service generator;
(e) Distilling plant;
(f) Hotel services (heating, cooking, laundry, etc.)

1102. MAIN TURBINE REQUIREMENT.

The thermal energy requirement for the turbine will be computed at near full power. The following data are assumed to be available for this condition.

Horsepower at propeller:	33,600 SHP
Turbine inlet pressure, p_a:	1150 psig
Turbine inlet temperature, t_a:	950°F
Condenser pressure, p_e:	28″ Hg
Estimated internal efficiency, η_i:	82 per cent
L.P. turbine exhaust area:	13 ft^2

The quantity sought is the steam flow in pounds per hour to the turbine. With this, and a knowledge of the steam inlet conditions, which are given, the total heat flow to the turbine is determined. If the total energy output of the turbine and the energy output per pound of steam flow is found, the steam flow is readily determined.

(a) *Energy output.* The horsepower delivered to the propeller is given. If total line shaft losses are estimated at 2 per cent, and reduction gear efficiency at 98 per cent, then the overall transmission efficiency is

$$.98 \times .98 = .96.$$

And the power required at the turbine flange is

$$33{,}600 \div .96 = 35{,}000 \text{ SHP}.$$

Converting to BTU, this represents an energy output of

$$35{,}000 \times 2544 = 89{,}040{,}000 \text{ BTU/hr}.$$

(b) *Turbine performance.* The total energy available in each pound of steam at the turbine throttle is found from its isentropic enthalpy drop from the throttle inlet steam state to the condenser steam state. From the data given, and using the Mollier Chart:

$$h_a = 1472 \text{ BTU/lb}$$
$$h_e' = 902 \text{ BTU/lb.}$$

From equation (28), Article 437, and the value of η_i assumed, the actual enthalpy drop is

$$h_a - h_e = .82\ (1472 - 902)$$
$$= .82\ (570) = 467 \text{ BTU/lb.}$$

From an examination of equation (28), Article 422, it is known that the work produced per pound of steam will be less than the above figure by the amount of the leaving loss and the mechanical losses. A cut and try process is necessary to compute the leaving loss, since the total quantity of steam flow, which we seek, is necessary to compute this loss. A reasonable assumption may be made that the combined losses are about 5 to 10 per cent, say 7 per cent, which will be checked later. Then the work available at the turbine shaft, per pound of steam, is

$$wk = 467 \times 0.93 = 434 \text{ BTU/lb.}$$

The assumed throttle flow is then

$$89{,}040{,}000 \div 434 = 205{,}161 \text{ lbs/hr.}$$

From the value of $h_a - h_e$ found above and the condenser pressure, the state of the exit steam is determined, which is necessary in order to find its specific volume for the leaving loss calculation. This gives

$$h_e = h_a - 467 = 1472 - 467 = 1005 \text{ BTU/lb.}$$

With this value, and the 2″ Hg condenser pressure, the Mollier Chart shows a moisture content of 9.7%. The specific volume, from the Steam Tables, is then 306 ft^3 per lb. From equaion (29), Art. 422,

$$V_e = \frac{201{,}000 \text{ lbs/hr} \times 306 \text{ ft}^3\text{/lb}}{3600 \text{ sec/hr} \times 13 \text{ ft}^2} = 1314 \text{ ft/sec.}$$

The leaving loss is then

$$\frac{(V_e)^2}{2gJ} = \frac{(1314)^2}{64 \times 778} = \underline{\underline{34.7}} \text{ BTU/lb.}$$

The mechanical losses are estimated at 3 BTU per pound [see Article 422 (c)]. The total losses are then

$$34.7 + 3 = 37.7 \text{ BTU/lb.}$$

This represents a total loss of 8.1 per cent of the actual enthalpy drop of 467 BTU/lb, which checks closely with the 7 per cent assumed.

Finally, the work delivered per pound of throttle steam is computed from equation (27), Article 422, as follows:

$$wk = (h_a - h_e) - \frac{V_e^2}{2gJ} - L_m$$

$$= 467 - 37.7 = \underline{\underline{429.3}} \text{ BTU/lb.}$$

(c) *Steam flow rate.* Since 89,040,000 BTU/hr are necessary, the throttle flow is

$$\frac{89{,}040{,}000}{429.3} = \underline{\underline{207{,}000}} \text{ lbs/hr.}$$

It is noted at this point that in the data given, an internal efficiency of the turbine was *assumed,* while all other data are susceptible to direct measurement or calibration. This estimate can be made very closely for a given power from the turbine condition curves provided by the manufacturer for different powers. In the absence of specific information, modern turbines will have an internal efficiency of from 78 to 82 per cent in the range of one-fourth power to ull power, so that an estimate of 80 per cent involves little error anywhere in this range.

1103. MAIN STEAM CYCLE PUMPS.

The pumps in the main steam cycle, taken together, must raise the pressure of the condensate in the main condenser from the condenser pressure to the boiler pressure. In addition, they must raise the pressure of additional condensed drains which flow into the main feed system from other sources. The quantity of turbine steam, and therefore main condensate, has been computed. Pending an accurate determination of the quantity of water condensed from other sources, an estimate must be made.

At higher turbine powers, the steam required by the main turbines is nearly a constant proportion of the total steam required by the plant, varying from about 80 per cent at half-power to about 85 per cent at full power. This gives a means of estimating the total steam flow, and therefore the total boiler feed flow.

In our example, then, an estimate of the total main feed flow is

$$207{,}000 \div .85 = 244{,}000 \text{ lbs/hr.}$$

This will be checked later.

It is necessary, then, to raise the pressure of a part of this water from condenser pressure to the DFT inlet pressure, and *all* of it from the DFT tank pressure (15 psig) to the boiler inlet pressure (1500 psig). Negligible error is involved in neglecting the first portion, and considering that the entire work is performed by the main feed and booster pumps.

From Article 602, equation (2), the pump head is

$$H = \frac{p_2 - p_1}{w} = \frac{(1500 - 15)144}{62.4} = 3430 \text{ ft.}$$

From equation (3) in the same article, the horsepower required is

$$p_0 = \frac{244{,}000 \text{ lbs/hr}}{3600 \text{ sec/hr}} \times \frac{3430 \text{ ft}}{550 \text{ ft-lbs/sec-HP}}$$
$$= 423 \text{ HP}.$$

If the pump efficiency is 70 per cent, the power required by the pump is

$$p_i = 425 \div .70 = 604 \text{ HP}.$$

which must be delivered by the auxiliary turbines.

The steam available to the auxiliary turbines is boiler pressure auxiliary steam in this case, reaching the throttle at 1100 psig and 920°F. It exhausts to a 15 psig auxiliary exhaust steam line, and therefore has an energy available, (see the Mollier Chart) of

$$1457 - 1100 = 357 \text{ BTU/lb}.$$

But the turbine has an overall efficiency of about 50 per cent, so only one-half of this appears as work at the turbine shaft flange. The power required by the turbine is

$$604 \times 2544 = 1{,}537{,}000 \text{ BTU/hr}.$$

Therefore the throttle steam required is

$$1{,}537{,}000 \div .5(357) = 8610 \text{ lbs/hr}.$$

Also, available for further use is the thermal energy in the exhaust steam, which amounts to approximately

$$1457 - .5(357) = 1299 \text{ BTU/lb}.$$

This will be used later in feed heating.

1104. BOILER AUXILIARIES.

The boiler auxiliaries have not been discussed in detail. The principal units to consider are the forced draft blowers, the fuel oil service pumps, the boiler circulating pump (for forced circulation installations) and the fuel oil heaters. These will not be taken up separately as individual units here. Data on each is available in the manufacturer's instructions provided with each installation. The total boiler auxiliary thermal energy requirement will be estimated here as 2.7 per cent of the main turbine energy requirement The range of this proportion varies from about 2.5 per cent at full power to about 3.5 per cent at one-half power. The main portion of this energy is used in the forced draft blowers.

The available energy per pound of steam is about 357 BTU/lb, as computed for the pumps in the preceding article. The efficiency of the forced draft blowers is low, in the range of 30 to 40 per cent, so an efficiency of 45 per cent will be assumed for all boiler auxiliaries. This means that the thermal energy finally converted to work, per pound of steam, is

$$357 \times .45 = 161 \text{ BTU/lb}.$$

The total energy required is, as discussed above, about 2.7 per cent of the main turbine energy, or

$$89{,}040{,}000 \times .027 = 2{,}400{,}000 \text{ BTU/hr}.$$

Therefore, the steam required is

$$2{,}400{,}000 \div 161 = 14{,}900 \text{ lbs/hr}.$$

This steam enters the boiler auxiliaries with an energy content of 1457 BTU/lb and leaves with an energy content of 1457−161 or 1296 BTU/lb. This will be utilized later in feed heating.

1105. SHIP'S SERVICE GENERATOR.

The electrical requirements will vary widely, depending on the climate, the battle condition set on the ship, and other non-engineering activities in which the ship is engaged. At high powers, ordinary engineering requirements, plus *normal* ship requirements will run roughly about one-third of main generator installed capacity. In our example it will be assumed that 2000 KW capacity is installed and that 650 KW are in use under the steaming conditions of this problem. Since we are analyzing one plant, one-half of this will be used here.

From Examples 81 and 82, it will be seen that a water rate of about 12 lbs. per kilowatt-hour is reasonable. Using this considerably simplifies the generator turbine calculations for heat requirement. Thus the throttle steam required is

$$325 \text{ KW} \times 12 \text{ lbs/KW-hr} = 3900 \text{ lbs/hr}.$$

The turbo-generator inlet conditions are the same as for the main turbines, and condensate leaving the auxiliary condenser is near saturation.

1106. DISTILLING PLANT.

About 24,000 gallons per day of distilling plant capacity is installed in a modern destroyer.

At 8.33 lbs. per gallon, this amounts to an hourly rate of

$$\frac{8.33 \times 24{,}000}{24} = 8330 \text{ lbs/hr.}$$

One-half of this will be associated with one propulsion plant. From Article 901, a steam rate of about 0.4 pounds per pound of evaporator capacity is reasonable. The steam requirement is then

$$4165 \times 0.4 = 1666 \text{ lbs/hr.}$$

if the evaporator is producing at full capacity. However, it is recalled that this steam is obtained from the auxiliary exhaust steam, and so does not add directly to the boiler requirement.

1107. HOTEL SERVICES.

Steam required for hotel services is difficult to estimate on a rational basis. Only by familiarity with a given ship can the engineer determine when the consumption ceases to be a reasonable value and becomes uneconomical. It does not vary appreciably with the ship's power, and is probably affected more by climate than any other single factor.

In our example, we will estimate a consumption of about 600 lbs/hr of saturated steam for hotel services, including laundry, ship's heating and galley requirements. This steam is produced at boiler pressure and throttled to the required pressure. However, its energy content per pound is equal to the enthalpy of saturated steam at boiler pressure, since the enthalpy before and after throttling is equal. In our plant, with a 1200 psig boiler, this is approximately 1183 BTU/lb.

1108. BOILER STEAM AND HEAT REQUIREMENT.

We are now ready to estimate the boiler steam requirement for one propulsion plant. This will be done in a tabular form, using previously calculated quantities.

a. Main Turbine	207,000 lbs/hr
b. Main Steam Cycle Pumps	8,610 lbs/hr
c. Boiler Auxiliaries	14,900 lbs/hr
d. Ship Service Generator	3,900 lbs/hr
e. Hotel Services	600 lbs/hr
Total boiler steam required	235,010 lbs/hr

(Distilling plant requirement omitted since it does not come directly from boiler.)

This compares with the 245,000 lbs/hr estimated in Article 1103. Inclusion of several minor units omitted, such as fire and flushing pumps, drain pumps, auxiliary air ejectors, etc., would increase the figure. In any case the original estimate gives the pump work within 3 or 4 per cent.

The next step is to compute the total heat added in the boiler. This is done by taking the difference between the thermal energy leaving the boiler, and that entering. The energy leaves as 234,410 pounds of steam per hour at the superheater outlet conditions, 1200 psig and 960°F and 600 pounds of steam per hour at 1200 psig saturated. The enthalpy of the superheated steam is 1477 BTU/lb and of the saturated steam is 1183 BTU/lb. The total is

$234{,}410 \times 1477 =$	346,000,000 BTU/hr
$600 \times 1183 =$	negligible BTU/hr
Total	346,000,000 BTU/hr.

The energy returned to the boiler by the steam cycle is the energy remaining in the auxiliary exhaust steam (except for that portion used to operate the distilling plant) and the energy in the condensate. It will be assumed that the main and ship service generator condensate is saturated, at a temperature of 105°F, and therefore it has an enthalpy of 73 BTU/lb. The auxiliary exhaust steam and the condensate are mixed in the DFT before entering the boiler, but unless feed temperatures are required, it is convenient to assume that they all flow into the boiler individually in this case. The effect of pump work on feed temperature was shown to be small in Chapter 6, and will not be considered. The tabular form given at the top of the next page shows the energy flow into the boiler.

The boiler balance is then

Energy output of boiler	348,000,000 BTU/hr
Energy in boiler feed	43,700,000 BTU/hr
Energy to be supplied by fuel	304,300,000 BTU/hr.

A well designed boiler operates at from 78 to 83 per cent efficiency in the conversion of the chemical fuel energy to thermal energy in the steam. Assuming an 80 per cent efficiency, it is necessary to supply sufficient fuel to provide the following amount of energy,

$$304{,}300{,}000 \div .80 = 380{,}000{,}000 \text{ BTU/hr.}$$

For a fuel having a heating value of 18,500

a. Main Turbine Condensate:	207,000× 73=15,100,000
b. Main Steam Cycle Pumps:	8680×1279=11,000,000
c. Boiler Auxiliaries: (Distilling plant requirement subtracted):	13,234×1296=17,200,000
d. S.S. Generator:	3900× 73= 285,000
e. Hotel Services:	600× 73= negligible
f. Distilling Plant Condensate:	1666× 73= 122,000
Total	43,707,000

BTU/lb, this requires a fuel flow of

$$380{,}000{,}000 \div 18{,}500 = 20{,}540 \text{ lbs/hr.}$$

The final performance figure is obtained by computing the fuel consumed in terms of pounds of fuel per shaft horsepower hour, taken at turbine shaft. Thus this figure, in our example, is

$$20{,}540 \div 35{,}000 = .587 \text{ lbs/SHP-hr.}$$

This is probably somewhat low, due to our simplified analysis. However, it is within the range of consumption for a modern ship, which lies between .550 and .600 lbs/SHP-hr at powers above one-half of that designed.

INDEX

T

U

V

W

X

Z

TRIMMED SIZE: $8 \times 10\frac{1}{2}$ inches

TYPE PAGE: 37×54 picas

TYPE FACE: Monotype 8A

TYPE SIZE: 10 point on 12

CHAPTER TITLE: 14 point Franklin Gothic Caps

PAPER: 50 lb. White Printone

CLOTH: Columbia Tanotex 320 Color 2531 Skiver Brite